Practical Human Factors for Pilots

Practical Human Factors for Pilots

David Moriarty

AMSTERDAM • BOSTON • HEIDELBERG • LONDON • NEW YORK • OXFORD
PARIS • SAN DIEGO • SAN FRANCISCO • SINGAPORE • SYDNEY • TOKYO
Academic Press is an imprint of Elsevier

Academic Press is an imprint of Elsevier
32 Jamestown Road, London NW1 7BY, UK
525 B Street, Suite 1800, San Diego, CA 92101-4495, USA
225 Wyman Street, Waltham, MA 02451, USA
The Boulevard, Langford Lane, Kidlington, Oxford OX5 1GB, UK

Notices
Knowledge and best practice in this field are constantly changing. As new research and experience
broaden our understanding, changes in research methods, professional practices, or medical
treatment may become necessary.

Practitioners and researchers must always rely on their own experience and knowledge in
evaluating and using any information, methods, compounds, or experiments described herein. In
using such information or methods they should be mindful of their own safety and the safety of
others, including parties for whom they have a professional responsibility.

To the fullest extent of the law, neither the Publisher nor the authors, contributors, or editors,
assume any liability for any injury and/or damage to persons or property as a matter of products
liability, negligence or otherwise, or from any use or operation of any methods, products,
instructions, or ideas contained in the material herein.

ISBN: 978-0-12-420244-3

British Library Cataloguing-in-Publication Data
A catalogue record for this book is available from the British Library

Library of Congress Cataloging-in-Publication Data
A catalog record for this book is available from the Library of Congress

For information on all Academic Press publications
visit our website at http://store.elsevier.com/

Typeset by MPS Limited, Chennai, India
www.adi-mps.com

Printed and bound in the United States of America

Working together
to grow libraries in
developing countries

www.elsevier.com • www.bookaid.org

Dedication

This book is dedicated to four outstanding scientists who have had a profound influence on my thinking

Daniel Kahneman and Amos Tversky
Erik Hollnagel
John Anderson

Short contents

Acknowledgements		xv
Author biography		xvi
Preface		xvii

1	**Introduction to human factors**	1
2	**Information processing**	11
3	**Error management and standard operating procedures for pilots**	77
4	**Error management and standard operating procedures for organizations**	119
5	**Personality, leadership and teamwork**	133
6	**Communication**	163
7	**Fatigue risk management**	189
8	**Stress management and alcohol**	227
9	**Automation management**	243

Conclusion	269
Index	273

Full contents

Acknowledgements	xv
Author biography	xvi
Preface	xvii

1	**Introduction to human factors**	**1**
	1.1 The start of modern human factors	1
	1.2 What is human factors?	3
	1.3 A picture of human factors in aviation	4
	1.4 Human factors and non-technical skills	7
	Chapter key points	9
	References	9

2	**Information processing**	**11**
	Introduction	12
	2.1 Introduction to brain structure and function	13
	2.1.1 Types of neuron	17
	2.2 Overview of information processing	18
	2.3 Sensation	19
	2.3.1 Sensory thresholds	20
	2.3.2 Sensory habituation	22
	2.3.3 Somatogravic sensory illusions	24
	2.3.4 Strategies for dealing with sensory limitations	25
	2.4 Attention	26
	2.4.1 Selective attention	27
	2.4.1.1 Strategies for selective attention	29
	2.4.2 Sustained attention (passive monitoring/vigilance)	30
	2.4.2.1 Strategies for sustained attention	32
	2.4.3 Divided attention and multitasking	32
	2.4.3.1 Strategies for divided attention	33
	2.5 Perception	33
	2.5.1 Mental models	35
	2.5.2 Perceptual difficulties with sensory-induced spatial disorientation	38
	2.5.3 Strategies for dealing with perceptual limitations	41
	2.6 Decision making	42
	2.6.1 The anatomy of decision making	43
	2.6.1.1 Goal module	43
	2.6.1.2 Imaginal (mental manipulation) module	43

		2.6.1.3	Production (pattern-matching) module	**44**
		2.6.1.4	Summary of the anatomy of decision making	**45**
	2.6.2	The two systems of human decision making		**47**
		2.6.2.1	System 1	**47**
		2.6.2.2	System 2	**48**
		2.6.2.3	Summary of Systems 1 and 2	**49**
		2.6.2.4	How Systems 1 and 2 interact and the role of workload	**49**
	2.6.3	Heuristics and biases		**50**
		2.6.3.1	Evolutionary origins of heuristics and biases	**51**
		2.6.3.2	Heuristics and biases in aviation	**52**
		2.6.3.3	Decision making and memory retrieval heuristics	**52**
		2.6.3.4	Social heuristics	**55**
	2.6.4	Strategies for decision making		**59**
		2.6.4.1	Managing high task-load	**60**
		2.6.4.2	Managing high time-pressure	**60**
		2.6.4.3	Managing problem underspecificity	**61**
		2.6.4.4	Managing System 1 effects	**61**
	2.6.5	An operational decision-making strategy: TDODAR		**61**
2.7	Response			**65**
2.8	Fear-potentiated startle and freezing			**66**
	2.8.1	Strategies for managing fear-potentiated startle and freezing		**69**
2.9	A note about the models used in this chapter			**69**
Chapter key points				**70**
Recommended reading				**72**
References				**72**
3	**Error management and standard operating procedures for pilots**			**77**
Introduction				**77**
3.1	Performance levels			**78**
3.2	Errors and violations at different performance levels			**79**
3.3	Detection of errors			**82**
3.4	The Swiss Cheese Model			**84**
3.5	Threat and Error Management 2			**86**
	3.5.1	Threat management		**86**
		3.5.1.1	Threat management opportunities	**88**
		3.5.1.2	Types of threat	**88**
		3.5.1.3	Threats associated with serious negative outcomes	**90**
		3.5.1.4	Threat identification framework	**95**
		3.5.1.5	Generic threat management strategies	**97**
		3.5.1.6	Threat management tool for briefings	**97**
	3.5.2	Unsafe act (error and violation) management		**97**
		3.5.2.1	Unsafe act prevention where no threat has been identified	**98**

		3.5.2.2	Unsafe act detection: checklists and crew communication	**99**
		3.5.2.3	Unsafe act management strategies	**105**
	3.5.3	Undesired aircraft state management		**106**
	3.5.4	Summary of TEM2		**108**
	3.5.5	History of Threat and Error Management and differences between original TEM and TEM2		**108**
3.6	TEM2 and unstabilized approaches			**110**
	3.6.1	Threat management for unstabilized approaches		**111**
	3.6.2	Unsafe act management for unstabilized approaches		**113**
	3.6.3	Undesired aircraft state management for unstabilized approaches		**113**
Chapter key points				**115**
Recommended reading				**116**
References				**116**

4	**Error management and standard operating procedures for organizations**		**119**
	Introduction		**119**
4.1	Beyond human error		**120**
4.2	Systems thinking		**120**
	4.2.1	Normal Accident Theory	**121**
	4.2.2	The Old View versus the New View of human error	**122**
4.3	Resilience engineering		**124**
4.4	Safety culture		**127**
	4.4.1	Just culture	**127**
	4.4.2	Moving from Safety I to Safety II	**129**
4.5	Principles of managing organizational resilience		**130**
Chapter key points			**131**
Recommended reading			**131**
References			**132**

5	**Personality, leadership and teamwork**		**133**
	Introduction		**133**
5.1	Personality		**134**
	5.1.1	Personality structure	**134**
	5.1.2	Personality and behavior	**138**
	5.1.3	Personality management strategies	**139**
5.2	Leadership and command		**140**
	5.2.1	Leadership and personality	**141**
	5.2.2	Non-technical skills for leadership	**143**
	5.2.3	Leadership, personality and flight safety	**144**
		5.2.3.1 The toxic captain	**146**
		5.2.3.2 The toxic first officer	**150**

5.3	Flight deck gradient	**152**
	5.3.1 Culture and flight deck gradient	**152**
5.4	Cooperation and conflict solving	**156**
	5.4.1 Individual differences and conflict-solving strategies	**157**
	5.4.2 Situational variables and conflict-solving strategies	**158**
	Chapter key points	**159**
	References	**160**

6 Communication **163**
Introduction **163**
6.1 The Sender–Message–Channel–Receiver model of communication **163**
  6.1.1 Barriers to communication **166**
  6.1.2 The NITS briefing **167**
6.2 Communication between pilots **168**
  6.2.1 Introduction to transactional analysis **168**
  6.2.2 Ego states **169**
  6.2.3 Complementary and crossed transactions **170**
  6.2.4 Ulterior transactions **176**
6.3 Establishing a positive team atmosphere **177**
6.4 Communication strategies for effective briefings **178**
6.5 Communication strategies for assertiveness **180**
  6.5.1 First officer assertiveness at the threat management stage **182**
  6.5.2 First officer assertiveness at the unsafe act management stage **183**
  6.5.3 First officer assertiveness at the undesired aircraft state management stage **185**
  6.5.4 How captains can encourage assertiveness **185**
Chapter key points **186**
References **187**

7 Fatigue risk management **189**
Introduction **189**
7.1 Introduction to sleep **190**
7.2 Fatigue **192**
  7.2.1 Cognitive effects of fatigue **193**
7.3 Role of sleep in managing fatigue **195**
  7.3.1 Physiology of sleep **196**
  7.3.2 Sleep inertia **199**
  7.3.3 Homeostatic sleep drive and sleep need **200**
  7.3.4 Biological clock, circadian rhythms, chronotypes and sleep urge **201**
    7.3.4.1 Biological clock and circadian rhythms **202**
    7.3.4.2 Chronotypes **203**
    7.3.4.3 Melatonin **203**
    7.3.4.4 Sleep urge **204**
  7.3.5 Sleep debt **206**

	7.4	Fatigue risk management strategies	210
		7.4.1 How to achieve sufficient high-quality sleep	211
		7.4.1.1 Planning your sleep	211
		7.4.1.2 Sleep hygiene	212
		7.4.2 How to nap effectively	213
		7.4.3 How to deal with insomnia and sleep disorders	214
		7.4.3.1 Chronic insomnia	215
		7.4.3.2 Obstructive sleep apnea	215
		7.4.4 How to mitigate risk if you find yourself fatigued	216
		7.4.5 How to use sleep medications	218
		7.4.6 How to deal with jet lag	219
		7.4.7 Organizational strategies for fatigue risk management	222
	Chapter key points	222
	Recommended reading	223
	References	223

8   Stress management and alcohol   **227**

	Introduction	227
	8.1 Chronic stress	227
		8.1.1 Prevention of allostatic load due to chronic stress	230
		8.1.2 Management of allostatic load due to chronic stress	230
		8.1.2.1 Management strategies aimed at the stressor	231
		8.1.2.2 Management strategies aimed at the individual	232
		8.1.3 Critical incident stress management	233
	8.2 Alcohol	234
		8.2.1 Alcoholism in aviation	238
	Chapter key points	240
	Recommended reading	241
	References	241

9   Automation management   **243**

	Introduction	243
	9.1 Systems of aircraft automation	245
	9.2 Flight control laws	246
	9.3 Levels of automation and their uses	249
	9.4 Flight mode annunciators	252
	9.5 Automation, perception and Newton's laws of motion	253
	9.6 The ironies of automation	254
	9.7 Skill fade and automation dependency	258
	9.8 Automation complacency	259
	9.9 Automation bias	260
	9.10 Automation surprises	262
	Chapter key points	265
	References	266

Conclusion   **269**
Index   **273**

Acknowledgements

Even sole authors do not really write books by themselves and I've been very lucky to have had a huge amount of support along the way. I'd like to thank Emily Ekle, Barbara Makinster, Charlotte Pover, Edward Taylor and all the team at Elsevier for their support and guidance throughout this project. I'm also very grateful for the support and encouragement of many of my colleagues, especially Sarah Skelton, Steve Jarvis, Paul Harris and Keith Fryer. Finally, I'm hugely grateful for the support of Matt Hutchinson and my parents and for their unceasing patience and understanding while I've been focussed on researching and writing this book.

Author biography

David Moriarty is a Captain and Chief CRM Instructor with a European airline and is based in London. Formerly a doctor, he left neurosurgical training to become a pilot and has a Masters degree in Human Factors and Aeronautical Safety. He runs Zeroharm Solutions www.zeroharmsolutions.com, a consultancy specialising in aeronautical and medical safety, particularly with regards to training, safety management, accident/incident investigation and procedure design (cognitive systems engineering). He is a member of the Royal Aeronautical Society and the Resilience Engineering Association. Comments and feedback about this book are very welcome and can be sent to david@zeroharmsolutions.com.

Preface

As many of you will know, every year, pilots and cabin crew have to receive training in crew resource management (CRM). This is essentially human factors training but specifically for aviation. Because this is a legal requirement, it has to be completed within a specified time-frame. Occasionally, because of illness or scheduling difficulties, we have to run a one-to-one CRM course so that the crew member remains legal to operate flights. As a freelance CRM instructor, I also do training for a variety of different airlines as well as my own airline. A while back, I carried out some CRM training for a business jet operator based in Europe. Unfortunately, one senior training captain was unable to attend the CRM training. Because he was due to start a work trip within the next couple of days, we had to arrange for him to complete his CRM training prior to leaving. In the event, the easiest way to achieve this was for him to fly to the UK and get him to undertake a one-to-one CRM course with me. If there are any CRM instructors reading this, you will know that it can be incredibly difficult to run a successful course with just one person. One of the key benefits of CRM training is the discussion that occurs between different crew members. Opinions and experiences are shared, situations are analyzed from a variety of different perspectives and, hopefully, consensus is reached. Needless to say, this becomes a lot more difficult when there is just one student in the room. The pilot in question was a highly experienced training captain who had served for many years in the military. He had flown for both commercial and business jet operators and had been exposed to CRM training since its development in the 1980s.

Given that there were just the two of us, I thought I would start the session by asking a candid question so I asked him what he thought of CRM. At first he gave me the answers that I have come to expect. He talked about how important it was, how communication is important, teamwork is important, and so on. He also said that it was relevant because it is the human rather than the machine that leads to aviation accidents these days. I pressed him a little further on this. These were answers that I had heard on just about every other CRM course where I had been an instructor or a participant: well-rehearsed, timeworn phrases that I could imagine hearing from any pilot, cabin crew member, aviation instructor or training department manager. Given that his answers were pretty standard, I took a different tack and asked him how CRM training has affected his behavior during day-to-day operations. This was clearly a more difficult question as he was unable to give me any examples of how his 20 years of CRM training had affected his performance in any way whatsoever. Based on this, I repeated my first question. "What you think of CRM?" Finally, he gave me his honest answer: "It's a waste of time".

Unfortunately, I think he is right.

When CRM training came into being back in the 1980s, it was generally felt that aircraft were becoming safer and safer and that it was now the human element that led to accidents and incidents. The old image of the gallant, swashbuckling captain, wrestling the burning aircraft on to the runway, while all around him panic, was beginning to fade. It was becoming increasingly apparent that successful management of abnormal situations was best achieved by utilizing the resources of all the crew members. This became the basis of modern CRM training. It was a profound step and, indeed, many other industries looked and continue to look to aviation as the pioneers of this new and exciting field. But what is CRM today? Has it lived up to the promise of revolutionizing aeronautical safety? Or is it, as I have heard from so many students, just common sense by a different name? I suspect that many crew members feel the same as that business jet training captain does: a good idea in principle; a waste of time in reality.

I think that it is important that from the outset that I give you my answers to these questions as this explains why I thought it was important to write this book. I think that we, as an industry, have failed. Although our initial analysis was correct, insofar as the human element had become the key to understanding the accidents and incidents happening in aviation, we have not followed this up in either our training philosophy or our legislation. A common statistic that is often repeated is that 70% of aviation accidents are due to human error. While I take some exception to this figure (show me an accident that does not include an element of human error!), it leads to an interesting observation. When we consider how we train pilots and cabin crew, the vast majority of the time allotted is spent in training them how to deal with technical failures. We spend comparatively little time looking at human factors. Why should this be so, given that we know that the majority of accidents are due to human error? For me, the answer is simple: it is easier to teach technical subjects. As will become apparent to you as you work through this book, human factors is a truly huge subject. It encompasses just about every aspect of human performance. It is far easier to teach crew members how to deal with technical failure than it is to equip them with accurate, research-based human factors strategies and behaviors that they can use in the workplace to enhance safety and efficiency.

Another major drawback of human factors in aviation is that it is rarely assessed. Although it is common for national regulators to stipulate that crew members' "non-technical skills" are assessed in the simulator and during line checks, we do not routinely assess an individual's *knowledge* of the human factors. I am reminded of a conversation I had with someone who worked for the national regulatory authority. We were talking about the importance of CRM training and I expressed my view that it was strange that it was the only subject that was not formally assessed and that this may, inaccurately, suggest to students that the subject is not important. The response I received was, "If you can get the subject across to just one or two people in the room, then you're doing well". Would we have the same, laissez-faire view if the subject was "hydraulics" or "safety and emergency procedures"? Of course not! We require all crew members to learn and be able to recall the details of these technical subjects. Failure to do so may have a negative impact on an individual's career. Why then, if we have all agreed on the importance of human factors with regard to aviation safety, do we not take the same approach to teaching and assessing this subject?

The only answer that occurs to me is that the science of human factors is so broad and so "messy" that we have not, as an industry, agreed on what is important and what is not. For example, the concept of "situational awareness" is still hotly debated. There are some human factors academics who feel that this is not a useful term.[1] There are others who think that it is a useful term. Who is right? The sad answer is that we do not know. Unlike hydraulics or safety and emergency procedures, the subject of human factors is not black and white. While national legislation specifies that human factors must be taught in the form of CRM and lists various topics that must be covered in such training, it does not, as yet, specify which approach should be taken to each of these topics. As you will soon see, there are multiple approaches that can be taken with each topic. If there is no uniformity in how different operators determine how a topic should be taught, is it fair to assess people on it? The unfortunate side-effect of this dilemma is that when crew members are scheduled to attend CRM training they are well aware that their knowledge of the subject will not be assessed at the end of the day. I believe that this, tragically, leads to the impression that the subject is not important. If it is not important enough to assess, then where is the motivation to pay attention and to learn?

Sadly, for the time being, the legislation is what it is although there do seem to be some changes on the horizon in Europe with recent proposed changes put forward by the European Aviation Safety Agency. When I look back on my experience of teaching this subject, I have come to the conclusion that the term "CRM" is tainted. This generation of crew has been brought up in a system that has not made up its mind on which is the most effective way to teach this safety-critical subject. Although some people are willing to engage with the subject, I believe that the majority of people are unconvinced about its importance and, for that, instructors like me must bear some responsibility. If there are any other CRM instructors reading this now, I am sure you will, as I have, have had the heartbreaking experience of standing up in front of a group that you have taught the year before and asked them to recall anything of what they have learned previously, and been met with blank stares. If the information is not being retained or, more importantly, not being put into practice, then what is the point of CRM training? This is a question that troubles me deeply.

Unfortunately, I am not a legislator. If I were, I would drop the term "CRM" and replace it with "human factors". I would completely revise and expand the training syllabus and give specific objectives to be achieved for each subject, and I would make it compulsory for each subject to be assessed formally. As it stands, I am a humble instructor trying to communicate the importance of these topics to front-line crew. However, there is one thing I can do. In writing this book, I hope that I can equip both instructors and crew members with accurate, science-based and, most importantly, practical knowledge of the science of human factors. My hope is that we can raise the quality of human factors training within our industry to ensure that it benefits from the huge advances that have been made in this fascinating, but sometimes complex field.

What this book is not

- **This book is not trying to reinvent the wheel** – One of the problems with human factors is that people are constantly trying to reinvent it. There are new avenues and approaches and, sometimes, there is a genuinely new discovery that explains something that we could not explain before. The drawback of this is that people are at risk of forgetting the past. There is a mountain of solid, elegant, beautifully thought-out and, most importantly, evidence-based science developed by genuinely deep thinkers in the field of human factors. Rasmussen, Reason, Hollnagel, Anderson, Kahneman, Tversky, Parasuraman, Helmreich, Bainbridge, Wickens, Hofstede, Woods, Berne, Costa, McRae, Simon … the list goes on. The more that modern science uncovers the workings of the human brain, the more the work of these visionaries makes sense to me. Many of them had insights about human brain function and its effect on behavior long before the technology was there to prove them correct. In short, nothing that you will read in this book is truly my work. All I have done is to try to draw together these pieces of outstanding research from these people whose abilities and knowledge far outstrip my own, and present them to you in an integrated, practical way. Human factors, although broad in scope, is an integrated discipline. These scientists may have been focusing on a small area of human behavior but their discoveries, when combined, create a model of human performance that we can use to our advantage. All I want to do with this book is to bring their work together into something that you can use day-to-day. The concluding chapter in this book will give you a summary of how everything you have learned in the individual chapters fits together to create a picture of how human beings interact with all the other elements in the aviation world.
- **This book is not a textbook** – The aim of this book is not cover the entire science of human factors. Rather, this book aims to give the reader a basic grounding in some of the key theories within the field of human factors and show how they relate practically to day-to-day aviation operations. I am fully prepared for the fact that I am going to receive a great deal of grief from my human factors colleagues saying that I have been too selective in what I have included in this book. Unfortunately, the reason that CRM is so badly taught at the moment is that, for many instructors, navigating the complexities of the scientific literature can be almost impossible. While scientists working in the field of human factors are excellent at developing and testing theories to explain human behavior, they often encounter difficulties in communicating these results to non-scientists. For example, in the field of decision making, there is a plethora of theories and approaches attempting to explain this phenomenon in subtly different ways. The choice I made before writing this book is that the reader would be better served by considering one of these approaches rather than several.
- **This book has not been chiseled in stone** – The science of human factors is both rapidly changing and controversial. Many of the subjects in this book are still being actively researched and, as with any scientific endeavor, debate about them continues. I am anticipating being asked why I have made some of the choices I have in selecting material for this book in light of the fact that a consensus has not yet been reached in the scientific community. My response is that just because a particular aspect of human behavior is still being researched, there is no reason why we should not attempt to put some of the current thinking into practice. Human factors, more than any other science, is about practicality. We are attempting to find ways of improving safety and efficiency through scientific research. The risk of taking a conservative approach and waiting until there is universal consensus about a particular aspect of human performance is that the consensus never arrives and the research goes unused. I would much rather make use of the science that is out there already to bring about changes in how we do things.

- **This book is not a complete reference tool** – This book is meant to be a starting point for both instructors and people who work in aviation. Throughout the book, guidance will be given about further reading material and other resources that readers may draw on to expand their knowledge of a particular subject.
- **This book is not meant to be abstract** – Although many of the concepts that will be covered in the chapters that follow may appear, on the surface, to be unrelated to day-to-day life, every effort will be made to relate the concepts to practical guidance, strategies and behaviors that can enhance safety and efficiency in real-world operations.
- **This book is not for the faint-hearted** – If you are looking for a really easy book that will explain how and why humans behave in the ways that they do, I'm afraid that this is not the book for you. In fact, I'm not sure that any book will be the book for you. The myth that pervades the aviation industry is that CRM and human factors are just common sense, that CRM is a "soft subject" and that it is entirely non-technical. The truth is quite the opposite. Human behavior is hugely complex and the mechanisms that underlie it are complex too. Of any subject related to aviation – aerodynamics, air law, structural engineering – human factors is, by far, the broadest and the most difficult. It is a highly technical subject that requires serious study if you hope to make best use of it in your operation. This is not meant to scare you or put you off reading, but I think it is only fair to make you aware of this at the start: you will find this book challenging but, if you stick with it, you will learn a huge amount about how and why people behave as they do, how such behavior can be predicted, why things go wrong in aviation, and how you can use this knowledge to your advantage to keep you and others safe.

How to use this book

- **Spend time reading Chapter 2 on information processing** – It is the toughest one in the book but, if are willing to spend some time understanding the basic structure and function of the brain, everything else in the book will slot into place. Brain structure and function lie at the very core of human factors, and once you have read this chapter you will be in the privileged position of understanding how that incredible three-pound mass of tissue between your ears, the most complex thing in the known universe, makes you who you are and allows you to interact with the world around you. After nearly 15 years of being interested in the human brain, of studying it, operating on it and conducting brain-based research, I still find this weird-looking organ utterly incredible and I hope that you will soon share in my fascination.
- **This book is for you** – If you are a pilot, there are things in this book that you might like to know. There are also things in this book that might be able to help you out. If you are a CRM instructor, the stakes are higher. My first desire when writing this book was to give CRM instructors a source of material that they could use in their training. Unfortunately, this means you are going to have to be more dedicated than other readers because, essentially, your job is to teach this stuff. If we are going to raise the standard of human factors knowledge and training across the industry, we need to get our hands dirty with the hardcore science behind it. It will be worth it, though, as you will be able to stand in front of a class and speak with the genuine authority of someone who knows the subject.
- **This book, like the science of human factors itself, is meant to be practical** – Ironically, for a science that is all about practicality, human factors can sometimes prove to be rather inaccessible. Every year, tens of millions of dollars of research funding are spent looking at how human behavior can be adjusted to improve safety and efficiency. Unfortunately,

because the science can be so impenetrable, the fruits of this well-funded research often fail to make any difference to the safety-critical industries that they are meant to benefit. This book is envisaged to be a bridge between the science of human factors and day-to-day aviation operations. As you will see in the chapters that follow, the general approach taken to each subject is that we will first look at the phenomenon and the science behind it. We will then look at the particular limitations of the phenomenon (e.g. the negative effects of fatigue on information processing) and illustrate them with real-life examples of incidents and accidents where they have been a contributing factor. Finally, we will consider some more of the scientific research and the recommendations that it makes to address these limitations. In other words, the goal of each chapter is to equip the reader with knowledge of the strategies and behaviors that can be used to mitigate the risks that have been highlighted earlier.

- **This book assumes no prior knowledge of the science of human factors** – In writing this book I have considered the reader to be a blank slate with regard to human factors. While I am assuming that the reader will have some knowledge of commercial aviation, my goal is to be able to give a reader who is new to the field as complete a grounding as possible in this safety-critical science.
- **This book assumes enthusiasm** – If we take a moment to consider the science of human factors, a science that relates to just about every aspect of human performance, especially in the workplace, you can begin to see how huge this field actually is. If you want to get to grips with this field, you really need to be motivated. Because there is so much information out there, would-be CRM instructors have two options: they can either take the easy route and assemble training material based on Google searches and Wikipedia or they can take the more difficult, but more rewarding route of acquainting themselves with as much of the science as possible and then making a reasoned judgment about which parts are best suited to their particular training needs. As I have said, this book is meant to be a starting point. There will be ample guidance for readers to expand their own knowledge in the form of "Recommended Reading" at the end of most of the chapters. To give the reader some idea of the accessibility of the various references mentioned, I have rated each recommendation according to its difficulty level. A reference rated as "Easy" is one that is written having assumed that the reader will have no prior knowledge of the subject. A difficulty rating of "Intermediate" suggests that the book assumes the reader has some basic knowledge of the subject in question (knowledge that you will likely have by the end of the relevant chapter of this book). A difficulty rating of "Advanced" means that the reference is probably aimed at someone who is already well acquainted with the subject of the book. Although these will require a lot more concentration, these references will give the reader a much more complete perspective of the subject.
- **Get used to lists** – There are some parts of this book that require the reader to understand some potentially complex concepts. In order for the reader to reach a position where he has an understanding, for example, of how the brain works, it might be necessary to learn a little bit of biology. In my experience, concepts such as biology or physics can be explained by building knowledge up from small, simple facts to progressively more difficult ones. The idea is that the reader works through the list step-by-step, moving on to the next point when the previous one has been understood. In this way, an in-depth knowledge of the subject can be built up from the simplest foundations.
- **Do not expect too much psychology** – Traditionally, human factors training has been based around applied psychology. As psychology developed as a science, it attempted to understand the structure and function of the human brain by analyzing behavior. Experimental psychologists would design scenarios to assess how humans function under a variety of different conditions. From these initial observations, they would construct models of how

they believed the human brain functioned and then develop experiments to test these models. Over a period of years or even decades, these models would be continually refined as more experimental data were collected. The achievements of psychological science are profound. Even before the advent of modern brain imaging and other direct measures of brain function, psychology had already produced a basic framework of how the brain works. During the past 20 years, neuroscience (the science of the brain) has approached human behavior from a different direction. Rather than trying to deduce structure and function based on behavior, neuroscience is now so advanced that we can begin to predict human behavior based on our knowledge of the structure and function of the brain. I am making this distinction early on because this book will not contain as many references to psychological theories as you might expect. While psychology and neuroscience remain bedfellows, neuroscience is progressively taking more of the blankets! One of the major chapters in this book (Chapter 2) covers information processing. I have decided to take a neuroscience-based approach to the subject as it will make a lot more sense to readers and will also equip them with tools to be able to understand many of the other subjects that are covered in this book. As the scientists would say, we are going to take a bottom-up approach to human behavior (looking at behavior based on the structure and function of the brain) rather than a top-down one (looking at behavior in isolation and then inferring structure and function based on that). Hence, I stress again, read Chapter 2 early on: it is the key to everything else.

- **You do not have to know it all** – There is a lot of material in this book, much of which will probably be new to you. If you are a pilot reading this book out of interest or an instructor reading this book to find some material that you can use for creating a human factors or CRM training course, the most important parts of each chapter are the strategies and behaviors that can be used to enhance safety and efficiency. If you are going to take something away from this book, take away the practical steps that can be used to improve safety and efficiency. The other information in the chapters is there to justify why these strategies and behaviors are useful.

A few other things to note

- **Use of gender pronouns** – For clarity, the masculine gender pronoun is used throughout this book.
- **Pilot roles** – Much of this book is based around the human factors of a two-crew flight deck. For normal operations, the duties of the pilots can be described using a variety of terms. For the purposes of this book, the pilot who is primarily responsible for operating the aircraft and maintaining the desired flight path is known as the "handling pilot". The other pilot, who is primarily responsible for monitoring the actions of the handling pilot, communicating with air traffic control and running checklists, is known as the "monitoring pilot".
- **The problem with human factors research** – For a science that prizes clarity of expression, human factors research can be notoriously confusing because of the multiple different approaches that researchers may take towards one subject. This leads to problems with terminology because two different approaches may use different language to describe the same thing. To avoid confusing the reader, I have had to make a choice about which terms I include in this book and which I exclude. Some of these choices may appear confusing, but if I have ignored a particular term, it is usually because the alternative term fits in much better with the rest of the material. For example, many readers may be surprised that there is no chapter or section on "situational awareness". As has already been mentioned, there

is still debate about whether situational awareness is a useful term. Many readers would expect it to have a place in the chapter on information processing, and my justification for leaving it out is based on the following: the subject is already amply covered by the sections on attention, perception and decision making. This goes back to the idea that human factors professionals sometimes think it necessary to reinvent the wheel. When you consider the definition of situational awareness, "the perception of elements in the environment within a volume of time and space, the comprehension of their meaning, and the projection of their status in the near future",[2] there is nothing there that cannot be explained by research in the fields of human attention, perception and decision making. The risk is that creating a new term suggests that it is also a new phenomenon. The term "situational awareness" only really came into being during the 1980s and was usually used in relation to the aviation environment. However, research into human attention, perception and decision making has been carried out for over 100 years across a wide range of environments. The human brain uses the same mechanisms no matter what environment it is in and yet we seem to equate situational awareness only with certain jobs, such as flying an aircraft. We do not see research into the "situational awareness" of children playing with toys, even though they will have to perceive the elements in their environment, comprehend their meaning and project their status in the near future. This is because there is already plenty of research that looks at children's attention, perception and decision making. If we assume that situational awareness is a subject in itself, we will probably miss out on the huge volume of research that has already been carried out into the phenomena that it tries to replace. However, the term "situation awareness" does make an appearance in Chapter 1 (Introduction to Human Factors) as it forms part of an assessment framework used to assess pilot's non-technical skills. In this context, the term refers to the operational behaviors that a pilot might exhibit as a result of his information processing abilities.

- **Case studies** – In the section on decision making, part of Chapter 2 on information processing, we look at a couple of phenomena that affect how we judge the behavior of others: hindsight bias and fundamental attribution error. These phenomena describe how we are generally programmed to blame other people for their misfortunes and convince ourselves that this would not happen to us. Unfortunately, this is certainly not the case. This book is full of case studies and it is important to note that, from the very beginning, these case studies are not given to be critical of the people or organizations involved. They are given as illustrations of some of the subject areas covered and to provide some real-world context.
- **The use of models** – Many of the chapters in this book rely on neuroscience-based or psychological models. In some cases, to make a chapter fit together, it has been necessary to try to uncover some common ground that can link different models together. As has already been stated, human factors is a science that continues to evolve and so it may turn out that there are different or more complex connections among different phenomena. However, the aim of this book is to provide practical guidance for pilots in their day-to-day operations and any linking of different theoretical approaches has only been done with this aim in mind.
- **This book is for guidance only** – Nothing in this book takes precedence over company procedures, national or international legislation or guidance from medical practitioners.

References

1. Dekker S, Hollnagel E. Human factors and folk models. *Cognition, Technology & Work* 2004;**6**(2):79–86.
2. Endsley MR. Toward a theory of situation awareness in dynamic systems. *Hum Factors: J Hum Factors Ergon Soc* 1995;**37**(1):32–64.

Introduction to human factors

Chapter Contents

1.1 The start of modern human factors   1
1.2 What is human factors?   3
1.3 A picture of human factors in aviation   4
1.4 Human factors and non-technical skills   7
Chapter key points   9
References   9

1.1   The start of modern human factors

In March 1979 an argument took place in an aircraft hangar in Pennsylvania.[1] The argument was heated and was between two experts working in a highly specialized field, each one arguing that his interpretation of events was the correct one. The stakes of this argument were about as high as one could imagine. If one of the experts was proved right, the nuclear reactor nearby was about to explode, spreading radioactive devastation across the USA in the same way as Chernobyl would spread radiation across Europe and Russia a few years later. To make the situation even more critical, the President of the United States, Jimmy Carter, and his wife were shortly to land at the airport before traveling to the plant to carry out an inspection. An explosion would then not only cause catastrophic radioactive fallout but would also kill the President of the USA.

The problems had started a few days earlier at Unit 2 of the Three Mile Island nuclear power plant near Harrisburg, Pennsylvania. The problem seemed simple enough at first. Under normal circumstances, water passing through the core of the reactor in the primary cooling system is heated to high temperature by the nuclear reaction. The pipework carrying this heated, radioactive water then comes into contact with the pipework of the secondary cooling system. Heat passes from the hot water in the primary cooling system to the cold water in the secondary cooling system, but no water is transferred between the two. For this reason, the water in the primary cooling system is radioactive but the water in the secondary cooling system is not. The water in the secondary cooling system must be extremely pure because when it turns to steam, it must drive carefully engineered turbine blades to generate electricity. Any impurities in the water would make this inefficient. A filter system keeps this water as pure is possible, although the filter in Unit 2 was known to be problematic and had failed several times recently. At 4 a.m. on 28 March 1979, a small amount of non-radioactive water from the secondary cooling system leaked out through the filter unit. Many of the instruments that are used to monitor the nuclear power plant rely on pneumatic pressure, and it seems that some of the water got into this air system

Practical Human Factors for Pilots. DOI: http://dx.doi.org/10.1016/B978-0-12-420244-3.00001-7

and led to some instruments giving incorrect readings. One of the systems affected by this leak was the feedwater pumping system that moved the hot radioactive water in the primary cooling system into proximity to the cool, non-radioactive water in the secondary cooling system. The water contaminating the feedwater pumping system caused it to shut down, and with no heat being transferred from the primary to the secondary cooling system, an automated safety device became active and shut down the turbine.

As well as circulating water in the primary cooling system so that heat can be transferred to the secondary cooling system, the constant flow of water caused by the feedwater pumps was what kept the temperature of the core of the reactor under control. With these pumps shut down, emergency feedwater pumps activated to keep the core temperature under control. For some reason though, the valves in both of the emergency feedwater systems had been closed for maintenance and had not been reopened. The emergency pumps were working and this was indicated in the control room, but the operators did not realize that although the pumps were working, the pipes were closed and so feedwater could not reach the reactor to cool it down. There was an indication in the control room that the valves were closed but one of these was obscured by a tag hanging from the panel above, and the default assumption of the operators was that these valves would be open as common knowledge had it that the valves are only closed during maintenance and testing.

With heat building in the core, the reactor automatically activated a system that would stop further heat from being generated. Graphite control rods dropped into the core to stop the nuclear chain reaction by absorbing the neutrons that would normally sustain it. Even though the chain reaction had stopped, this huge stainless steel reactor was still incredibly hot and required vast quantities of water to cool it to a safe level. Without this cooling, the residual high temperatures would continue to heat the remaining water in the core and could lead to a significant build-up of pressure, enough to critically damage the reactor. To avoid this, an automatic pressure relief valve opened to allow some of the hot, radioactive water to leave the core and so reduce the pressure. The pressure relief valve is only meant to be open for a short period to regulate the pressure. Unfortunately, the pressure relief valve failed to close after the pressure in the core had been reduced. Unbeknownst to the operators, the core now had a huge hole in it (the open relief valve). To add to the confusion in the control room, even though the valve was now stuck open, the system that would normally close it sent a signal to the control room saying that it had ordered the relief valve to close and this was interpreted as confirmation that the relief valve was *actually* closed.

When it comes to problem solving in complex systems such as nuclear power plants and modern aircraft, it is often the highly automated nature of such systems that makes it a lot more difficult for the human operators to keep up with an evolving situation. Within 13 seconds of the initial failure in the filter unit, the automated safety devices had initiated a sequence of major changes in how the reactor was being controlled. Not only was the plant in a very different configuration than it had been 13 seconds previously, but the problems with the various valves being in the incorrect position and the fact that this was not being clearly signaled in the control room meant that the operators were completely in the dark regarding the true nature of the problem that they had to deal with. Problems continued to mount over the following days to the extent that a mass evacuation was ordered and churches in the local area offered general absolution to their congregations, something that is normally only offered when death is imminent. As all levels of local, state and federal government tried to solve the problem, the situation continued to deteriorate. The argument in the aircraft hangar was about whether a dangerous concentration of hydrogen gas was accumulating at the top of the reactor to such high pressures that it could detonate, blowing open the reactor core and spreading tons of nuclear material all over the USA. Fortunately for the population, it seemed that while hydrogen was accumulating, it was not at sufficient pressure to detonate spontaneously. By the time President Jimmy Carter had landed and was taken on a tour of the plant, an improvised system of pipes was feeding water to the core, thus cooling it, and the unexpected positions of the various valves had been discovered and corrected. Three Mile Island did not explode but the reactor was critically damaged and a substantial amount of the nuclear material had melted in the core, rendering it unusable. There had also been some release of nuclear material, but not enough to pose a significant health risk.[2]

This case illustrates many of the technological, procedural and human limitations that can be seen in just about every accident or incident in any industry. The human operators had to deal with a highly automated, highly complex nuclear power plant that could radically change its own configuration at a speed that they could not keep up with. The indicators that they relied on to form an adequate mental model of the situation were unreliable, and many assumptions were made during the decision making process that not only failed to fix the problem but actually made it worse. This accident started in the early hours of the morning when humans are not best suited to complex problem solving. Because of a limited number of phone lines running into the plant, it was almost impossible for people with vital information (such as the company that built the plant) to communicate this information to the people making the decisions. Information processing, decision making, error management, communications, leadership, fatigue and automation management were all impaired, and it was the subsequent investigation of this accident that highlighted the vital importance of understanding human factors in making safety critical industries as safe as possible.

1.2   What is human factors?

The scope of human factors can make it a hard discipline to define. There are many definitions, but most share some key elements:

- It is a multidisciplinary science that includes research from the fields of psychology, biology, sociology and engineering.
- Human factors research is aimed at improving the safety and efficiency of a system, a system being a collection of components such as humans, procedures, and/or machines that are designed to achieve something.
- By optimizing these components, especially the human component, and by optimizing the interfaces among components, the system can be made to work as safely and efficiently as possible.

As we will see in Chapter 4 (Error Management and Standard Operating Procedures for Organizations), as systems become more and more complex, it can be difficult to predict where failures will occur and how they will evolve and affect the operation of the system. It is also important in human factors to respect that there needs to be a balance between safety and efficiency and that by trying to maximize safety with more checks, rules, and procedures, the efficiency of the system can be compromised to the extent that it no longer works. Where the optimum balance is will be different for every system. Human factors is about finding this balance point and then optimizing the system by optimizing the function of the components of the system and the interfaces among them.

1.3   A picture of human factors in aviation

A commonly used statistic is that 70% of aviation accidents are a result of human error. This statistic has been accepted for the past few decades but conceals a more troubling truth. If 70% of aviation accidents are caused by human error, what are the other 30% caused by? The most common answer is that these are due to technical failures. However, machines do not fail by themselves. If part of an aircraft breaks, it is either because it has been designed incorrectly (the design is insufficient for its intended purpose), it has been built or maintained incorrectly, or it has been used in the wrong way. When you probe into "technical accidents", human error will be involved somewhere. With the possible exception of a completely freak accident that no one could possibly predict (e.g., a plane being hit by a falling meteorite), every accident, incident, and near miss will have some element of human involvement in the sequence of events leading up to it. If human error is everywhere, why do we insist on classifying accidents and incidents as human related or technology related? In fact, this classification is something of a fallacy. Once an accident cause has been determined as human error, the reaction of the people with a stake in determining its cause is that an isolated, one-off, human-mediated event occurred and that this can be easily corrected by dealing with the human. As you will discover in Chapters 3 and 4 on error management, nothing could be further from the truth. The quality and success of the complex interactions between humans and other humans, humans and machines, and the associated procedures that are used to achieve this are at the root of aviation safety. Hence, an understanding of human factors is the first and most important step in understanding where success and failure come from.

As the name suggests, for human factors to be applicable there must be a human in the system somewhere. If not, it becomes an engineering problem. For example, the design of a small component that will be integrated into a larger machine might not need any human factors input unless that component has some part to play when the human interacts with the machine (e.g., an indicator light or a dial). For most other systems, however, there will be a human operating within the system. The early work that considered types of systems that combined human and mechanized components was carried out in the 1940s in the British coal-mining industry and led to the concept of a

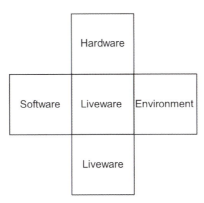

Figure 1.1 The SHEL model.

"sociotechnical system."[3] During the 1970s, this idea of a sociotechnical system was expanded to cover human factors in the aviation industry and led to the development of the SHEL model by Edwards, later developed further by Hawkins.[4] The SHEL model, SHEL being the acronym of its four components (software, hardware, environment and liveware), considers all the elements of an aviation system and is shown in Figure 1.1.

The "environment" component is self-explanatory, liveware refers to the human components, hardware to the machine components and software to any procedures. As you can see from Figure 1.1, there is liveware at the center of the model. This liveware is, essentially, the human operator. That human operator will need to inter-act with machines (hardware), procedures (software) and other people (liveware). All of this is done in some sort of environment, perhaps on the ground, in an office, or in a flight deck at high altitude. Although the SHEL model has been around for decades, it has fallen slightly out of favor in human factors. It is not mentioned as much as it used to be and, as suggested in the Preface to this book, may be one of the "old" concepts in human factors that people are all too willing to try to replace. The reality is that the SHEL model is as relevant today as it ever was, perhaps even more so given the highly automated nature of modern aviation. Aside from the individual components of the model, the real insight that the original designers of this model had was to focus on the *interfaces* among components, that is, the points where the squares touch. As well as wanting to optimize the components themselves, we need to consider how we can optimize the interfaces among components. Although the SHEL model can serve as the foundation for our understanding of human factors, to make it more relevant to aviation it can be adapted in a few ways:

- First, to give us a point of focus, let's make the human in the center of the model more prominent.
- The environment affects all the components of the model and can be shown surrounding all the other components.
- The software (procedures) are one way that the central liveware interacts with other liveware and with the hardware. As well as procedural interaction, there can also be direct, non-procedural interaction between liveware and hardware.

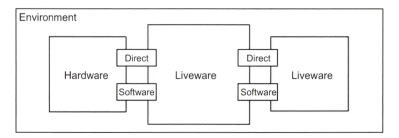

Figure 1.2 Adapted SHEL model.

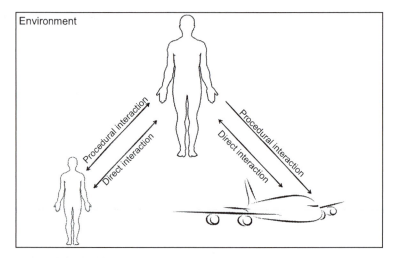

Figure 1.3 Aviation-specific human factors model.
Human image copyright vadimmmus and aircraft image copyright mmarius (both from www
.shutterstock.com).

With these adaptations, the new model looks something like Figure 1.2.

We can now make this model more pictorial and more specific to aviation, as shown in Figure 1.3. For the purposes of this illustration, the interaction between the second liveware component and the hardware is not shown.

We can divide the human up into two main sections, the body and the brain, each of which has characteristics that would be of interest when considering that human's involvement with the system:

- the body – strength, nutritional state, muscle fatigue, general health and disability
- the brain – information processing capabilities, personality, general intelligence, communication skills, fatigue levels, mental health and cognitive disabilities

Because the human is at the center of our model of human factors, optimizing the human's performance is key to optimizing the entire system.

The other component of the system that the human may have to deal with on a regular basis is other humans (other liveware). Now that we are looking at interactions between two components of the system, the human factors practitioner must start to consider how these interactions occur and how they can be optimized. Humans can interact with each other in two ways:

- directly
- procedurally, using standard operating procedures (SOPs)

The distinction is subtle but important. In aviation, the two pilots will interact with each other procedurally by making standard callouts during take-off and approach and the procedural interactions are important for maintaining safety. However, if the interaction between the two pilots was limited to procedural communication only, the atmosphere in the flight deck may end up being quite cold and unfriendly. We therefore have direct interaction and this can be important in establishing rapport, detecting errors and solving novel problems that are not covered by procedures.

We next have to consider the technology that the human operator must interact with for the system to be successful, in this case, the aircraft. The range of technologies that humans may have to interact with is vast, but a general consideration for the human factors practitioner would be how easily the human can interact with the technology and how clearly the technology can give feedback and information to the human. A machine, no matter how clever it is, is useless if it is unusable; for example, a powerful computer is useless if it is too complicated for the human operator to use. In this book, we will be looking especially at the interaction between pilots and the aircraft that they fly. In the same way that two humans can interact either directly or procedurally, humans can interact with technology either directly or procedurally. The direct interaction between humans and technology is two way as the machine will usually give indications of its status to the human. Procedural interaction tends to be one way, from human to machine, as there are likely to be SOPs that determine how the pilots control the aircraft.

The rest of this book will focus on optimizing the pilot, optimizing the direct interaction between the pilot and the other pilot (and any other humans), optimizing the interaction between the pilot and the aircraft and, finally, optimizing the procedures through which the pilot can interact with the other pilot (or other humans) and the aircraft. Crew Resource Management (CRM) training is intended to achieve the same aim but does so with only variable success. There is considerable variability in the quality of CRM training, a worrying fact given that this is the only dedicated way of delivering human factors knowledge to crew working in this high-stakes environment.

1.4   Human factors and non-technical skills

While having a knowledge of the science of human factors is a good first step, the science must be put into practice in order to be useful. A captain who understands the science behind intercultural communication and conflict-solving strategies should also know how to use this knowledge in practice. As the aviation industry puts increasing emphasis on good CRM skills, a framework has been introduced to formally assess these skills in both the simulator and the aircraft environments. In Europe, assessment of non-technical skills forms part of operator proficiency checks, license proficiency checks and line checks. The Federal Aviation Authority in the USA has similar processes to assess non-technical skills. Although an operator can elect to use any framework for their non-technical skills assessment (provided it is acceptable to their national authority), a commonly used framework is the NOTECHS system.[5] NOTECHS divides non-technical skills up into four major domains, each with three or four subdomains. These are shown below:

1. Cooperation
   a. team building and maintaining
   b. considering others
   c. supporting others
   d. conflict solving
2. Leadership and managerial skills
   a. use of authority and assertiveness
   b. providing and maintaining standards
   c. planning and coordination
   d. workload management
3. Situation awareness
   a. awareness of aircraft systems
   b. awareness of external environment
   c. awareness of time
4. Decision making
   a. problem definition and diagnosis
   b. option generation
   c. risk assessment and option selection
   d. outcome review

Formal assessment of non-technical skills using any framework such as NOTECHS relies on the examiner detecting and recording specific behavioral markers that indicate the presence or absence of particular skills. A behavioral marker is a specific, observable behavior that demonstrates the presence of a particular non-technical skill or, in the case of an expected positive behavioral marker being absent or the presence of a negative behavioral marker, that the non-technical skill is deficient. For example, a crew member who actively seeks the input of other crew members when solving a technical problem will have demonstrated good option generation skills. On the other hand, the crew member who makes no effort or disregards the input of other crew members could be said to be lacking in option generation skills. A crew member cannot fail a simulator check or line check based solely on deficient non-technical skills unless there is an associated technical failure. For example, if a crew member elects to land at an inappropriate airport, that could be seen as a technical failure. If the decision to go to that particular airport was reached without discussion with other crew members or any sort of risk assessment, the technical failure is associated with failure of a particular non-technical skill, and this information allows for appropriate remedial training to be planned in order to correct the deficiency. The purpose of non-technical skill assessment is not to pass or fail people purely on their CRM abilities but to provide a better framework for understanding where people's strengths and weaknesses are in this regard and to allow weaknesses to be addressed when they may have an impact on flight safety. Based on feedback from examiners, it seems that the vast majority of technical failures during checks are the result of problems with non-technical skills.

In developing guidance for implementing and using the NOTECHS system, five principles were identified to ensure that the system is used fairly and reliably[6]:

1. Only observable behavior is to be assessed – It is not appropriate to make judgments on the crew member's personality or attitude. The aim is to be as objective as possible.
2. Technical and non-technical skills are associated – For non-technical skills to be rated as unacceptable, flight safety must be actually or potentially compromised.
3. Repetition is required – Repetition of an unacceptable behavior must be observed to conclude that there is a significant problem.
4. Acceptable or unacceptable rating is required – The result of the check should include a rating of whether the overall non-technical skills were acceptable or unacceptable.
5. Explanation is required – In the event that a skill is rated as unacceptable, the examiner must be able to defend this assessment using examples of negative behavioral markers and how these led to safety consequences.

To ensure maximum objectivity in assessing behavioral markers in the context of non-technical skills, studies were carried out to determine the validity of the framework. It was found that 80% of examiners were consistent in their ratings of non-technical skills and 88% were satisfied with the consistency of the method.[5]

Some topics covered in this book refer directly to one specific non-technical skill domain or subdomain, while others, such as information processing and communication, will have an impact on several domains. Reference will be made to the relevant non-technical skills that a chapter or section refers to.

Chapter key points

1. Human factors is about optimizing the components of a system – the humans, the hardware, and the procedures – in order to improve safety and efficiency.
2. The humans in the system (the "liveware") interact with machines and with other humans either directly or using procedures.
3. This book contains guidance about optimizing the human components, the procedures and the interfaces among components.
4. Crew Resource Management training is the way that most pilots are exposed to human factors.
5. A pilot's human factors (non-technical) skills are assessed during simulator checks and line checks.
6. These non-technical skills are assessed against some sort of framework such as NOTECHS.
7. Non-technical skills assessment relies on observing behavior and looking for positive and negative behavioral markers.

References

1. The American Experience: Meltdown at Three Mile Island. PBS; 1999. [TV Program].
2. Perrow C. *Normal accidents: Living with high risk technologies*. Princeton, NJ: Princeton University Press; 1999.

3. Trist E. *The evolution of socio-technical systems.* Toronto: Ontario Quality of Work Life Centre; 1981. Occasional paper 2.
4. Civil Aviation Authority. *CAP 719: Fundamental human factors concepts.* West Sussex, UK: CAA; 2002.
5. van Avermaete JAG. *NOTECHS: Non-technical skill evaluation in JAR-FCL.* Amsterdam: National Aerospace Laboratory NLR; 1998.
6. Flin R, O'Connor P, Crichton M. *Safety at the sharp end: A guide to nontechnical skills.* Farnham, UK: Ashgate; 2008.

Information processing

2

Chapter Contents

Introduction   12
2.1 Introduction to brain structure and function   13
   2.1.1 Types of neuron   17
2.2 Overview of information processing   18
2.3 Sensation   19
   2.3.1 Sensory thresholds   20
   2.3.2 Sensory habituation   22
   2.3.3 Somatogravic sensory illusions   24
   2.3.4 Strategies for dealing with sensory limitations   25
2.4 Attention   26
   2.4.1 Selective attention   27
         2.4.1.1 Strategies for selective attention   29
   2.4.2 Sustained attention (passive monitoring/vigilance)   30
         2.4.2.1 Strategies for sustained attention   32
   2.4.3 Divided attention and multitasking   32
         2.4.3.1 Strategies for divided attention   33
2.5 Perception   33
   2.5.1 Mental models   35
   2.5.2 Perceptual difficulties with sensory-induced spatial disorientation   38
   2.5.3 Strategies for dealing with perceptual limitations   41
2.6 Decision making   42
   2.6.1 The anatomy of decision making   43
         2.6.1.1 Goal module   43
         2.6.1.2 Imaginal (mental manipulation) module   43
         2.6.1.3 Production (pattern-matching) module   44
         2.6.1.4 Summary of the anatomy of decision making   45
   2.6.2 The two systems of human decision making   47
         2.6.2.1 System 1   47
         2.6.2.2 System 2   48
         2.6.2.3 Summary of Systems 1 and 2   49
         2.6.2.4 How Systems 1 and 2 interact and the role of workload   49
   2.6.3 Heuristics and biases   50
         2.6.3.1 Evolutionary origins of heuristics and biases   51
         2.6.3.2 Heuristics and biases in aviation   52
         2.6.3.3 Decision making and memory retrieval heuristics   52
         2.6.3.4 Social heuristics   55
   2.6.4 Strategies for decision making   59
         2.6.4.1 Managing high task-load   60
         2.6.4.2 Managing high time-pressure   60
         2.6.4.3 Managing problem underspecificity   61
         2.6.4.4 Managing System 1 effects   61
   2.6.5 An operational decision-making strategy: TDODAR   61

Practical Human Factors for Pilots. DOI: http://dx.doi.org/10.1016/B978-0-12-420244-3.00002-9

2.7 Response   65
2.8 Fear-potentiated startle and freezing   66
   2.8.1 Strategies for managing fear-potentiated startle and freezing   69
2.9 A note about the models used in this chapter   69
Chapter key points   70
Recommended reading   72
References   72

[The brain is] a machine for jumping to conclusions.

Daniel Kahneman[1]

Introduction

This chapter is the longest and most technically challenging one in the entire book. Information processing is the subject that causes the most confusion during Crew Resource Management (CRM) classes a lot of difficulties. CRM instructors tend to deal with information processing in one of two ways. The instructor either makes the effort to cover the subject comprehensively and risks losing the class because of the technical detail involved or, alternatively, glosses over the subject quickly and moves on to other topics which are deemed to be more practically relevant. The latter strategy, although understandable, means that the students miss out on learning about the most fundamental area of human factors.

Information processing is at center of what we do every day as human beings. Humans encounter many stimuli during the day (anything that triggers any of our senses) and come up with many responses (actions, inaction, words, etc.). The basic nature of information processing is that we encounter a stimulus and this leads to a response. If we look at this process more deeply, we can see that it involves the acquisition, processing and analysis of information and also the process of deciding on actions based on that analysis. From our first waking moment, our entire existence centers around information processing. Later on in this chapter, we will look at the several different stages of information processing and, importantly, the limitations of each of these stages together with examples of aviation incidents and accidents connected with failures at each stage. All the other subjects in this book are only important because of their effect on information processing.

Information processing covers some very large subjects that are normally covered separately in CRM training. This is a legacy of the psychology-based approach that is normally taken to the subject. The reason this chapter may appear more technical than information processing material that you may have encountered previously is that we are going to try and understand what is actually going on in the brain, rather than just labeling a phenomenon with the name of some theoretical psychological concept. By approaching information processing using neuroscience, it becomes a lot easier to link many of the major human factors subjects together. Rather than seeing these concepts as distinct, when we take a neuroscience-based approach it is immediately clear that these concepts need to be considered together. Once you have recognized this, you will have a deeper understanding of how the science of human factors fits together and, indeed, how pretty much everything is based around information processing. Trying to understand these concepts in isolation means missing out on many crucial inter-relationships that have a massive impact on the effectiveness of human performance.

To get the most out of this subject, it is necessary to have a basic understanding of how the brain works. I would encourage readers to devote some time to this particular area as it will form the basis of not just this chapter, but also several of the chapters that follow. Although it may seem quite daunting, there is absolutely no reason why, by the end of Section 2.1, which covers brain structure and function, the reader should not be able to achieve a basic understanding of how this incredible organ works. Everything that you are and everything that you will ever do is down to this three-pound mass of tissue between your ears. It is well worth understanding how it works.

2.1   Introduction to brain structure and function

The human brain is composed of brain cells. These cells are called neurons and there are approximately 86 billion of them in the human brain. Before we look in more detail at how these neurons work, let's consider the brain as a whole. The human brain is broadly divided into functional areas. Figure 2.1 shows the brain as seen from the left side and Figure 2.2 shows the midline structures of the brain (as if the brain has been cut lengthwise down the middle).

The upper part of the brain, the cerebral cortex, is highly developed in most mammals but may be considerably smaller or even absent in other types of animal life. In most mammalian brains, the cerebral cortex is divided into two halves, known as hemispheres and is further divided into lobes. The left and the right hemispheres are linked and communicate through a bridge of nerve tissue called the corpus callosum.

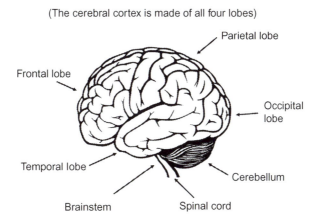

Figure 2.1 Brain anatomy as seen from the left side.
Image copyright takito (www.shutterstock.com).

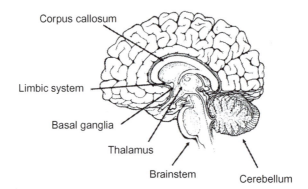

Figure 2.2 Midline brain anatomy.
Image copyright Denis Barbulat (www.shutterstock.com).

Most animals with a nervous system have part of the brain dedicated to regulating basic physical functions; for example, heartbeat, respiration rate and sleep–wake cycles. Higher mammals such as human beings went on to develop the more complex cerebral cortex in order to achieve more advanced functions; for example, more advanced forms of communication, imagination and creativity, the ability to plan and execute complex behaviors (known as executive function) and the ability to carry out all manner of social interactions. Explanations for the major areas shown in Figures 2.1 and 2.2 are given below.

- The brainstem, located towards the rear of the brain and leading out into the spinal cord, is the part that we share with most forms of animal life. It is this part of the brain that regulates our most basic physical functions such as heart rhythm and breathing. Damage to this area is usually fatal as it disrupts these essential functions.
- The thalamus, located at the top of the brainstem and surrounded by the cerebral cortex, acts as the junction box for the brain. Because the brain is divided into several different functional areas, the various inputs to these areas must be routed through the thalamus and their outputs integrated as part of information processing.
- The basal ganglia, a collection of brain structures located towards the center of the brain, is thought to act as the central decision maker; that is, it will choose responses from several possible options. It also has a role in learning procedural motor skills.
- The cerebellum, located towards the rear of the brain but separate from both the cerebral cortex and the brainstem, is more highly developed in mammals that possess fine motor skills. This part of the brain is involved in adjusting our motor outputs (e.g. hand and finger movements) to achieve a high level of dexterity.
- The cerebral cortex is more highly developed in humans than in other mammals. It is the largest part of the brain and helps us to achieve all the functions that distinguish human beings from other animals. Other mammals do possess a cerebral cortex and the size and organization of it dictates the intellectual capabilities of the animal. It is divided into four main regions, known as lobes:
  - The occipital lobe, located at the rear of the brain, mainly processes visual information. As the vast majority of the sensory information we receive is visual, a correspondingly large area of the brain is devoted to processing it.
  - The temporal lobe, located on each side of the brain (remember, the brain is divided into two hemispheres and so there is a temporal lobe on both the left and the right sides), is

involved in processing auditory information, particularly speech. It also has a crucial role in the formation and storage of memories.

- The parietal lobe, located at the top of the brain just forward of the occipital lobe, processes sensory information coming from the body and combines this with information from other brain regions. For example, it would combine the sensation of a cold object in the palm of your hand with the visual information derived from the occipital lobe to conclude that the object is probably an ice cube.

- The frontal lobe, as the name suggests, is located at the front of the brain and has a variety of functions. A section of the frontal lobe sends instructions to the muscles in the body to allow them to move in a coordinated manner. The frontal lobe is also responsible for executive function, which is to do with planning complex behavior as well as interpreting facial expressions and emotions in other people. It also has a role in moderating human behavior based on socially acceptable norms. Damage to this area can result in loss of inhibitions, a failure to interpret the emotions of others and an inability to plan ahead.

- The spinal cord – This is the main route for information coming into the brain and instructions flowing out. It connects the brain with the rest of the body and can be seen as the main communication network that allows information to be gathered and instructions to be distributed.

- The cranial nerves – Although the majority of the information comes into the brain through the spinal cord there are 12 nerves that feed directly into the brain. The main ones that we will consider later on in this chapter are the optic nerves (visual information) and the vestibulocochlear nerves (both auditory and balance information).

- The limbic system – This, again, is one of the brain areas that we share with many other animals. It is responsible for processing emotion and is closely associated with the thalamus and the cerebral cortex.

Now that we have covered the basic function and location of the main brain regions we need to go a step further and attempt to understand how these regions actually *achieve* their functions. In order to do that, we need to understand the basic building block of the human brain: the neuron.

A neuron is a cell within the brain. Think of the neuron like a wire insofar as information passes from one end to the other. Sometimes the distances can be tiny, perhaps a fraction of a millimeter, and sometimes the distances can be large, perhaps up to a meter. Despite their small size, if all the neurons in your brain were laid end to end, they would reach 180,000 km, enough to circle the Earth four times. Unlike most wires that we are familiar with in our homes, a single neuron may have up to 10,000 links with other neurons. These links are called synapses and although the adult brain has about 86 billion neurons, the huge number of links between neurons means there are about 500 trillion synapses. Figure 2.3 shows a diagrammatic representation of a network of neurons and an enlarged diagram of a synapse showing the release of neurotransmitters, the chemicals that enable neurons to communicate with each other.

When people are told that the complexity of the human brain exceeds even the most complex computer ever created, they often find it difficult to understand why humans are not capable of performing billions of calculations per second like a computer can. To understand the reason for this, we need to look at how information is passed along these different types of "wires". Although we learned at school that the brain is an electrical organ, this is not entirely correct. If it were correct,

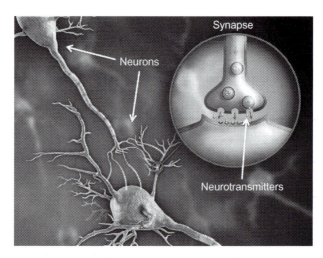

Figure 2.3 Neurons, a synapse and neurotransmitter release.
Image copyright Andrea Danti (www.shutterstock.com).

we would be able to exceed the capabilities of even the most advanced computer by virtue of the fact that brains are wired up far more intricately than even the most modern supercomputers. The difference between the wiring in a computer and the neurons in our brain is that information transmitted through computer wires is *electrical* in nature (traveling at between 66% and 97% of the speed of light, depending on the nature of the wire) and the information transmitted through neurons is *electrochemical*.

Electrochemical transmission in a neuron involves positively and negatively charged chemicals moving in and out of the surface of the neuron. As this happens in one section of the neuron, it triggers the movement of charged chemicals in the next section. In this way, if a neuron is activated, the impulse will travel from the point of activation throughout the neuron and to all the points where it might connect with other neurons. This activation of the neuron is known as "firing". When a neuron fires, the impulse will reach a synapse (the area where it connects with other neurons) and the activated neuron will release chemicals known as neurotransmitters into the tiny space between the two neurons as shown in Figure 2.3. These neurotransmitters will then have an effect on the second neuron and may cause it to activate as well. The speed of electrochemical transmission is a lot slower than that of electrical transmission. If we consider that electrical transmission in a normal copper wire occurs at about 97% of the speed of light (about 291,000,000 m/s), the equivalent speed in an average neuron is only 100 m/s. That is only 0.000034% of the speed in an electrical wire. The ability of the brain to do all the amazing things it does is by virtue of the sheer *number* of connections between neurons (synapses), a number estimated to be about 500 trillion. The brain has roughly 120,000 times more connections than the most advanced microchip that is commercially available.

We can summarize this contrast between complexity and speed with the following two facts:

• Because each neuron may have up to 10,000 connections with other neurons, far more interconnectedness than even the most advanced microchips, the complexity of the wiring of the human brain is about 120,000 times greater than that of the most advanced microchip that is commercially available.
• However, the speed of transmission in the human brain is only 0.000034% of that within a computer.

This means that although the human brain is capable of achieving far more complex tasks than a computer, a computer can perform many more simple tasks within a short space of time.

2.1.1 Types of neuron

There are several different types of neuron that allow the brain to carry out its functions. Sensory neurons bring information into the brain and motor neurons carry instructions from the brain to all the muscles of the body. Surprisingly, only a very small percentage of the 86 billion neurons in the human brain are sensory or motor neurons. Between 95% and 99% of the neurons in the human brain are of the type known as interneurons. As the name suggests, these neurons form the links between other neurons and so are the basis for the highly complex networks that make the brain capable of carrying out complex functions.

The sheer number of interneurons in the brain means that it can be wired in a vast number of different ways. To understand how this complexity allows us to learn new skills, make sense of complex, rapidly changing situations and decide on courses of action, we need to understand how the human brain learns. Consider the following:

1. A single interneuron may be connected to as many as 10,000 other neurons.
2. Imagine a child learning to read. As you have learned already, visual information is processed in the occipital lobe.
3. When a child first encounters letters, the brain has no stored information that allows sense to be made of these letters. All they are to the child is a new visual stimulus.
4. When first learning how to recognize the letters of the alphabet, the child will initially be presented with an unfamiliar symbol.
5. As the visual information is processed in the occipital lobe, the brain will identify two sloping lines that join at the top as well as a horizontal line linking them midway along their length. This, unsurprisingly, is the letter "A".
6. If you have ever seen a young child learning the letters of the alphabet, you will know that this does not tend to happen in silence. It is usual for parents or teachers to sound out the letters as they are presented to the child. The associative nature of the brain means that the child quickly learns that this arrangement of slanted and horizontal lines is associated with a sound – "ah".
7. Through repetition, this knowledge becomes deeply embedded in the child's memory. The symbol presented to them is no longer just an arrangement of lines but now represents the first letter of the alphabet and is associated with the "ah" sound.

8. Once the child learns to recognize such a letter, he no longer needs to analyze the arrangement of the lines every subsequent time because the neurons that are activated in the occipital lobe when the letter "A" is presented are now accustomed to being activated at the same time. This is the crucial feature that allows human beings to learn. If multiple neurons are activated at the same time it becomes easier and quicker to activate them in the future. As we said earlier, neuron activation is also known as "firing". When it was discovered that it becomes increasingly easy to activate multiple neurons if they are frequently being activated at the same time, a renowned neuropsychologist summarized this concept with the phrase "cells that fire together, wire together".[2]

9. This very simple idea, when applied to different functions in the brain, explains how human information processing occurs. Once the child recognizes the arrangement of lines as the letter "A", he will soon associate this arrangement of lines with a corresponding sound. Once the child has learned to recognize more letters and associate them with different sounds he will soon associate the lines going to form the letters of "A P P L E" as corresponding to the sound of that particular word as well as to a fruit that can be eaten.

10. As adults, we do not need to analyze the lines that form the letters of a particular word because we are capable of recognizing the word as a whole. Not only that, but from our experiences, if we were to see the phrase "apple pie" we would not think of two separate objects: an apple and a pie. We have learned that these words in combination refer to a specific item of food, namely a pie made from apples.

11. It goes further though. If we are used to having apple pies at home, chances are that we are also used to the smell of an apple pie cooking in the oven. If we should arrive home and smell that particular odor as we walk through the front door, the neurons that detect smell will recognize and remember the association between the odor and the apple pie. All this has been learned through association. We associate the lines with the letters, the letters with the sounds, the letters and the sounds with the word, the words with the phrase, the phrase with the object, the smell with the object, and so on. Association is key to how the brain learns and also, as we shall see later, how the brain thinks. One of the best examples of this is when you are trying to recall a song or a poem, but are struggling. Once someone tells you the first line, all of a sudden there is a cascade of neural network activation that summons up the next line. That line then triggers the following line, and so on.

2.2 Overview of information processing

Let's consider how human beings interact with the world around them. At its most basic level, we sense what is going on around us (stimuli) and then we respond to them. What happens in between sensation and response is known as information processing. To complete the picture, we tend to include sensation and response as part of information processing although, in reality, they are the inputs and outputs of the process.

The model of human information processing that we are going to look at has five stages:

1. sensation
2. attention
3. perception
4. decision making
5. response.

We are going to look at each one of the stages in turn, consider what is going on in the human brain during each of them, and look at what can go wrong. Most importantly, we will look at what practical steps we can take to stop the limitations of each of these processes affecting the safety and efficiency of our day-to-day performance. At the risk of being repetitive, these five processes are what allow us to interact with the world. Every single type of unsafe act that has led to any accident in the history of aviation can be traced back to a problem at one of these stages. This entire book is aimed at optimizing each one of these stages to improve the integrity of our information processing capabilities.

This process of learning by association as laid out in Section 2.1.1 has a role in attention, perception and, especially, decision making. The neural mechanisms of attention and perception are complex and are beyond the scope of this book. However, the neural basis of decision making is extremely important and this knowledge about how neurons connect together is crucial to understand how humans make decisions based on cognitive processes known as "production rules". This then forms the basis of our understanding of human error. In summary, although this knowledge will not be directly applied until we cover decision making, it is important that you fully understand the concept of learning by association before you move on.

2.3   Sensation

Ask anyone in the street how many senses we have and 95% of them will give the same answer: five. There is, in fact, a sixth sense: balance (the sensation of how our body is oriented in space) so, in total, we have six sensory modalities:

*   sight
*   sound
*   touch
*   smell
*   taste
*   balance.

These six senses are our only means of receiving information from the world around us. Sight is clearly the most important sense and the one we rely on most to receive information about the world around us. It can sometimes override seemingly contradictory information coming from other senses and this can be advantageous or disadvantageous depending on the situation. Conflicting sensory information is usually resolved at the perception stage and this will be considered in Section 2.5.

We like to think that our senses give an accurate picture of our surroundings. Unfortunately, this is not always the case. The human sensory system is intrinsically limited by biology. Other animals may be more sensitive to stimuli than we are (e.g. a dog's sense of smell is far more sensitive than a human's) and some animals can detect certain types of stimuli that we simply do not have the ability to detect (e.g. a kestrel can see ultraviolet light and uses this to track its prey by the urine trails it leaves behind on the ground). We know that only a small spectrum of radiation is visible to us. We cannot detect infrared light, X-rays, microwaves or many of the other wavelengths that surround us and that which we know exist because we have developed scientific instruments that can detect them.

2.3.1 Sensory thresholds

This idea of sensory sensitivity is important. Many readers of this book will have had the unfortunate experience of being bitten by a mosquito. More often than not, we only know that we have been bitten after it has already happened. Our first sign may be a red, itchy lump at the site of the bite. Why did we not feel the mosquito land on us and, so, be able to flick it off or squash it before it had the chance to bite us? Fortunately for the mosquito (but unfortunately for us), nature has developed a way to allow the mosquito to bite animals without being detected: a mosquito is so light that we are not sensitive enough to detect it when it lands on us. If a mosquito were the size of a hummingbird then we would always feel it landing and be able to do something before it bit us. From this example, we can see that our sense of touch has a threshold below which a stimulus will not be detected. By this we mean that in order for a sensory stimulus to be detected it has to be of a great enough magnitude to exceed our sensory threshold.

Through various experiments, scientists have worked out the sensory thresholds for all of our six senses, thresholds below which we would not be able to detect the stimulus at all:[3]

- Sight – the flame of a match from 50 km away on a pitch black night. The eye can be 10,000 times more sensitive to light once it has become accustomed to the dark. This was why convoy ships had to travel without any lights and with blacked out windows at night during the Second World War. Any submarines scanning the horizon, especially if the submariner looking through the periscope had allowed his eyes to become accustomed to the dark, could spot a flame, such as would be used to light a cigarette by a careless sailor on deck, from a great distance. If the ship was any more than 50 km away, the submariner would not be able to detect the flame.
- Sound – the tick of a watch from 6 m away in a quiet room.
- Touch – the wing of a bee falling on our cheek from a height of 1 cm (there is a huge variations in skin sensitivity depending on which body part it covers. For example, the skin on your fingertips is a lot more sensitive than the skin on your elbow).
- Smell – one drop of perfume diffused throughout a three-room apartment.
- Taste – 1 g of salt dissolved in 500 liters of water.
- Balance – the average human being can detect a roll rate of 2 degrees per second.[4]

If we took an average person and assumed that all of these sensory thresholds were accurate, it would mean that that person would be able to tell the difference between a glass of water taken from the 500 liters that had had 1 g of salt dissolved in it and a glass taken from a vat of 501 liters that it had 1 g of salt dissolved in it. The first would taste ever so slightly salty and the second would not. They would also be able to hear the watch ticking from 6 m away but would not be able to hear it from 6.1 m away. In this way we can see that sensory thresholds denote the "all-or-nothing" nature of detecting stimuli. If we take it back to the level of the neuron, if the stimulus is below the sensory threshold then the neuron will not fire but if it is above the sensory threshold then it will fire. The number and intensity of the neurons that fire will give an indication of the intensity of the stimulus.

We now come to one of the confounding difficulties of the science of human factors: we are all different. The previous list gives sensory thresholds for the average person. However, there are certain individual factors that may make a person more or less sensitive to stimuli than the average person. An obvious example is someone who has worked in a noisy environment for many years and is suffering from noise-induced hearing loss. There is a high chance that their sensory threshold for sound will be considerably higher than someone who has worked in a quiet environment and doesn't have noise-induced hearing loss. The person who has worked in the noisy environment may only be able to detect the tick of the watch from 1 m away.

If we ignore factors such as illnesses or specific injuries that affect our sensory threshold, age is probably the factor that affects our sensory thresholds the most. For example, the average person can detect a roll rate of 2 degrees per second. However, the possible range of this sensory threshold when adjusted for a variety of ages is from 0.2 to 8 degrees per second. Before we go on to consider how these facts affect us in aviation, let's look at an accident that was related to sensory thresholds (see Box 2.1).

The nature of the technical failure meant that two out of the three instruments that the pilots had available to them to indicate the aircraft attitude were not functioning. When flying the aircraft without an autopilot in conditions where it is not possible to see out of the window, these instruments are essential in helping the pilot to maintain his orientation. The pilots of this flight may have been attempting to ascertain why the autopilot disconnected and why they had lost some of their instrumentation. It is unfortunate that when the autopilot disconnected, the aircraft entered a gentle roll to the right instead of a roll that was more pronounced. If, for example, when the autopilot disconnected the aircraft had jerked violently to the right, it is likely that one of the pilots would have noticed this, looked at the standby attitude indicator and corrected the roll. In the event, it appears that the roll was below both pilots' sensory thresholds and so it was not detected until it was too late.

Box 2.1 Accident Case Study – Sensory Thresholds

Adam Air Flight 574 – Boeing 737-400
Polewali, Indonesia, 1 January 2007[5]

While flying in bad weather between two cities in Indonesia, the pilots of this Boeing 737-400 noticed a problem with the navigation system. During the subsequent troubleshooting, one of their actions caused the autopilot to disconnect. After the autopilot disconnected, the aircraft began a very gentle roll to the right. While both pilots were attempting to resolve the technical problem, it appears that neither of them detected that the aircraft was in a roll. The roll was not corrected and the aircraft reached a bank angle of 100 degrees with 60 degree pitch nose down. The aircraft entered a dive, exceeded its maximum speed and underwent major structural failure before crashing into the sea, killing 102 people.

Roll rates that occur below our sensory threshold can lead to spatial disorientation. This is defined by the Flight Safety Foundation as:

> when a pilot fails to properly sense the aircraft's motion, position or attitude relative to the horizon and the Earth's surface. Spatial disorientation can happen to any pilot at any time, regardless of his or her flying experience, and often is associated with fatigue, distraction, highly demanding cognitive tasks and/or degraded visual conditions.[6]

It is important to note that this definition does not require the pilot to be conscious of any feelings of disorientation. A pilot can be spatially disoriented without recognizing that anything is wrong. Their reaction when they do recognize that something is wrong is considered in Section 2.5. There are other sensory limitations that can lead a pilot to be unable to detect the actual movement of an aircraft, another major one being sensory habituation.

2.3.2   Sensory habituation

As well as the problems of sensory thresholds, our senses are limited by the phenomenon of habituation. To illustrate this, all you need is a coin. Place the coin in the palm of your hand and close your eyes. In a few seconds you will no longer be able to detect the pressure that the coin is exerting on your palm. This is sensory habituation. When a stimulus is continuous and remains unchanged in nature, our body does not continue to process that stimulus. Perhaps you have had to try and sleep in a room which is subject to a lot of traffic noise. If the traffic noise is relatively constant, after a short period you do not really notice it any more. However, if an emergency vehicle were to pass by with its siren on, you would once again be aware of the traffic noise. In the first instance, because the volume, intensity and regularity of the sound are fairly constant, your brain is able to filter the sound out. When the stimulus changes, your attention is drawn back to it. Sensory habituation allows us to filter out continuous, relatively unchanging stimuli so that they do not constantly occupy our attention. For example, when getting dressed, we may initially sense the texture and temperature of the fabric of our clothes but once we have been dressed for a some time, we are no longer aware of these sensations despite the fact that approximately 90% of our skin's surface is being constantly stimulated by our clothing. The body accomplishes this by altering the way that neurons work when subjected to constant stimulation. As we described earlier, a neuron "fires" (is activated) when it is stimulated to a sufficient extent that electrochemical changes occur along the surface of the neuron in a cascading way that activates the rest of the neuron and this may stimulate other neurons that are connected to it. There is normally a very small latency period while the electrochemicals move back to their original locations before the neuron can fire again. If a neuron is being constantly stimulated, it progressively decreases the frequency and intensity of its firing by means of a process known as sensory adaptation.[7] In this way, a hot bath may feel very hot when we get into it, but eventually, the thermoreceptors in our skin will adapt and it will not feel so hot. This sensory adaptation could be described as "getting used to something". Unfortunately, there are some stimuli that we do not want to get used to because their persisting nature is telling us something very important about how we are moving relative to the Earth's surface.

In the same way as with sensory thresholds, it is our sense of balance that is most susceptible to sensory habituation. If an aircraft is at a constant angle of bank for 20 seconds or more, the body will get used to this new position because the neurons that are transmitting this information begin to adapt to the constant stimulation and so give progressively weaker signals to the brain. Once this adaptation has occurred, we have become habituated to this new orientation and it will feel like we are not in a banked turn at all. When we then go to roll the aircraft wings level, it will feel as if it is now banked in the opposite direction. This can be illustrated with the following example. An aircraft is performing a climbing turn with a bank angle of 20 degrees. The aircraft is in the turn for approximately 30 seconds, long enough for habituation to occur to the pilot's sense of balance. Although the aircraft remains in the banked turn, for the pilot the sensation will be that the aircraft is flying wings level. Under normal circumstances, this feeling can be counteracted by visual reference to the instruments or, if conditions allow, to the outside world. Our visual sense is the most powerful and can override other senses when there is any sort of conflicting information. In the event of an instrumentation failure or a sudden, unexpected decrease in the conditions outside, the pilot's instinct may be to rely on his sense of balance to orient himself in space.

If, once sensory habituation has occurred, the pilot wishes to roll the wings level, decreasing the angle of bank will give the impression that the aircraft is now moving from the wings level position towards a bank in the opposite direction. Without any other references, the pilot will now feel that he is going from the wings level position into an uncommanded left bank. To correct this, the pilot will then adjust the control surfaces to bank the aircraft to the right in order to regain the imaginary wings level position. Unfortunately, the effect of this is simply to increase the angle of bank to the right, potentially to the stage where the aircraft enters a spiral dive. Instrument-rated pilots are trained to rely on their instruments but, in cases where spatial disorientation occurs because of sensory habituation, the situation is made far more difficult when the output of these instruments is in doubt (Box 2.2).

Other effects associated with sensory habituation are[6]:

- The leans – Once habituation has occurred in a banked turn, after the wings have been rolled level, a pilot flying on instruments will be consciously aware that the aircraft is now in level flight. However, the act of rolling level gives a secondary impression that the aircraft is banked in the opposite direction. The pilot is able to avoid severe spatial disorientation because the visual information presented on the artificial horizon confirms level flight. In some cases, this powerful visual information may not be enough to completely override the erroneous sensory input from the pilot's balance-sensing system and so in order to resolve this conflict, the pilot moves his head or entire body to try and orient it with the imagined vertical axis. For an observer, this will have the appearance of the pilot leaning in his seat. This highly disconcerting effect can last for 30 minutes, although there is some suggestion that a quick shake of the head can reset the correct datum for balance. Unfortunately, a shake of the head may cause the second effect that we need to consider: coriolis.
- Coriolis – Even when flying on instruments, if sensory habituation has occurred, any movement of the head (e.g. to look up or down in order to make switch selections) can give confusing sensory data to the brain and produce the illusion of rolling or yawing.

Box 2.2 Accident Case Study – Sensory Habituation

Air India Flight 855 – Boeing 747-200
Arabian Sea near Mumbai, India, 1 January 1978[8]
Shortly after take-off, the captain, who was the pilot flying, reported a possible problem with his attitude director indicator (ADI). The aircraft was in a climbing turn to the right, but when the captain rolled with wings level, his ADI still showed that the aircraft was in a right bank. The departure routing had taken them out over the Arabian Sea and it was night-time, meaning that there were no external visual references. The first officer did not seem to note the captain's concern. The captain seems to have reacted to the ADI and attempted to correct the indicated right bank by commanding left bank. Unfortunately, the wings were already level and so the aircraft entered a bank to the left. This roll continued until the aircraft reached 40 degrees angle of bank to the left. The flight engineer tried to alert the captain to a disparity in the ADIs but the aircraft continued to roll and reached a left bank angle of 108 degrees, descended from its low altitude and crashed into shallow water 3 km offshore, killing all 213 people on board.

2.3.3   Somatogravic sensory illusions

Sensory illusions, unlike sensory thresholds and sensory habituation, are where the biological sense mechanisms that we use to maintain our orientation give incorrect information to our brains rather than no information (in the case of sensory thresholds) or diminished information (in the case of sensory habituation). The explanation for why these sensory illusions happen is very simple: man did not evolve to fly aeroplanes. Our sensory organs evolved to support an organism that stays relatively close to the ground, can reach a maximum speed of about 44 km/h (achieved by Usain Bolt during the 2009 International Association of Athletic Federations World Championships in Berlin) with a maximum acceleration of 9.6 m/s^2 or about $0.98\,g$ (calculated based on Bolt's record-breaking performance). We have now taken those sensory organs and put them into vehicles that can achieve speeds 3530 km/h (achieved by the SR-71 Blackbird in 1976) with a maximum acceleration of 88 m/s^2 or about $9\,g$ (the limit that military pilots with protective suits can endure without loss of consciousness). Even the lowest performing aircraft can achieve speeds and accelerations far in excess of those that our sense organs have evolved to deal with. This mismatch between what our bodies have evolved to do in the 7 million years since humans have existed and what we are now forcing them to do using technology that has only been in existence for the last 100 years leads to many of the information processing problems that this chapter explores.

Because aircraft can accelerate and decelerate at rates beyond those which our senses have evolved to deal with, our bodies interpret these sensations in an incorrect way leading to somatogravic illusions. For example, rapid acceleration can give the illusion of climbing as fluid in the inner ear is pushed backwards and stimulates neurons that would normally be stimulated by tilting the head upwards or orienting the entire body upwards. The reaction of the pilot to this seemingly uncommanded pitch-up can be to reduce the pitch, a situation that could be extremely serious if low to the ground. Deceleration has the opposite effect as fluid moves forward in the inner ear and stimulates the neurons that would normally only be activated by tilting the head or the body downwards. The response of the pilot would be to pitch-up, a response that could be fatal in low-speed states. The bigger the magnitude of the acceleration or deceleration, the stronger the effect.

2.3.4 Strategies for dealing with sensory limitations

The three limitations we have discussed are sensory thresholds, sensory habituation and somatogravic illusions. As you might expect, the best strategy for preventing these sensory limitations affecting the safety of our operation involves a reliance on flight deck instrumentation. We have demonstrated and seen through examples how our senses can be tricked and there is very little we can do to change this aspect of our biology. Fortunately for us, aircraft designers have provided us with highly sensitive "technical sensory organs" such as ring laser gyroscopes. These can provide us with visual information via the various instruments in the flight deck that will inform us about changes in our speed and attitude that we may be too insensitive to detect ourselves. The instruments should be considered not just as tools to use when flying in instrument conditions, but rather as highly sensitive sensory organs that give us far more information about the world than we are capable of detecting ourselves. With this in mind, here are some specific strategies for dealing with our sensory limitations:

- Manual flight in conditions where there are no external visual references must be carried out with sole reference to the instruments. In the event that the output of these instruments is in doubt, or when there is any doubt about aircraft attitude, the crew's first action should be to cross-reference the available instrumentation and decide which is most likely to be giving accurate information. It may be necessary to hand over control at this point.
- In certain scenarios, it may be possible that a pilot flying with reference to his own attitude indicator may be unaware that there is a technical malfunction with that attitude indicator. Through a combination of sensory habituation and roll rates below the pilot's sensory threshold, it is possible that he may detect nothing abnormal and will feel that the aircraft is correctly oriented in space. The only warning may be an alert from another crew member suggesting that the aircraft is in an abnormal attitude. The difficulty here for the pilot flying the aircraft is that for him, the situation feels normal even though the aircraft may be at a completely different attitude from that indicated on the instrumentation. To prevent this situation deteriorating to the point where the pilots lose control of the aircraft, any suggestion from any crew member that there is any doubt as to the aircraft's attitude must be followed up with cross-checks of all instruments to ascertain whether a technical failure has occurred and, if so, which instruments can be trusted.
- If it is impossible to determine which, if any, instrument is giving an accurate reading, and the aircraft is flying in instrument meteorological conditions, this is an extremely serious situation. The strong temptation for the pilot will be to rely on his senses to keep the aircraft stable. Unfortunately, this is almost guaranteed to fail because of our sensory thresholds and the effects of sensory habituation. Depending on the stability characteristics of the aircraft

being flown, it may be safer to avoid making any control inputs and rely on the natural stability of the aircraft until such time as a reliable source of attitude information can be identified or the aircraft leaves instrument meteorological conditions.

- In the event of expected or unexpected autopilot disconnection, the designated handling pilot must now concentrate solely on flying the aircraft with reference to the flight deck instruments. The monitoring pilot must do everything possible to preserve the handling pilot's focus on the instrumentation. This may involve taking on more tasks than normal to allow the handling pilot to focus solely on flying the aircraft with reference to the instruments.

- During manual flight, it may be necessary for the handling pilot to participate in cross-checking switch selections or checklist items that may be actioned by the monitoring pilot. This may require the handling pilot to look away from the flight deck instruments to carry out these tasks. The time looking away from the instruments should be minimized whenever possible and, in cases where several switch selections must be confirmed at the same time, the pilots should return his attention to the flight deck instruments in between each selection or checklist item. Given that the range of sensory thresholds for balance between individuals is from 0.2 to 8 degrees per second depending on age, if the handling pilot is approaching retirement age, looking away from the instruments for 4 seconds may result in the aircraft undetectably rolling to a bank angle of 32 degrees. As you can see, it would only take a few seconds more for the aircraft to reach an unsafe bank angle. This risk is particularly pronounced in the event of two older pilots flying together.

- Visual flying should only be performed in full visual conditions. When transitioning from visual to instrument flight, there will come a point when the visual cues start to degrade. The temptation is to continue flying visually while there remain visual cues that may seem to be reliable. As we will see later on in this chapter when we discuss perceptual illusions, this poses a significant risk. This, in combination with the limitations caused by sensory thresholds and sensory habituation, means that as soon as there is any degradation in visual conditions outside, flight should be continued solely with reference to the instruments.

- Other people may have more finely tuned senses – a younger member of the cabin crew may be able to detect sounds or odors that an older pilot cannot.

- Just because you cannot sense something, does not mean something is not happening – trust your instruments and any alarms that they trigger. Hardware is usually far more sensitive than the liveware.

- If experiencing the leans, a shake of the head may resolve the situation.

2.4 Attention

The brain receives a huge amount of information through the senses. It has been calculated that the human brain receives about 120 megabytes per minute of information from the eyes alone.[9] This is the equivalent of having an ethernet cable plugged into your head. That volume of data is too great to be able to process and so humans have evolved a capacity to filter out data that are less important in order to be able to focus on data that are more important. This is known as attention.

There are three types of attention that we will look at:

- selective attention
- sustained attention
- divided attention.

Attention is a controversial topic and is closely associated with strategies for automation management. There is currently a drive in the aviation industry for pilots to become better at monitoring. Accident reports frequently cite a lack of monitoring as being a causal factor. As we will briefly see here, but will look at in more detail in Chapter 9 (Automation Management), this is a far more complex problem that it initially seems. Before we go into the neuroscience of attention, its limitations and how these limitations can be overcome, it is worth considering the language we are going to use. Aside from the three types of attention listed already, this paragraph also includes the word "monitoring". Although this word seems to be used as a catch-all for any task that requires attention, we need to break it down further so that we can understand it in the context of what follows. Below are descriptions of two tasks defined according to the attentional system in use and how one would describe them in terms of monitoring:

- Cross-checking a colleague making a switch selection – This requires selective attention; that is, there is a defined stimulus (the actions of your colleague) that requires you to direct your attention towards it to provide short-term supervision. This could also be called "active monitoring".
- Observing an automated system to ensure that it continues to function correctly – This requires sustained attention because the stimulus that is being attended to is relatively unchanging. The reason for sustaining the attention is to be able to detect any abnormal occurrence even though this may never occur. This could also be called "passive monitoring" or "vigilance".

Although every type of active monitoring requires selective attention, not every case of selective attention could be categorized as active monitoring. In this way, active monitoring is an example of our ability to selectively attend to a stimulus, rather than being a term that is interchangeable with that of selective attention. On the other hand, sustained attention and passive monitoring are interchangeable terms. However, for clarity, "attention" will be used in preference to "monitoring".

2.4.1 Selective attention

Selective attention is one of our key attributes that allows us to function in a complex, dynamic and highly variable world. This ability is used whenever we need to focus on one particular part of our environment. The advantage of having a mechanism that allows us to focus on, for example, a single technical problem is that it allows us to dedicate a substantial portion of our brain's processing power to solving the problem. As we saw earlier in the chapter, what the brain lacks in speed it makes up for in complexity and capability. Human beings are excellent problem solvers compared to computers and selective attention is the faculty that allows us to do this.

As with just about any other evolved capability, selective attention also has its drawbacks. One of these is illustrated in Box 2.3.

Why did the crew not notice the altitude deviation warning chime? This chime had been designed to get the crew's attention but failed in this particular case. The evidence suggests that all three crew members were completely focused on addressing

---

Box 2.3 Accident Case Study – Selective Attention

Eastern Airlines Flight 401 – Lockheed L-1011 TriStar
Florida Everglades, USA, 29 December 1972[10]
On approach to Miami International Airport, the indicator light that signified that the nose gear was fully down and locked did not illuminate. The crew discontinued the approach, climbed to 2000 feet (600 m) and entered a holding pattern over the Florida Everglades. All three crew members in the flight deck became preoccupied with trying to solve the technical problem, namely ascertaining whether this was a problem with the indicator light or a problem with the landing gear itself. Even though the autopilot was engaged, an inadvertent action by the captain led the aircraft to start a slow descent. As with most commercial aircraft, when an aircraft leaves its preselected altitude a warning chime sounds in the flight deck to alert the crew to this fact. In this case, the warning chime sounded but it appears that this went unnoticed by all three crew members. The plane continued descending until it crashed into the Everglades, killing 101 people.

---

the technical problem that was delaying their landing. The cockpit voice recorder indicates that the captain, who was the designated handling pilot, was leading the efforts to solve the technical problem. Like many other cognitive abilities, selective attention comes at a price and one of the disadvantages is that once selective attention is brought to bear on a particular element of the environment, there can be significant changes in the rest of the environment that go unnoticed. This is what is known as change blindness (or inattentional blindness).

Change blindness is a phenomenon where we don't notice large changes in our visual environment often because our selective attention is focused somewhere else. A classic example of this, which is often shown to students during CRM courses, is a video of basketball players. There are two teams of players, one team dressed in black and one team dressed in white. Each team has a basketball and the people watching the video are asked to count how many passes one of the teams makes between its members. In this scenario, the visual environment is very dynamic; the players move around, the ball is passed in a variety of directions and the team members weave it in and out between one another. Even a sophisticated computer program would have difficulty completing this task but, for our highly evolved brains capable of focusing selective attention on a small portion of the environment, most people are able to track the number of passes with a reasonable degree of accuracy. But there's a catch! Shortly after the video starts, something unexpected appears among the basketball players. In some cases it is someone dressed as a gorilla or, perhaps, someone dressed as a bear "moon-walking" from one side of the screen to the other. Once people have given their answers as to the number of passes that a particular team has made, they are asked if they noticed anything else unusual about the video. The vast majority of people do not notice the unexpected appearance of the gorilla or bear, even though it was clearly presented within their field of vision.[11]

In another experiment carried out by the same team, students and teachers at a university were asked to participate in an experiment. They entered a room and approached the desk, where an experimenter asked them to fill in a consent form. Once they had signed a form, the experimenter stooped down behind the desk as if to collect an information pack. Unbeknownst to the participant, once the experimenter had stooped down behind the desk, he remained there and a different experimenter stood up and handed the information pack to the participant. The second experimenter was of the same gender and approximately the same age but had a different appearance and wore different clothes. In 75% of cases, the participants did not notice that the person who handed them the information pack was not the same person who stooped down behind the desk initially, even though the change occurred within a matter of seconds.[12]

When discussing change blindness with students in a CRM course, I find it useful to ask them how they would design a warning system for pilots such that there would be little or no chance of the pilot not noticing the warning. After all, the purpose of an alarm system is to get our attention. It is there to suggest that no matter what we are doing at the time there is something more urgent that requires us to intervene. Examples of this in the flight deck are master warning lights that alert pilots to certain technical problems, fire bells and ground proximity warning system (GPWS) alerts. GPWS alerts in particular are designed to get a pilot's attention. Among other things, they are there to alert the pilot that he may be in close proximity to the ground or a mountain and needs to change course immediately to avoid collision. It is very rare that the GPWS is activated during a flight and most pilots will go their entire career without ever hearing an alert. A fairly common feature in accident reports where an aircraft has ended up in a situation where the GPWS was activated and started issuing warnings is that the pilots subsequently reported that they did not hear them. When we discuss workload, some of the reasons for this will become clear, but for the time being it seems that selective attention links a pilot to a particular stimulus in a way that is difficult to break.

2.4.1.1 Strategies for selective attention

When an individual is highly focused on the task, it can be incredibly difficult to break their selective attention in order to alert them to something else that requires their input. When I ask students to suggest designs for warning systems that would have a better chance of getting a pilot's attention, the answers are often very interesting. Some people suggest electric shocks (although I think this might increase the stress of a particular situation!). Other people suggest using a different sensory modality, for example vibration and, indeed, some aircraft control columns vibrate when an aircraft is near its stalling attitude. Perhaps the best answer I heard was from a male pilot. He suggested that a guaranteed way to get most pilots' attention would be to present erotic images on all the screens in the flight deck. Interestingly, he has identified a potential route for breaking through someone's selective attention.

Our brain is wired up to pay more attention to things that are emotionally charged or biologically relevant. Imagine a warning system where, instead of saying in a disembodied robotic voice that there is terrain ahead, it spoke to us in the voice of our partner, child or parent and warned us that we were about to die. It is far more likely that this would break through our selective attention and allow us to properly evaluate the deteriorating situation that we find ourselves in. In the same way, the example of using erotic images to get a pilot's attention isn't as strange as it sounds because of the high biological relevance that sex has to humans. There are other strategies that can be used by one crew member to alert another crew member of something that may be going wrong, and we will cover these in Chapter 6 (Communication).

Other strategies to consider when dealing with selective attention are:

- If one pilot is focusing on a task, the other pilot should be especially vigilant as the pilot focusing his selective attention will be susceptible to missing large, seemingly obvious changes in the environment.
- Flying an aircraft requires a lot of attention. The handling pilot may need to be reminded of this should they attempt to divide their attention, particularly when flying manually.
- Crew engaged in safety-critical tasks should announce this to other crew members to limit the chance of being distracted and allow them to preserve their selective attention.
- Selective attention is required for critical tasks such as performing checklists. If something breaks through this selective attention (such as the dispatcher coming to the flight deck), important steps can be missed. Any distraction should be immediately followed up with rechecking the situation, for example reassessing where you were in the task before the distraction occurred. For any in-flight distraction, particularly a technical malfunction, before selective attention is brought to bear on the problem, checking the flight mode annunciators may reduce the risk of not noticing a major change in the situation because of change blindness.

2.4.2   Sustained attention (passive monitoring/vigilance)

Sustained attention is our ability to focus on a particular section of the environment in order to ascertain whether any relevant changes occur that might require our intervention. It could be argued that sustained attention is simply selective attention that is extended over a long period, but the difference lies in the nature of the stimulus. When we selectively attend to something, it is because that stimulus has changed or is changing in some way that requires our attention. If we attend to an unchanging stimulus for a prolonged period in the hope of being able to detect any relevant changes (even though they may not occur), this is sustained attention. This concept is very relevant when we come to cover automation management and is the source of many of the problems that occur during pilot–automation interactions, particularly automation complacency. Navigating the scientific literature on this topic can be difficult because this one phenomenon is investigated under three different names (sustained attention, passive monitoring and vigilance). What follows is an amalgamation of the relevant research.

Sustained attention can be defined as "a fundamental component of attention characterized by the subject's readiness to detect rarely and unpredictably occurring signals over prolonged periods of time".[13] Despite being told that as pilots we are responsible for monitoring the automation, what follows, in conjunction with

Chapter 9 (Automation Management), will demonstrate why this is not a realistic strategy to prevent automation-related accidents.

The ground-breaking research into sustained attention was carried out during the Second World War. Norman Mackworth wanted to discover why airborne radar and sonar operators on antisubmarine patrols would miss weak signals that suggested the presence of submerged enemy vessels and why this tended to happen more towards the end of their shifts.[14] His experiments showed that when subjects were tasked with sustaining attention on one task, after 30 minutes there was a decline of 10–15% in the accuracy of signal detection and this decline continued as the time on task increased. This decrease in performance was called the "vigilance decrement" and has been extensively studied since then. More recent research has shown that a measurable decrement can occur within 5 minutes, particularly if the task demands are high.[15] It was originally thought that this decrease in the ability to sustain attention was due to boredom and underarousal. However, since the introduction of functional brain imaging as an investigative tool in attention research, the evidence is suggesting the opposite. A task requiring sustained attention is mentally hard and imposes a high workload on the brain. This is seen by increased activity in certain parts of the brain together with increased blood flow. The progressive increase in the vigilance decrement is matched by a decrease in blood flow to the brain, suggesting that sustaining attention is an extremely difficult task.[16] If sustaining attention on a single task results in a vigilance decrement, this effect is exacerbated when other tasks have to be carried out as well. The brain has a finite capacity and attentional resources are limited. When these resources must be shared, the quality of the attention decreases.

The implications of this for aviation are profound. Human beings cannot be relied on to sustain attention on an area of the environment in order to try and detect rare and unpredictably occurring changes. If a system is designed that requires this, the system is prone to failure because it has missed several of the fundamental findings of research on attention. There is absolutely no point in suggesting that pilots must constantly monitor automation because this task is precisely the kind that is exceedingly hard for humans; watching a relatively unchanging system in case of some sort of subtle, insidious failure (which very rarely happens with modern automation). This dilemma will be explored further in Chapter 9.

Aside from the nature of the task, there are several other factors that affect an individual's ability to sustain their attention:

- Fatigue – If the brain is already fatigued, it is likely that that vigilance decrement will occur more quickly.[17]
- Personality – There is some research to suggest that certain personality traits predispose someone to be better or worse at tasks requiring sustain attention. For example, with reference to the Five-Factor Model that will be covered in Chapter 5 (Personality, Leadership and Teamwork), high scores in the extraversion factor (a factor where pilots tend to score highly) resulted in poorer performance on tasks requiring sustained attention.[18]
- Stimulants – One of the earlier studies into managing the vigilance decrement found that amphetamines delayed it[19] and this effect was also seen with some other stimulants such as caffeine and nicotine.[20] It should be noted that only one of these three should be used on the flight deck and that the vigilance decrement may be delayed rather than negated.

2.4.2.1   Strategies for sustained attention

The strategies for improving sustained attention are quite limited. The best strategy for avoiding the problems associated with the vigilance decrement is to redesign the task/system to avoid this requirement. However, there may be occasions where some sort of technical malfunction means that a pilot needs to sustain his attention on a defined area of the environment, perhaps an instrument, in order to detect any changes. One situation like this would be an automation failure during cruise. The handling pilot now has to manually control the aircraft. Once the aircraft is appropriately trimmed and the thrust is set, there may be very little input required by the handling pilot, particularly if the flying conditions are smooth. In this case, he may have to sustain his attention on the primary flight display and correct any deviations. Over a prolonged period, his sustained attention may begin to decrease and he may not notice small deviations that he would have picked up had they occurred earlier on in the task. Without the assistance of the monitoring pilot or any deviation alerting systems, if he does not identify these deviations himself, they may exceed the acceptable tolerances and the aircraft could deviate from its assigned lateral or vertical path. Some strategies for best achieving a task that requires sustained attention are given below:

- Ensure you are well rested before carrying out a task requiring sustained attention.
- Hand over the task as frequently as possible – the more challenging the task, the faster the vigilance decrement will occur. It may be necessary to hand over the task to someone else after a relatively short period, say 5–15 minutes, depending on what other tasks are being performed concurrently.
- Caffeine may temporarily delay the vigilance decrement. More information on the use of caffeine can be found in Chapter 7 (Fatigue Risk Management).

2.4.3   Divided attention and multitasking

Divided attention is required when multiple tasks require selective attention, sustained attention or a combination of both. The problem with dividing attention is caused by the brain mechanism that allows us to switch between multiple different tasks. The theory of threaded cognition suggests that we use the same neural mechanisms for different tasks but switch between the tasks rapidly, thus giving the impression that we are carrying out the tasks concurrently.[21] Problems occur when a task that is using a neural mechanism needs to use it *in the same fashion* as another task. For example, trying to carry out two tasks based on visuospatial processing may lead to interference.[22] An example of this is driving and manipulating a phone to send a text message. Even if the driver keeps his eyes on the road, the processing required to work out the movements of his fingers may cause him to deviate from his intended track. However, if two tasks either use different mechanisms or use the same mechanism but in a different fashion, there is more chance that both tasks can be carried out without any significant impairment to either. For example, a driver may have no problem staying in his lane while talking at the same time. Older theories suggested this interference but could not account for why practice can improve the ability of a person to carry out multiple tasks that have the potential to interfere with each other.

Threaded cognition explains the effect of practice as a speeding up of the switching between tasks and the ability to process more of the task in the short period between switches. This is down to the formation and use of more complex production rules, a phenomenon that will be discussed further in Section 2.6 on decision making.

2.4.3.1   Strategies for divided attention

- Workload management is a lot more important when attention is divided – the overall cognitive workload that a pilot is under is not just the sum of the workload generated by the individual tasks. It is also increased by the *number* of tasks as this will relate to the number of switches that the brain has to make while processing these tasks concurrently. For example, carrying out two tasks with an equal workload will generate a lower overall workload than carrying out four tasks that are each half as hard as the two original ones. This is because switching between four tasks generates more workload in itself than switching between two.
- As well as minimizing the number of tasks, consider the nature of the tasks – because different tasks that require the same neural mechanisms can interfere with each other, multitasking will be more successful when the tasks are quite different. For example, it will be easier to manually fly the aircraft and talk on the radio than to fly the aircraft and have to make manual switch selections to deal with a technical failure.

2.5   Perception

If we assume that the sensory information that is being conducted to our brain is an accurate reflection of our environment and that our attention is focused in the right place then we come to the third stage of information processing: the formation of a perception. Perception is a slightly slippery concept but can be best described as the mental model the individual forms that contains a representation of themselves and their environment. The importance of this idea of a mental model cannot be overstated and understanding how we use mental models is at the root of understanding human information processing.

It may surprise you to learn that we do not, in fact, make decisions or perform actions directly as a response to our surroundings. We make decisions and perform actions based on our *mental model* of our surroundings and it is this model that is built at the perception stage of information processing. As humans, we are capable of forming very accurate mental models based on our surroundings and we are then able to use these models as the basis for our decision making. The advantages of mental models are that they can act as simulations of how potential decisions may play out; that is, based on our knowledge of how the physical world works, we use our mental model to trial various options that we are considering. For example, "If I put this ball on the table, it will probably roll off. If I put it on the floor, with all these people around, someone will probably slip on it. However, if I put it in this drawer, it will stay in the drawer and there is no chance of people slipping on it". I did not need to physically put the ball on the table to know that there was a high likelihood that it would not stay there, in the same way as I did not need to put it on the floor to know that it would pose a risk to others. By applying my knowledge of the physical world to my mental model, I can play out a variety of scenarios to help me to decide on the best course of action. We will talk more about this process when we discuss decision making. The neural basis of perception is derived from the following:

- Our knowledge of the physical constituents of the world; that is, the sensations that would suggest the presence of a particular object – Our perceptual system has access to a store of information about particular objects and can integrate this information when determining what objects are. For example, we know the following about the sensory experience we would expect from an ice cube and may use some or all of this information in perceiving that the object we are encountering is, in fact, an ice cube:
  - vision: it will look like a transparent/translucent cube
  - hearing: it will be silent but may emit a cracking noise as it melts
  - smell: it will be odorless
  - touch: it will be cold and it will feel like a cuboid with smooth edges. It will also be slippery
  - taste: it will taste like water.
- The biological/emotional importance of the physical constituents of the world – As well as being able to recognize one or several of the physical characteristics of a stimulus, we integrate our biological and emotional knowledge of that object. For example, while the appearance of two snakes may be similar, if we are sufficiently well versed to know that one is highly poisonous and the other harmless, we will also have a biological or emotional reaction to the poisonous one. This aspect of our perceptual system is eloquently proven by patients with Capgras syndrome. As well as being able to recognize a friend or family member by their appearance, voice or even odor, there is a component of emotional recognition caused by a neural link between our visual processing system and our emotional processing system. This is why we feel attached and have a biological drive to protect our offspring and family members. Patients with Capgras syndrome have sustained damage to the neural link between the visual and emotional processors, meaning that they do not experience any emotional recognition to accompany the sensory recognition. This emotional recognition is so important when it comes to integrating sensory and emotional information to determine the identity of the person being looked at, the tragic outcome of this illness is that the sufferer believes that all their friends and family have been replaced by identical looking imposters.[23]
- Our knowledge of how physical laws affect physical objects – We know how familiar objects react to friction, gravity and other forces. Based on this knowledge, we can predict what will happen if we let go of the ice cube: it will fall. If an object reacts in an unexpected way, we may re-evaluate our perception and correct our mental model.

A mental model based on visual sensory information that is being attended to is formed in the occipital lobe and is partly transmitted to the parietal lobe, where it is combined with other sensory data including auditory information. The key elements of the mental model are what the stimulus is and where it is relative to the individual. Most of this information is processed in the parietal lobe. The following section on decision making (Section 2.6) will explain how these mental models are used.

The key point about perception is that it is based on integrating multiple pieces of information to make sense of the world. This integration of processed sensory information from all over the brain (e.g. visual information from the occipital lobe, auditory from the temporal lobes) takes place in the parietal lobe. If all the required information is there, our parietal lobe will create a robust mental model. If some bits of information about a particular stimulus are missing, it may form a mental model based on the most likely object that stimulus represents. If there is insufficient information, we may find ourselves experiencing the highly disconcerting feeling of cognitive dissonance, the result of either having contradictory perceptions of an ill-defined object or being unable to form a meaningful perception at all. Imagine that a highly unusual alien artifact suddenly appeared in your home. Its completely novel appearance combined with the fact that it defied the physical laws we know by materializing out of nowhere would be mean that you have no framework upon which to form a perception of what this object is or what it will do. This impenetrability and the lack of any familiar characteristics that would allow any sort of understanding will result in a state of fear and confusion. Wherever possible, we prevent this by filling any gaps in our mental model with plausible, experience-based "guesses", given that we do not have reliable, evidence-based information from our environment. If we have struggled to form a mental model, perhaps because of the complexity and dynamic nature of the unfolding situation, the brain is reluctant to revise it either partially or completely as that would lead to further cognitive dissonance.

If, on the other hand, we have formed a mental model but it transpires that something about that mental model is wrong because we encounter something that we were not expecting, we update our mental model with the new information. We also have the ability to take our mental model and manipulate it mentally to project what is going to happen as the physical laws and behavior of objects take effect over time. Because of the many variables that we encounter in the physical world, we are not always able to do this reliably and this can lead to significant problems when it comes to aviation.

2.5.1   Mental models

Because our mental model forms the basis of our decisions, it is vitally important that our mental model is accurate. Running a simulation based on bad data is likely to give an inaccurate outcome. If our sensory information is correct and our attention is in the right place, how could our mental model possibly be wrong? To understand this, look at Figure 2.4. Looking at the squares labeled A and B, would it surprise you to know that squares A and B are the same color? This can be quite difficult to see, for reasons we will discuss shortly. To help you see that squares A and B are the same color, I would suggest that you take a piece of paper, make two small holes in the paper that correspond to the positions of squares A and B, and place the paper over the image so that only squares A and B are visible. Bizarrely, it will now be obvious that squares are the same color. Despite its benign appearance, this is a profoundly unsettling picture. What it shows is that even though our sensory information is correct and our attention is focused in the right place, our perception of the world can still be wrong. The next question we need to address is why this is so.

If you showed this image to a baby who, for the sake of argument, we will say has no real experience of the world but is able to talk, they would have no difficulty in saying that the two squares are the same color. The reason that their assessment is accurate and ours is not is that our knowledge of the world is acting against us. Most

Figure 2.4 A checkerboard.
© William H. Adelson (http://web.mit.edu/persci/people/adelson/checkershadow_illusion.html)

people who see this picture have two bits of information in their memory that affect how they perceive the colors of the squares:

• We know that checkerboards have alternating light and dark squares.
• We know that objects in shade appear darker than objects that are not in shade.

The terrifying thing is that our expectation that the world will conform to our previous experiences means that our perception of the image is altered by our prior knowledge. We do not necessarily see things as they really are, but we often see them as we *expect* them to be. There are many examples of these perceptual illusions where our prior experience adjusts our mental model and makes it inaccurate. Why this should be the case is unclear. It may be that adjusting our mental model based on previous experience decreases the overall workload for the brain in forming the mental model. Instead of having to examine each aspect of the environment, blanks can be filled in using prior knowledge. An analogy for this phenomenon is that of the visual blind spot. There is a section of our visual field that we are unable to detect because of how the optic nerve connects to the retina. This is usually unnoticeable because the brain "fills in" the gap using information from the rest of the visual field. The implication of this phenomenon of altering mental models based on experience is profound. If, as we have seen, it happens automatically, how can we ever be sure that our mental model of the world is accurate? The short answer is that we can't and this makes it very difficult to come up with effective strategies for preventing this phenomenon from affecting safety.

Perceptual illusions, or the adjustment of our mental model based on previous experience, have been known to cause problems in aviation. A classic example involves landing on unfamiliar runways. For pilots who operate in and out of major international airports, the visual appearance of a runway when flying an approach is fairly consistent. These runways are usually of similar dimensions, are normally flat, and the approach angle that the plane takes towards them is usually the same. For

---

Box 2.4 Accident Case Study – Perceptual Illusion

Canadian Airlines International Flight 48 – Boeing 767-300
Halifax, Canada, 8 March 1996[24]

During the final stages of the approach to Runway 06 at Halifax Airport, Nova Scotia, the crew of this Boeing 767-300 perceived that they were high on the approach and so reduced thrust on both engines in an effort to correct the perceived deviation from the expected approach path. This resulted in the aircraft crossing the threshold much lower than would normally be expected and with a reduced thrust setting. The aircraft touched down 200 feet (60 m) along the runway, with the tail striking the runway, causing substantial damage. An investigation subsequently showed that the first part of the runway was upsloping by approximately 1%. This upslope meant that the crew perceived that they were higher than they actually were during the final stages of the approach and so reduced thrust and increased their rate of descent to correct this. In fact, they had been on an appropriate approach and these "corrective" actions based on the incorrect mental model of their height relative to the runway led to this tail-strike event.

---

these reasons, the pilot is normally presented with a fairly consistent visual picture of how the approach should appear at various stages. The visual pictures are stored in the pilot's memory and form the basis of his expectations. As we have seen previously, expectation has a significant effect on our mental model.

If the pilot continues to fly between these similar looking runways, then the expectation becomes more reinforced. The difficulty arises when he must then operate to a runway that does not conform to the "normal picture". This runway may be longer or shorter than normal, wider or narrower, or maybe sloping up or down. The pilot is now faced with a mismatch between what he actually sees and what he expects to see. The risk is that the mental model formed at the perceptual stage of information processing is adjusted to conform with expectations. In the event that the approaching runway is wider than the pilot is used to, the pilot will feel that he is lower than normal. The reason for this is that the visual appearance of a wider runway from 2000 feet (600 m) above it may be the same as a narrower runway from 1500 feet (450 m) above. In this way, a pilot making a visual approach may adjust his approach profile to regain the "normal" picture that they were expecting to see. The risk now is that the pilot lands too far along the runway, with the possibility of running off the end. A similar thing occurs for downsloping runways.

Conversely, when making an approach to a runway that is narrower than the pilot is used to, the tendency will be to fly a lower approach than normal to regain the expected picture. The risk here is that the pilot lands short of the runway. A similar thing occurs for upsloping runways (Box 2.4), and these illusions have been implicated in many general aviation crashes as well as incidents and accidents in commercial aviation.

Misty conditions can also make runways look more distant, thus giving the impression of being higher on the approach than you actually are. Ground mist makes it difficult to assess speed in the final stages of the approach. We visually assess our velocity by noting the speed at which the environment moves through our peripheral vision. Ground mist has the effect of making environmental features appear less distinct, thereby decreasing the "visual texture flow" of the world around us. Objects appear to be moving through our peripheral vision less quickly and so our perception is that we are slower than we actually are. The risk here is that we inappropriately increase our speed when we are low to the ground. This may cause difficulties on shorter runways. Any reduction in visibility also makes ground features, including the runway, more distant. As with a narrow runway, pilots may perceive that they are further away or higher than they actually are and their reaction will be to descend. This is clearly a significant risk in low visibility as there is a chance of impacting the terrain.

Another less well understood perceptual illusion is the black hole effect. This perceptual illusion has been implicated in multiple crashes in general aviation, military and commercial aircraft. Although the precise reasons why this illusion occurs are still unclear, there does seem to be a set of conditions that put a pilot at increased risk of experiencing it:[25]

- night-time with little or no light from the moon or stars
- visual approach
- featureless terrain around the airport with no ground features or lights visible; the sea or some other large expanse of water would have the same effect
- sometimes a bright feature beyond the end of the runway, such as a town.

It seems that under these conditions, pilot attempts to maintain a constant visual angle between the landing threshold of the runway and the lights designating the far end of the runway. The effect of this maneuvering to keep a constant visual angle between the start and the end of the runway is that the pilot performs a steeper approach than normal and may either impact the ground short of the runway or end up flying a level segment at very low altitude. Other theories have been suggested about why the black hole illusion occurs but the motivation to keep a constant visual angle to the runway may, subconsciously, feel like the only way for the pilot to maintain his orientation with reference to the ground. Unfortunately, this attempt to maintain orientation gives a false perception of being too high during the approach and so increases the risk of crashing short of the runway (Box 2.5).

2.5.2   Perceptual difficulties with sensory-induced spatial disorientation

Spatial disorientation that occurs as a result of the limitations of our sensory systems may go unnoticed. A pilot who does not notice a gentle bank below his sensory threshold may collide with another aircraft performing an approach to a parallel runway. In this situation, spatial disorientation has clearly occurred but, crucially, the pilot is unaware of it until impact. A more complex issue is how we manage spatial disorientation and the confusion that arise when we do become aware that something

---

Box 2.5 Accident Case Study – Perceptual Illusion

Federal Express Flight 1478 – Boeing 727-200
Tallahassee, Florida, USA, 26 July 2002[26]

During a night-time visual approach to Runway 09 at Tallahassee Regional Airport in Florida, the crew of this Boeing 727 descended below the normal approach path, struck trees in front of the runway threshold and crashed short of the runway, seriously injuring all three crew members. Although vertical guidance was available in the form of Precision Approach Path Indicators (PAPIs – four lights adjacent to the runway that indicate whether the aircraft is too high, too low or at the correct height during the approach), the approach was flown without reference to them and the descent path matched that which you would expect for a pilot experiencing the black-hole illusion: a steep descent followed by a shallower descent closer to the ground.

---

is wrong. We are now aware that something is seriously wrong with our mental model but because we have conflicting sensory data, it may seem impossible to determine what is wrong with our mental model and so we are unable to fix it. The pilot is now experiencing that highly stressful feeling of cognitive dissonance, specifically, a form of perceptual dissonance, and will be looking for any possible way to resolve this as quickly as possible. Consider the case described in Box 2.6.

One of the key questions that investigators tried to answer was why, when abruptly given control of the aircraft that was in a steep left bank, did the captain increase the bank angle further on several occasions. He was clearly trying to build a mental model of the situation and despite having his first officer telling him to bank in the opposite direction, he continued to bank left. The surprising answer to this seems to be how he perceived the attitude of the aircraft based on the visual information he was receiving from the attitude direction indicator (ADI) in the flight deck.

There is a difference in ADI design philosophy between Western and Russian aircraft manufacturers. Western ADIs work on the principle that the aircraft symbol remains fixed and it is the artificial horizon behind it that moves when the aircraft is in a banked turn. The perception of the pilot is based on being "inside" the aircraft symbol and seeing the artificial horizon as shifting in his visual field in the same way as the real horizon would if he was flying visually. Russian ADIs work on the principle that the horizon remains fixed but it is the aircraft symbol that moves. The perception of the pilot is based on being "outside" the aircraft, watching it move with reference to the horizon, much like a video game. These two design philosophies lead to two very different ADI presentations of a left banked turn depending on what type of aircraft you are flying; Western built, or Russian built. Figure 2.5 shows what a 30 degree left turn would look like on a Western-built ADI and what a 30 degree right turn would look like on a Russian-built ADI.

Under the high-workload conditions of flying the approach, it would appear that the captain, who had spent far more time piloting Russian-built aircraft, had misinterpreted

Box 2.6 Accident Case Study – Perceptual Dissonance

Aeroflot Flight 821 – Boeing 737-500
Perm Airport, Russia, 14 September 2008[27]

The first officer was the handling pilot and the captain was the monitoring pilot during this approach to Perm Airport. Owing to a variety of factors, including problems managing the thrust on both engines and inexperience of the effects of this in aircraft with more widely spaced engines that both pilots were more familiar with, the crew experienced very high workload during the approach. During the approach, the autopilot was disengaged. The autothrottles had not been working at all during the flight. At one point, the captain instructed the first officer to make all the necessary configuration changes himself as he was occupied with talking to the tower. Despite being designated as the monitoring pilot, it seems that the captain made a series of control inputs that surprised the first officer. As the situation deteriorated, the first officer put the aircraft in a 30 degree left bank but concluded that he could no longer maintain adequate control. Although the captain was engaged in a radio communication with the tower, the first officer said "Take it, err ..., take it". It seems that the captain was not ready to take control and was most likely out of the loop regarding the attitude of the aircraft. He replied "Take what, [expletive]? I can't do it either!" The captain took control at this point and, despite the aircraft being at a bank angle of 30 degrees to the left, applied an abrupt left control wheel input and so increased the bank angle to 76 degrees. At this point, the first officer called out "On the contrary! In the opposite direction!" The captain was now totally disoriented as he made another, almost full, left control input. The aircraft made an almost full barrel roll and crashed into the ground with a left bank roll rate of approximately 35 degrees per second.

 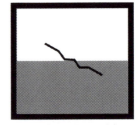

Western ADI: 30 degree left turn   Russian ADI: 30 degree right turn

Figure 2.5 Western attitude director indicator (ADI) showing 30 degree left turn and Russian ADI showing 30 degree right turn.

the left bank presented on the Western-built ADI for a right bank owing to how a right bank would have looked on a Russian-built ADI, and had built this incorrect perception into his mental model of the situation. In the event, it seems that he tried to "fly the horizon"; that is, he became fixated on the moving element of the ADI. In a Russian

ADI, that moving element is the aircraft symbol. On a Western ADI, the moving element is the horizon. Even the first officer's warnings were not enough to counteract this incorrect perception and this led to a crash that killed all 88 people on board.

As we saw with the checkerboard image in Figure 2.4, the power of our previous experience to alter our perception of a situation can be incredibly strong and almost impossible to overcome. The perceptual problems caused by transitioning from Russian-built to Western-built aircraft with their different design philosophies have been implicated in several other crashes, most notably the crash of Crossair Flight 498[28] shortly after take-off from Zurich in 2000. The captain, who was also the handling pilot, had significantly more experience of Russian-built ADIs than the Western-built one in the Saab 340 and so progressively overbanked to the right despite correct information being presented to him and the first officer warning him that he needed to bank left. The aircraft entered a spiral dive and crashed, killing all 10 people on board.

If we look at cases like this in the context of our model of information processing, we can say the following:

- Sensation – accurate: The visual information being transmitted to their brains was correct; that is, they could see the ADIs.
- Attention – accurate: They were looking at the ADIs.
- Perception – inaccurate: The meaning of the visual information was misinterpreted, leading to an inaccurate mental model of the world.

It is unfortunate that in these cases, the perceptual confusion arises from how visual information is integrated into our mental model. Because we get the majority of the data that we need to construct our mental models from our sense of vision, when a perceptual error occurs that affects this information, it is almost impossible to correct. In both of the cases described in this section, the pilots had very strong auditory cues telling them that they were banking the wrong way but this was not enough to counteract their erroneous perception based on the visual information.

2.5.3   Strategies for dealing with perceptual limitations

The difficulty in dealing with perceptual illusions and other inaccuracies in our mental model of the world is that it is impossible to detect when our perceptions have been affected by illusions or adjusted by our expectations. Errors in our mental model are normally completely undetectable. If we resign ourselves to this fact, our only remaining option is to mitigate the negative effects that this may have on safety. As we will see in the section on Threat and Error Management in Chapter 3, we do have some options available to us when it comes to managing this risk:

- In circumstances that may lead to perceptual illusions, such as any of the meteorological conditions that have been mentioned in this section, considering the possibility of visual illusions in advance, particularly if this can be discussed with a colleague, can allow contingency plans to be formulated to mitigate the risk of such illusions affecting flight safety. For example, perceptual illusions due to ground mist can give the impression of being too slow just before landing. By considering this possibility in advance, the pilot may be sufficiently primed that before he acts on the impulse to increase thrust, the perceived airspeed is cross-referenced with the indicated airspeed. The monitoring pilot can also be asked to actively monitor the actions of the handling pilot in this regard during the final stages of the approach.

- Perceptual illusions predominantly occur during visual flight. Wherever possible, flight should be conducted with reference to the flight deck instruments; where this is not possible, visual information should be cross-referenced with the flight deck instruments to minimize the risk of perceptual illusions leading to inaccuracies in the pilot's mental model.

- Accidents that occur in the approach and landing phase of flight are three times more likely to occur at night.[29] A visual approach at night is a particularly risky maneuver and, in circumstances where an instrument approach is not available, the visual approach will require significantly more preparation than might ordinarily be expected but should be avoided wherever possible.

- A visual approach at night when there are no ground-based indicators (e.g. precision approach path indicators) to give the pilot guidance on the vertical approach path represents a significant threat to flight safety and should definitely be avoided.

- In multicrew operations, probably the most useful strategy for preventing the limitations of our perceptual system affecting flight safety is being able to compare our mental model with someone else's. If the possibility of perceptual illusions is discussed in advance, the handling pilot can ensure that the monitoring pilot highlights any unexpected flight conditions as soon as they occur. It may be that the monitoring pilot's mental model of the situation is correct and the handling pilot's mental model of the situation is wrong and, as we will see in Chapter 6 (Communication), ensuring that all crew members are confident enough to raise any issues they have is an essential step in maintaining a safe operation.

- Handing over controls to a pilot who is not prepared to accept them may be problematic. If the monitoring pilot is out of the loop because he is performing other tasks, being suddenly expected to take control of an aircraft when he may not have a fully updated mental model of the situation can lead to problems.

2.6   Decision making

[P]eople are not accustomed to thinking hard, and are often content to trust a plausible judgment that comes to mind.

Daniel Kahneman[30]

If our sensory information is accurate, our attention is in the right place and our perception (mental model) of the situation is entirely correct, we now have to decide what to do next. The science of decision making is worthy of a chapter in itself and, indeed, many thousands of books and scientific papers have been written on the subject. More than just about any other topic in human factors, decision making polarizes the scientific community and there have been many different approaches taken to understand this complex subject. More than any other stage of human information processing, decision making is most often implicated when things go wrong. In this section, we are going to look at the mechanics of decision making, what situations lead us to use various decision-making systems, why things commonly go wrong with decision making and how we can prevent this. In this context, problem solving is a form of decision making, in that once a problem has been identified, there is a need to decide what to do about it.

2.6.1   The anatomy of decision making

Once a mental model of a situation has formed, we may find that there is an element of the situation that requires our attention. In the previous section, we described the phenomenon of selective attention. We are selectively attending to a variety of different things for the majority of time that we are awake: driving, watching television, flying a plane, replacing an engine, and so on. Once we have decided to pay attention to a particular task, we become less aware of the rest of our environment and our mental model becomes focused on the task at hand. This allows us to enhance this aspect of our mental model in order, for example, to solve a particular problem. Our mental model of our environment is primarily driven by visual information, given that vision is the most relied upon sense. Other sensory information such as sound, touch and balance are also integrated into this mental model to give the most accurate representation of our environment and our relationship to it. This integration happens in the parietal lobe of the brain. Various other capabilities are required for decision making to happen, and there are brain regions (known as "modules" in the context of decision making) that carry out these tasks. Their arrangement is shown in Figure 2.6.

This model is known as the Adaptive Control of Thought – Rational (ACT-R) cognitive architecture and was developed primarily by John Anderson and his team at Carnegie Mellon University as a way of understanding the neural processes that underlie the phenomenon of decision making.[31,32] What follows is a description of each module and its purpose.

2.6.1.1   Goal module

This module, located in the anterior cingulate cortex towards the center of the brain, defines and keeps track of the particular thing we are trying to achieve (e.g. replace this windscreen wiper, add 45 to 76, or ascertain what a warning indication means). At each stage of problem solving, the goal module is referred back to in order to ensure that the decisions being made relate to the initial goal.

2.6.1.2   Imaginal (mental manipulation) module

This area of the brain, located in the parietal lobe (the same area primarily responsible for building mental models), allows elements of the mental model to be manipulated in order to attempt to solve a particular problem. For example, it may be necessary to mentally manipulate an object to see how it fits into another object. Before learning this ability, young children can only do this by trial and error. Imagine a child placing differently shaped pieces of wood into a series of differently shaped holes. There may be several failed attempts before he realizes that a square peg will not fit in a round hole. Once the imaginal module is more fully developed, the child will be able to

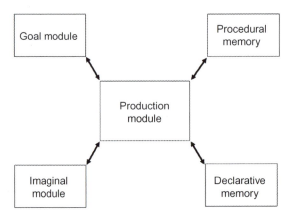

Figure 2.6 Arrangement of modules used for decision making [part of the Adaptive Control of Thought – Rational (ACT-R) cognitive architecture].

reach this conclusion simply by manipulating the relevant pegs and holes represented in his mental model. When attempting to solve more complex problems that require several steps, the problem representation in the imaginal model is updated as each step is completed.

2.6.1.3 Production (pattern-matching) module

This module is the main coordinator of decision making and is thought to be localized to the basal ganglia towards the center of the brain. It interacts with the imaginal model as well as the goal module to define the scope of the problem. It also interrogates the memory stores to try and match the problem to past experience and so be able to select the most appropriate response. This pattern-matching process is the very core of decision making and relies on retrieving information from various memory stores to find a suitable response. The two types of memory store used are declarative and procedural, and the number of trials the production module is willing to go through when trying to match the best response to the presented pattern of stimuli is determined by workload and stress. These elements are described below:

- Declarative memory – This comprises discrete bits of factual data such as the year the Second World War started, the name of your best friend at school or the knowledge that $2 + 2 = 4$. These bits of data are called "chunks".
- Procedural memory – This is the store of production rules used by the production module to solve problems. They normally take the form IF (this problem), THEN (this solution). For example, most people will be able to solve the equation $8 + A = 12$ to determine the value of A. The production rule says that IF (First number plus A equals Second number), THEN (A equals Second number minus First number). The production rule is generic and all the brain has to do is to fill in the values of the first and second numbers (the "slots" in the production rule). Some problems will require the use of several production rules and the more often they are used, the higher the chance that these multiple production rules will combine into a single, more complex production rule that can be activated more quickly in the future.

This is the key to learning and refers back to the material we covered earlier in the chapter regarding the importance of association in learning. Problems that could only be solved by using multiple, simple production rules can, with practice, lead to the formation of new, more advanced production rules. This is because, to repeat Donald Hebb's insightful words, "cells that fire together, wire together". The multiple neural firings required to activate multiple production rules lead to an association between these different networks so that the next time the stimulus is encountered, they can be quickly activated in a synchronous manner rather than individually. For example, a single technical failure on an aircraft may lead to multiple warning lights in the flight deck. An inexperienced pilot, unsure of what is really going on, can either deal with the different warnings individually or search his declarative memory and try to formulate a hypothesis about what could have gone wrong to cause this myriad of warnings. An experienced pilot who has encountered this scenario before may have formed a new production rule that says IF (Warning lights A, B, D and X are illuminated), THEN (Shut down the left engine because it is critically damaged). It is this ability to form new production rules that allows us to develop expertise in various complex tasks.

- Workload – Given that neuronal communication is limited both by the speed of electrochemical transmission along the neuron itself and by the chemical communication (using neurotransmitters) between neurons, the brain has a processing speed limit that means it can only perform a certain number of tasks in a given amount of time. This limit is a lot lower than that of a computer because the transmission speeds in the brain are a lot slower than the transmission speeds in a microchip. In the same way as a computer can be made to crash, or generate incorrect results in situations where the required calculations exceed the machine's computational capacity, the same thing can happen with the human brain. The brain has a certain number of tasks to perform during a fixed volume of time. It is estimated that to activate a production rule takes 50 milliseconds (ms), meaning that about 20 production rules can be selected and tested per second. This may seem like a lot but when you consider that humans often have to perform several tasks concurrently, each needing their own production rules, and that a novel task that requires pattern matching may require many of these rules to be activated and tried in the search for an adequate match, this speed limit has been described as the bottleneck of information processing.[32] There are two factors that can lead to this bottleneck becoming a problem and these are the main determinants of workload. They can cause high workload when they occur separately or together:
  - high task-load: multiple tasks having to be carried out within a limited period
  - high time-pressure: even a single problem, if it is complex, may generate a high workload if it has to be completed within a specified period.
- Acute stress – High workload can generate acute stress and anxiety. Other situational and emotional factors can generate acute stress and, unfortunately, it appears that having to deal with this acute stress impairs information processing. Attentional Control Theory suggests that an individual experiencing acute stress will have a reduced cognitive capacity because some of their information processing machinery is occupied with processing the feelings of stress.[33] Unfortunately for some human operators, such as pilots, when faced with complex, novel and dynamic problems, their own safety is also at stake. As well as the workload generated by attempting to solve the problem, cognitive resources may be partially occupied by processing these stress responses.

2.6.1.4   Summary of the anatomy of decision making

Given what we have covered about how neurons interact with each other (i.e. neurons that are repeatedly activated at the same time form an association and so are easier to activate in future), and given the incredible complexity and number of neuronal connections in the human brain (500 trillion), from the very basic idea of learning and practicing simple rules you can begin to see how we learn information and how our behavior, especially our amazing capacity to solve problems, is generated in the brain. Instead of one neuron being associated with another neuron, there may be a million neurons associated with a million other neurons in a highly complex way and yet still operating on the basis of that phrase that we encountered earlier: "cells that fire together, wire together". Certain subgroups of interconnected neurons go to form the various modules that have been covered in this section and it is the interaction of all of these modules that allows us to make decisions. Taken in the context of the other stages of information processing that we have already discussed, we can illustrate this using an example. Figures 2.7 and 2.8 show the process of solving an equation from the initial visual stimulation coming from the writing on the board to deciding on the correct answer.

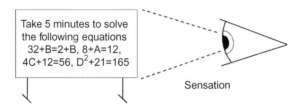

Figure 2.7 Visually sensing the stimulus of a whiteboard with mathematical equations.

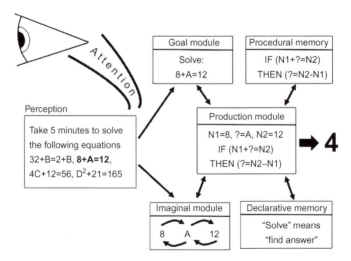

Figure 2.8 The attention, perception and decision-making process used to derive answers to the equations.

2.6.2   The two systems of human decision making

All animals are capable of making decisions in some form. Simpler forms of animal life are able to make decisions based on instinct. For example, for most organisms there is an instinct to withdraw from painful or unpleasant stimuli. Through this simple form of associative learning, animals can learn that a certain set of environmental conditions that led to pain in a previous encounter are also likely to lead to pain if those conditions are met again. In this way, animals learn to avoid situations that could potentially cause harm. In the language of the ACT-R model of decision making that we have just described, the animal is now equipped with some basic production rules that govern its behavior. Because of limitations in brain size, complexity and available memory storage space, it is impossible for simpler organisms to learn more complex behaviors that require combinations of multiple production rules.

There are, however, certain production rules and other hardwired programs that humans and some higher forms of animal life possess that allow decisions to be made rapidly under complex, uncertain or dynamic conditions. For example, as well as being able to recognize situations that are the same as ones previously encountered, humans and higher animals can recognize situations that are *similar* to ones previously encountered. At first glance, this may not appear to be a radical step forward in evolutionary terms but, in reality, it is unlikely that we will encounter a situation in our environment that is absolutely identical to one that was previously encountered. The ability to judge that a situation is similar enough to one that has been previously experienced enables the animal to activate the same production rules as before and so execute the same successful strategy to protect itself from harm. We will return to this concept of *similarity matching* later.

Modern research in the field of decision making has concluded that human beings possess two distinct systems that govern our decision making:[34]

- System 1 – This is the system that we have inherited from our animal ancestors and allows us to make decisions in high-workload situations.
- System 2 – The system has evolved more recently and allows us to make more rational decisions using all of the processing power of our highly evolved cerebral cortex.

2.6.2.1   System 1

System 1 is always active when we are conscious. It allows us to make rapid decisions when we are in complex, time-pressured, dynamic or high-workload situations. One such example would be firefighters having to deal with the fire inside a building. In such a situation there may be considerable pressure to look for survivors, concern about one's own safety and the difficulty with having to deal with a rapidly evolving situation. People who have been in situations like this often have difficulty in recalling how they made the decisions they did at the time. Unlike many other jobs, time-pressure is a critical factor when fighting a fire and it is rarely possible to make a slow, relaxed decision having carefully considered all of the variables. And yet, decisions do get made and so there must be some wiring in our brains that allows us to do this.[35] Of course, firefighting is not the only profession where people are faced with situations that require rapid decision making. People working in roles within military, healthcare and aviation regularly encounter such situations and one of the major research questions is: how can we make decisions without conscious and deliberative assessment of the nature of the problem? Imagine a word processor on a computer. If you press the letter "A" on the keyboard, you will expect it to appear on the screen. Input leads to response. What is not seen are the thousands of lines of computer code that allow this to happen. For someone who does not know how the system is set up, trying to work out the programming that translates the push of a button into a letter appearing on a screen is almost impossible. It was the same situation for researchers trying to work out why certain conditions led to certain, apparently instinctive and subconsciously triggered behaviors. Through the groundbreaking work of two scientists, Daniel Kahneman and Amos Tversky, the unseen, subconscious programming that System 1 uses has been revealed.[36]

The input to System 1 is the mental model formed at the perception stage and the output is some sort of decision. It would seem that the brain's decision-making system is still used during this process but that it works using some preprogrammed, hardwired production rules that we did not have to learn but that we seem to have been born with. The ability to make quick decisions has been passed on evolutionarily from our animal ancestors and so System 1 is always operational when we are awake in order to allow us to react quickly to threats. When describing System 1, it could be called the system of instinctive decision making. If all we had was System 1, we would still be able to function as an organism but we would behave very differently from how we behave with System 2 there to modify the outputs of System 1. If you consider the brain as a computer, the hardwired rules that System 1 utilizes are like the operating system, allowing the computer to carry out some basic functions even though no more advanced software has been installed yet. The good thing about this operating system is that it seems to allow us to make lots of decisions very quickly, like the parallel processing capabilities of an advanced computer. These hardwired programs that System 1 uses are very relevant to aviation, particularly as they are a major source of errors, and will be covered later in this chapter.

2.6.2.2 System 2

This system is what makes us human. If System 1 gives us our instincts, System 2 gives us our rationality. It is like our inner chess player; able to carefully assess the situation, consider the variables, calculate options and probabilities in a logical way but also apply abstract reasoning to derive elegant solutions. The cost of this is time, effort and energy. Trawling through memory stores, activating and testing production rules, manipulating sections of the mental model stored in the imaginal module, all takes time. Those individual blocks of 50 ms (the time taken to activate a production rule) all start to add up and, as well as that, every neuron that fires requires energy and every impulse that is transmitted across a synapse releases chemicals that will then need to be broken down. The huge burden that System 2 makes on our cognitive resources can be demonstrated with a very simple experiment. When you are walking with a friend, ask him a complex, multipart question such as "start with the date you were born, add all the numbers together, square that answer and work out what 15% of that answer is". Because this task requires a lot of activity between the goal, imaginal and procedural modules and the activation of multiple production rules, the cognitive workload is very high. What is very likely to happen is that the person will stop walking. His brain is now so occupied with the task that it is shedding less important tasks (such as navigating the footpath).

As if the higher cognitive workload were not problematic enough, we may be faced with several problems that require conscious and careful engagement but, unlike System 1, we cannot consciously solve multiple problems in a parallel fashion with System 2. System 2 works in a serial way so that we can only consciously consider one problem at a time. In the previous example where you gave your friend a multi-step mathematical problem, each step has to be carried out sequentially because the input for the next step relies on the output from the last step. If we were to be presented with a piece of paper with three unrelated mathematical problems to solve, we could only solve them one by one. These limitations of capacity and serial processing are why thinking really hard is actually very taxing and it is why, in the words of Daniel Kahneman quoted at the start of this section, "people are not accustomed to thinking hard, and are often content to trust a plausible judgment that comes to mind".[30] These "plausible judgments" are the outputs from System 1 but, as we will see shortly, they can be dangerously flawed.

2.6.2.3   Summary of Systems 1 and 2

Table 2.1 describes the different characteristics of Systems 1 and 2.

2.6.2.4   How Systems 1 and 2 interact and the role of workload

Each of these systems is not used exclusively and, in fact, they are normally both operational at the same time. When we have formed a perception (mental model) of the situation, it is System 1, the system that is always operational and is automatic, that first makes a decision based on a number of hardwired production rules. In high-workload conditions this decision will probably lead directly to a response, but in low-workload conditions there is an opportunity for System 2 to modify or override the output of System 1. Although System 1 can save us when there is no time

Table 2.1 Characteristics of Systems 1 and 2

System 1	System 2
Subconscious decision making	Conscious decision making
Automatic, always operational	Controlled, needs to be activated
Low effort	High effort
Rapid	Slow
Evolutionarily old	Evolutionarily recent
Non-logical	Logical
Parallel processing	Serial processing

to engage System 2, we tend to make much better decisions when System 2 has an opportunity to reconsider the problem.

Workload is the key determinant in whether System 2 will get a chance to modify the output of System 1. There is a high likelihood that the outputs from System 1 will lead directly to a response in any of the following three situations (or when a combination of them occurs). The first two are what determines workload:

- High task-load – Where multiple tasks have to be carried out within a limited period.
- High time-pressure – Even a single problem may be dealt with using only System 1 if there is significant time-pressure.
- Problem underspecificity – A problem that is underspecified is one where there is not enough information to be able to make a decision based on System 2. The situation that was unfolding in the control room at Three Mile Island nuclear power plant (as described in Chapter 1) was completely underspecified; there was no reliable information about what was going on in the core of the reactor and, even if there was, the problem was evolving too rapidly for the human operators to be able to keep their mental model updated.

When you consider the aviation environment and the highly technical nature of some of the problems we have to deal with, it is something of a perfect storm when it comes to preventing us from making rational decisions. Aviation is an environment where System 1 based errors are very likely.

2.6.3 Heuristics and biases

The hardwired programs in our brains that allow System 1 to operate and so let us make quick-fire decisions are know as *heuristics*. One of the psychologists who discovered the existence of these programs, Daniel Kahneman, defines a heuristic as "a simple procedure that helps us find an adequate, but often imperfect answer, to a difficult question".[1] It is derived from the same Greek word as eureka – *heuretikos*, meaning "inventive". It can be considered as a mental shortcut; that is, a quick way of coming up with a rough-and-ready answer to a difficult question. An additional aspect to System 1 programming is *biases*. These biasing programs lead us to overemphasize the importance of some information, often at the cost of another piece of information. Biases can profoundly affect decision making in a negative way.

Another famous researcher in the field of decision making, Gerd Gigerenzer, gives an excellent example that describes the principle of what a heuristic is meant to do and how it works.[37] Imagine a cricket player or baseball player attempting to get into a position to catch a ball in flight. The mental effort required to mathematically calculate the path of the ball and plot an interception point is far beyond what most humans are capable of without the aid of a calculator, let alone under the pressure of having to complete the calculation while the ball is still in flight. If we do not have time to calculate where the ball is going, how is it possible for the player to decide where to position themselves to catch it? Fortunately, we have a mental shortcut that allows us to achieve the same objective, albeit with less accuracy, in the limited amount of time available. The player, if he is experienced, will know that if he adjusts his running speed and the path he is taking in order to keep the ball at a constant visual angle (i.e. rather than moving his head to keep the angle of gaze towards the ball constant, he moves his entire body), his path will intersect with that of the ball. If the player reaches the predicted point of intersection while the ball is some way off, he may have time to refine his estimate and adjust his position accordingly. There are now more mental resources available for this task as the player is not having to devote some attention to running and to avoiding any obstacles. If, on the other hand, the player is still running when the ball reaches the intersection point, there is less chance of catching it as there has been no opportunity to refine the estimate of where the ball is going and the player is still having to devote some mental resources to the act of running. The mental shortcut that allows a player to do this is known as the gaze heuristic.

2.6.3.1   Evolutionary origins of heuristics and biases

Aside from the gaze heuristic, hundreds of other heuristics have been identified. Many of them have a profound impact on our decision making and affect our lives in ways we do not even realize. To do justice to this fascinating area of psychological science, let's look at the evolution and classification of these heuristics.

*   Heuristics evolved because animals needed to be able to make rapid, quick-fire decisions in complex, time-pressured situations and also to make decisions in situations where all the required information is not available, such as during social interactions.
*   These abilities needed some sort of hardwired programming that would permit rapid but approximate pattern matching of stimulus to response and perceptual gap filling when all the required information was not available.
*   While these heuristics are designed to improve our performance, some of them, particularly those whose names include the word "bias", can have a negative effect on the accuracy of our decision making because of how much the world around us has changed from the time when this hardwired programming first evolved.

In the same way as the gaze heuristic allows us to solve a complex problem by simplifying the calculation, many heuristics involve replacing a complex decision with a simpler one. The ones that are of most interest in aviation are those where we are in a complex, highly dynamic situation and have to satisfy our need to remain oriented with respect to our mental model of the world. This is just a convoluted way of saying that humans do not like uncertainty. We struggle in situations that are too complex to understand or where the situation is changing too rapidly for the brain to be able to keep track of it (such as in the Three Mile Island nuclear disaster summarized in Chapter 1). Remember, the human brain has evolved over millions of years. It is only in the last century that we have taken our brains, an organ that evolved to suit an organism with a top speed of perhaps 44 km/h, and put them in aircraft traveling at 900 km/h with a computer interface between us and the environment. Biologically speaking, we are still essentially cavemen. The programming in our brains is relatively unchanged from the days when our primitive ancestors walked the Earth. It is no wonder that we sometimes find ourselves behind the curve. Another unfortunate side-effect of the fact that the heuristic programming of System 1 is evolutionarily hereditary is that we cannot rid ourselves of it. Many experiments have been carried out to see whether people can be "debiased", but the fact is that this hardwired programming in our brains, for better or worse, is here to stay.[36]

2.6.3.2 Heuristics and biases in aviation

There are literally hundreds of heuristics that allow System 1 to function. Because System 1 is functioning whenever we are conscious, large numbers of these heuristics may be active at any one time. Out of the list of hundreds, there is a smaller subset that seem to have a direct impact on aviation, mainly because of the unique nature of the aviation environment. We can classify these heuristics according to the functions they are meant to support:

- decision-making and memory-retrieval heuristics
- social heuristics.

2.6.3.3 Decision making and memory retrieval heuristics

Aside from the specific heuristics that we will look at, there two general patterns that some of these heuristics seem to be based on. These patterns relate to how information is stored and retrieved in the brain and how production rules are employed:

- Similarity matching – the tendency to associate a new stimulus with a similar one that was previously encountered, resulting in the same response. This is accounted for by the decision-making mechanisms of the brain not requiring a perfect match between a stimulus and the appropriate response, but rather an adequate match.
- Frequency gambling – using a strategy that frequently works in novel situations. Production rules become more easily activated the more often they are used (a process known as "raising base-level activation"). In this way, a commonly used rule becomes more prominent than other, similar rules and so is more likely to be activated.

There are five heuristics (some referred to as "biases") that have an impact on aviation decision making, especially as a cause of systematic errors that occur under particular types of uncertain or high-workload situation. Because these are important sources of error, they will be covered briefly here and in more detail in Chapters 3 and 9.

Confirmation bias

This is the tendency to avoid cognitive dissonance by continuing to believe in a particular mental model and putting extra emphasis on environmental cues that support it while disregarding cues that would refute it. Having to abandon a mental model of a situation in light of contradictory evidence is very difficult because the brain then has to build a new mental model before decisions can be made based on it (Box 2.7).

Availability heuristic

This is the tendency to utilize a piece of information or a strategy that has been used recently. In a similar manner to frequency gambling, even if a production rule has not been used frequently, the recentness of its last usage can increase its prominence among other similar rules and so it is more likely to be activated (Box 2.7).

Box 2.7 Accident Case Study – Confirmation Bias and Availability Heuristic

British Midland Flight 92 – Boeing 737-400
Kegworth, Near East Midlands Airport, UK, 8 January 1989[38]
During a flight from Heathrow to Belfast, this Boeing 737-400 suffered a fan fracture in the left engine resulting in severe vibrations and smoke in the cabin. From the very start of this malfunction, the situation induced stress and the workload would have been very high, leading to an overreliance on the heuristic-based outputs from System 1. The crew elected to divert to East Midlands Airport but during the course of their handling of the technical malfunction, they shut down the incorrect engine. There were several factors that led the crew to think that it was the right-hand engine that was damaged, including that on the previous model of 737 they were accustomed to flying, the air conditioning came from the right-hand engine. Smoke in the cabin would suggest damage to that engine. Unfortunately, one of the differences with the newer model of aircraft that they were flying was that air was taken from both engines and so even damage to the left engine could cause smoke in the cabin. The fact that the crew activated a production rule that said "IF engine-induced smoke in the cabin, THEN it must be a problem with the right-hand engine" is an example of the availability heuristic. This rule was true for the previous type of aircraft that they had far more experience on and so it was activated in the high-workload situation that they were faced with.

The real tragedy was the extremely strong confirmatory cue that the crew got when they shut down the incorrect engine. Despite the damage to the left engine, the autothrottle system was still commanding fuel to be injected into the damaged engine core to try and maintain thrust. This was what was causing the vibration and smoke. In pulling back the thrust-lever for the right-hand engine, the autothrottle was disengaged and so stopped pumping fuel into the damaged left-hand engine. However, all the crew knew was that when they pulled back the right-hand thrust-lever, the smoke and vibration decreased significantly. This would have strongly reinforced their mental model that it had been the right-hand engine that was causing the problem and, based on the current understanding of stress, workload and decision making, I think it is very probable that another crew could have made the same mistake. There were several other factors that contributed to the outcome, including difficulties in interpreting the new instrumentation on the flight deck and a missed opportunity to pass on information from the cabin, that would have told the pilots that the wrong engine had been shut down. Unfortunately, the plane crashed just before the runway at East Midlands Airport, killing 47 out of the 126 people on board.

Box 2.8 Accident Case Study – Plan Continuation Bias and Representativeness Heuristic

Southwest Airlines Flight 1455 – Boeing 737-300
Burbank, California, USA, 5 March 2000[40]

In an effort to comply with air traffic control speed and altitude instructions, the crew of this Boeing 737-300 were left excessively high and fast on the approach and yet decided to continue to land, resulting in a runway excursion. The final approach was at an angle of approximately 7 degrees and a speed 50 knots higher than the normal approach speed. There were multiple ground proximity warning system alerts. The aircraft landed fast and ran off the end of the runway into a petrol station.

During a subsequent interview, the captain reported that he was aware that the aircraft was not in a stable condition at 1000 feet (300 m) above runway and was fully aware that this required him to execute a go-around (discontinue the approach). He indicated that he became fixated on the runway and could not explain why he did not perform a go-around maneuver. Forty-four passengers were injured.

Plan continuation bias

This evolutionarily inherited bias means that when we have settled on a plan of action, we are unlikely to change our minds, particularly when we are in a high-workload situation. For our more primitive ancestors, when energy was scarce, after making a plan to find food it was preferable to continue with that plan rather than switching to a different plan. Switching means that any energy expended up to that point was wasted and so natural selection favors the tendency not to keep switching plans. In aviation, this means that we are unlikely to change our plan when we are under high task-load or high time-pressure or have an underspecified problem (Box 2.8).

Representativeness heuristic

The underlying programming that leads to this heuristic is still unclear, although it does seem to be based on the idea of similarity matching. Representativeness in the context of aviation leads to pilots basing their beliefs about a current situation on those from their previous experiences despite the fact that there may be salient differences. In their book exploring the role of cognitive psychology in 19 crashes, Dismukes, Berman and Loukopoulos suggest that this heuristic can lead to risky behavior when there is no clear dividing line between what makes a situation safe and what makes a situation unsafe.[39] They give the example of an unstable approach as being susceptible to representativeness. If a pilot has performed unstable approaches in the past but has managed to land safely, they may be more inclined to allow a subsequent approach, one that is even more outside the stabilized approach criteria, to continue to a landing. Under high-workload conditions, a judgment is made about the safety of the approach based on prior knowledge (Box 2.8).

Automation bias

Very simply, we tend to trust what computers tell us. In an effort to decrease our own cognitive workload, we are overly willing to delegate decision making and diagnosis to automated systems and tend to believe what those systems tell us. This will be explored in more detail in Chapter 9 (Automation Management).

2.6.3.4   Social heuristics

Although some of these heuristics play a role in decision making, they also have an effect on how we regard the accuracy of our decisions, how we view others, particularly as resources, and how we interpret events that we are not directly associated with (e.g. accident case studies).

Overconfidence heuristic

The heuristic of overconfidence needs to be clearly defined from any personality trait that might lead to overconfidence. The heuristic of overconfidence is an internal process that relates to how we assess the accuracy of our stored knowledge and our perceptual models. Studies have repeatedly shown that humans have more confidence in the accuracy of their knowledge and actions than is justified.[41] The ability to assess the correctness of our knowledge and actions is known as calibration and humans have been repeatedly shown to have poor calibration. A normal process of feedback should lead to us refining our calibration ability and yet this does not seem to be the case. As we have described already, this heuristic of overconfidence biases our own judgment and heuristic-based biases cannot be overridden through training. The evolutionary basis for this is intriguing. It may seem counterintuitive that we have inherited a hardwired system that makes us overestimate our abilities. However, when we consider it in the context of our more primitive ancestors, it begins to make sense. Bullfrogs sometimes position themselves within hollow tree trunks and adapt their calls so that they resonate within the empty cavity. By doing this, they can amplify and deepen their characteristic mating call and so seem more dominant than they are in reality.[42] This is an example of an animal employing deception to achieve an objective. For more advanced organisms, an important social ability is being able to detect deception in others. For a social organism that is intent on convincing other members of its species that it is more dominant, strong or intelligent than it really is, the most reliable way of doing this without being detected is to deceive itself as well – to genuinely believe this to be the truth.[43] This is the reason that we are more confident in our abilities and knowledge than is justified (see Box 2.9).

Halo effect heuristic

The halo effect heuristic is a way for the brain to try to "complete" some elements of a perceptual mental model where no definitive information is available to fill in the gaps. The elements of the mental model that are most susceptible to the halo effect are those involving people. The halo effect uses information that we know about a person and infers other characteristics based on that information. It is the halo effect

Box 2.9 Accident Case Study – Overconfidence Heuristic

KLM Flight 4805 – Boeing 747-200
Los Rodeos Airport, Tenerife, 27 March 1977[44]
In heavy fog, this KLM Boeing 747-200 was backtracking along Runway 30 before take-off. A Pan-Am 747-100 was cleared to backtrack behind the KLM and vacate the runway towards the end. The KLM aircraft was due to depart first. The worsening visibility and the risk of the crew exceeding their flight duty limitations meant that there was some situational pressure to complete the take-off as quickly as possible. On turnaround, the captain evidently thought that the Pan-Am aircraft was clear of the runway as he advanced the thrust-levers. The first officer alerted him that they had not yet received their air traffic control (ATC) clearance. The thrust-levers were retarded and the captain said "I know that, go ahead, ask". After receiving ATC clearance (but not take-off clearance) the captain advanced the thrust-levers and began the take-off. Having heard some radio communication from the Pan-Am aircraft, the following exchange occurred between the KLM flight engineer and the captain (translated from Dutch):

> Flight Engineer: Is he not clear, then?
> Captain: What do you say?
> Flight Engineer: Is he not clear that Pan American?
> Captain: Oh, yes (emphatic – the Dutch word "jawel" was used, meaning an emphatic yes as opposed to "ja", that would just mean yes).

The KLM aircraft continued its take-off roll and 13 seconds later collided with the Pan-Am 747, killing 583 people. It remains the worst air disaster in history. The subsequent investigation highlighted that the KLM aircraft had taken off without clearance. Despite the evidence to the contrary, it is clear that the captain's mental model of the situation (a product of his sensory inputs, his attention and the data integration that occurred at the perception stage) was that the Pan-Am aircraft had vacated the runway. From what has been covered in the previous sections, once a mental model has been established, it is quite difficult to alter it in any significant way because of the feelings of cognitive dissonance that will occur during this process. The overconfidence heuristic adds to this by making us more sure of our knowledge and perceptions than is justified. The captain's response to the flight engineer's query about the Pan-Am aircraft still being on the runway suggests complete assuredness that he was clear of the runway. At this stage, it is important to differentiate the heuristic of overconfidence from the personality trait of overconfidence. As we will see in Chapter 5, we can choose to override personality tendencies using behavior. Heuristic programming in System 1 is almost impossible to overcome without engaging System 2.

that supports the view that first impressions are important. When someone knows very little about you except your appearance, their mental model of you will be "completed" based on the information that is to hand. Given that once a mental model is established it is difficult for it to be amended, it will then take incontrovertible evidence to adjust the model further. For example, if a pilot attends an interview for a job with a major national airline, if the interviewers have not had an opportunity to review his CV and so know nothing about him, the "gaps" in their mental model of him will start to be filled as soon as he enters the room. If he walks in wearing jeans, a torn T-shirt and a baseball cap, they may infer various things regarding his diligence, cautiousness, conformity and judgment. Remember, this is a function of System 1, and System 1 is automatic. The interviewers may have no control over this. They now have a mental model and it will be up to the candidate to demonstrate why these inferences about his attributes should be revised in light of his answers to the interview questions. If, on the other hand, the candidate arrives and is dressed conservatively with neatly cropped hair and presents the appearance of a pilot already employed by that company, he can use the halo effect to his advantage. All he has to do then is to make sure that none of his answers suggests that the interviewers should change their positive impression of him.

The halo effect has a role in crew interaction. If a crew are flying together for the first time or do not know each other particularly well, there are a great many boxes to fill in their mental models of each other. A first officer who is shabbily dressed and is wearing unpolished shoes and an unironed shirt may be assumed by a captain, especially if that captain is neatly dressed with highly polished shoes, to be inattentive to detail, unfocused and lazy. The reality is that these assumptions may or may not be true. It could be that he turns out to be an excellent first officer in every respect apart from his appearance. It could be that this first officer encountered a problem at home that meant he did not have an opportunity to iron his shirt or polish his shoes and this is an isolated event. The problem with the halo effect is that the captain may now have a fixed mental model that will affect how he views the first officer as a resource. He may be less likely to ask for or listen to suggestions, delegate responsibilities or trust the abilities of such a first officer unless that first officer can demonstrate that the captain's assumptions were unfounded. This is difficult to do if you are not given the opportunity. Conversely, a trainee first officer flying with a captain who happens to be the chief pilot may know only one thing: this person is the chief pilot. Based on that, he may assume that this captain has excellent flying and management skills and so be reluctant to question him when something does not appear to be going to plan, particularly early on in the flight before he has had a chance to gather enough information about the captain to make an accurate assessment of him.

Hindsight bias

This bias is hugely problematic in accident investigation and can severely limit the accuracy of the analysis and invalidate the recommendations. The evolutionary basis of hindsight bias is similar to that of fundamental attribution error (FAE) and will be covered in the next section.

Box 2.10 Accident Case Study – Hindsight Bias

TWA Flight 843 – Lockheed L-1011 TriStar
John F. Kennedy Airport, New York, USA, 30 July 1992[45]

After the aircraft had passed its V1 speed and began to rotate, the stick shaker activated a warning that the aircraft was about to stall. The first officer, who was the handling pilot, handed control over the captain. The aircraft was about 16 feet (5 m) off the runway when the captain elected to abort the take-off. He retarded the thrust-levers, and applied full reverse thrust and maximum braking. As there was insufficient runway remaining to stop and to avoid hitting a barrier at the end of the runway, the captain steered the aircraft left on to an area of open grass. During this maneuver, it became apparent that the nose wheel had collapsed and a fire had started. The aircraft came to a stop and all the passengers were evacuated.

The National Transportation Safety Board report was finalized on 31 March 1993, 8 months after the event. A fault was subsequently found in the stall detection system. The report stated that "the pilots should have concluded that the stick shaker was a false stall warning". It also stated that "the pilots' improper interpretations of information, their false perceptions, and their failure to evaluate all available information were major factors in the cause of this accident".

Evidence of hindsight bias, like FAE, is often found when the word "should" is used. In the case described in Box 2.10, the National Transportation Safety Board (NTSB) panel spent 8 months determining how the various clues both inside and outside the flight deck *should* have informed the captain's decision to abort the take-off. For the captain, he was faced with a truly awful situation: the golden rule says you must continue the take-off if you have passed V1, but the aircraft is telling you that it is about to stall. It is too late to stop but something is wrong that means you cannot safely continue. He probably had less than a second to make the decision to abort the take-off. If he had continued and the stall-warning had been a true one, the aircraft could have crashed shortly after take-off with a much higher likelihood of killing the passengers, much as happened to Air Florida Flight 90 that stalled on 13 January 1982 before crashing into the icy Potomac River, killing 78 people. There were no fatalities on TWA Flight 843.

The cause of hindsight bias is that it is very difficult to ignore knowledge. When investigators begin looking into an accident, they know what the final outcome of the situation was; for example, they know that the plane ultimately crashed. The tendency is to look at the sequence of events leading up to the accident with this knowledge in mind. This is completely the opposite situation that the crew would have been in at the time. They had no knowledge of what the outcome was going to be and so had to make decisions with the information that they had to hand. The more salient question that investigators need to ask is this: in the absence of knowing the outcome, what crucial information did not get integrated into the crew's mental model, and why? As we will introduce in Chapter 4 (Error Management and Standard Operating Procedures for Organizations), a crucial test is something called the substitution test. If another crew could make the same choice when given the same situation, this is probably not a liveware problem. The problem may be with the procedures or with the hardware.

Fundamental attribution error

In a similar manner to the heuristic of overconfidence, we have evolved mechanisms that allow us to see ourselves in a better, more positive light that is justified by our actual performance. These mechanisms of self-deception have evolved because the more positively we can present ourselves, the more likely we are to find a mate. We deceive ourselves because, in so doing, we do not have to consciously deceive others as to our imaginary enhanced capabilities. Humans are good at detecting lies and so we are less likely to be caught out if we actually deceive ourselves as well. FAE allows us to maintain this self-deception by attributing the negative outcomes experienced by others to their poor abilities while being able to attribute negative outcomes that we experience to situational variables outside our control. FAE is a major barrier to learning from case studies because it allows us to explain away accidents by attributing the cause to poor crew performance. In the same way as the word "should" is often symptomatic of hindsight bias, FAE is often characterized by "How could they miss that?", "How could they be so stupid?" or "I'd never do something like that". The startling fact about many of the accidents and incidents covered in this book is the experience level of the people involved. The vast majority of the pilots in these case studies are far more experienced than I am at the time of writing about them, and I have to accept the fact that it could easily be me in some of these situations. The only possible defense against FAE is consideration of a paraphrase that is attributed to John Bradford, a sixteenth century English reformer, when he saw prisoners being taken away to be executed: "There, but for the grace of God, go I". Once we accept that bad things can happen to any of us, we can start to look for ways to try to prevent them from happening. This is what this book is designed to do.

2.6.4   Strategies for decision making

The bad news for aviation decision makers is that the heuristic programming that System 1 uses, for all the benefits that it can give us, will lead to systematic errors under certain conditions. It is even more unfortunate that many of these conditions are exactly the same as those we must face in the flight deck when dealing with abnormal situations: high task-load, high time-pressure and underspecified, rapidly evolving, complex problems. Evidence has shown us that we cannot undo this programming. The nature of the hardwiring of these heuristics into the neuroanatomy of our decision-making mechanisms means that we cannot "deprogram" them. It would be like trying to use behavioral training to affect someone's brainstem so that they could stop and start their hearts at will. The body has a variety of functions, known as auto-nomic functions, that are beyond our conscious control. It is time to start considering System 1 as an autonomic function. There is only one way of preventing System 1 from causing us to make these errors and that is to do everything that we possibly can to engage System 2. The relationship between the two systems is such that the output of System 1 will lead directly to a decision, one that could be erroneous, unless that output is modified by System 2.

Our strategies for improving the quality of our decision making are based on allowing System 2 to function effectively. To do this, we need to address the three conditions that lead to an overreliance on System 1 outputs and consider how we can avoid the negative effects of System 1 if they do occur. We will look at these ele-ments individually and then consider how they can all be managed effectively in an operational setting.

2.6.4.1   Managing high task-load

System 2 cannot operative effectively if the task-load is too high. If a pilot is manually flying the aircraft, attempting to action a checklist or manipulating other systems, he will not be a position either to run or to participate effectively in a rational decision-making process. Given that someone who is walking needs to shed this relatively simple task when trying to solve a complex or multistep problem, a pilot who has a high task-load either has to shed some of these tasks to engage System 2 or is des-tined to make a System 1 based decision. Complex decisions are best made when the aircraft is in as stable a condition as possible, the relevant checklists are completed and the pilots have as low a task-load as possible. Of course, some emergencies do not permit this (e.g. US Airways Flight 1549, which ditched on the Hudson). In this case, System 1 can work to our advantage to allow us to make quick decisions when there is no opportunity to reduce the task-load.

2.6.4.2   Managing high time-pressure

A decision made under time-pressure will probably be made using System 1 heu-ristics, with all their potential drawbacks. In the same way as managing task-load, the pilot needs to consider what the actual time-pressure is and whether this can be reduced. Some technical malfunctions may appear very dramatic but incur no added time-pressure; for example, a generator failure causing multiple instrument failures. In this case, a rushed action may instinctively seem to be necessary to regain the lost instruments but could heighten the risk of error. Unless the nature of the emergency itself causes time-pressure, the question the pilot needs to consider is: "Is there any-thing that requires me to act quickly?" If there is a fuel leak, a pressurization problem or another worsening technical problem, quick action may be required, but if that is not the case, both pilots need to realize that they have got as much time as necessary to make the best decision that they can. It can also be advantageous to consider gen-erating more time. Slowing down to the maximum endurance speed gives more time to consider the decision as well as saving fuel. Entering a hold near an airport at this speed keeps the aircraft in a defined block of airspace, limits fuel consumption and can give the crew an opportunity to engage System 2.

2.6.4.3   Managing problem underspecificity

Some technical problems are complex and dynamic and may have features that are not readily apparent to the pilot during the decision-making process. System 1 will be frantically trying to match the perceived problem with any previous experience and may latch on to similarities between the current situation and previous situations without regard for salient differences. This is the representativeness heuristic at work and once this mental model becomes fixed, confirmation bias and the overconfidence heuristic will make it very difficult to amend. Actively searching for additional information is a high-workload task and is a function of System 2, which means that problem underspecificity can only be dealt with using a strategy that employs System 2. This strategy is given at the end of this section.

2.6.4.4   Managing System 1 effects

The automatic nature of System 1 means that as soon as master warnings are activated or gauges start showing strange readings, our heuristics will be rapidly trying to work out what is going on and devise a plan about what to do about it. This is how people can make decisions without even seeming to realize it. When we have not managed our task-load or time-pressure, that is, when we have not engaged System 2, we have one very important strategy that can prevent the inherent drawback of our System 1 programming from leading us down the wrong path: the System 2 of the person sitting next to us. This is a very important resource because it can highlight where perceptual differences occur between pilots and can prompt exploration as to which (if any) perception is correct. If two pilots have two different perceptions of a situation, they could both be wrong, one of them could be wrong, but they simply cannot both be right. The perceptual differences need to be analyzed and the correct perception shared before effective decision making can happen. In addition, if our own System 1 has made an error, it may only be the System 2 of our colleague that spots it. The next section introduces a decision-making strategy that integrates all these factors into a system that can be used operationally.

2.6.5   An operational decision-making strategy: TDODAR

There are several decision-making tools in use across the aviation industry and they are used with varying degrees of success. This section contains guidance for the most effective use of one of these tools, TDODAR (Time available, Diagnose, Options, Decide, Assign tasks, Review), but the principles are equally applicable to other formats. The recommendations for how this can be used to best effect in practice are based on what has been covered so far on information processing. For TDODAR to be effective, as many of the following conditions need to be established. These are designed to minimize the task-load and time-pressure for both pilots:

- Automation is engaged and flight mode annunciators are confirmed as appropriate.
- The aircraft is in level flight above the minimum safe altitude and preferably in visual conditions to avoid icing and allow perceptual cues from the environment to add to the pilots' mental models.
- The aircraft is in a flight management system (FMS) flown hold or under positive radar control (e.g. on vectors) and, if necessary, the speed and configuration are optimized to maximize endurance.
- All checklists required for initial management of the emergency are complete.
- Both pilots are ready and have an appropriately low task-load to participate in the decision-making process.

TDODAR should be led by the captain and both pilots should be trained in how to carry it out. The key features of a successful TDODAR are:

- Key steps are carried out independently. This gives the best chance of spotting System 1 errors.
- Although the captain leads the TDODAR, the first officer will be taking the lead for the early stages.
- It is carried out in a bullet-form fashion: facts are stated as briefly as possible in point form rather than prose.
- The attention-grabbing problem is not allowed to hijack the selective attention of both pilots. While procedures may differ between airlines and opinions may differ between captains, my recommendation is that the first officer takes or remains in control of the aircraft. Since the captain is more likely to be able to manage the situation and to ensure that his future decisions are made using System 2, his task-load should be minimized. In the case of a control problem, the captain may always resume control shortly before landing. However, it is more important in the early stages that he is able to take a broad view of the situation.
- The captain needs to make it very clear that the first officer now has the primary responsibility for the flight path of the aircraft, especially as managing the problem will inevitably mean that the captain will not be able to actively monitor the first officer as effectively as he would be able to normally.

The scenario used in this example is fictitious and is meant to demonstrate how the TDODAR process can work. The malfunction can be assumed to have occurred on a three-engine jet aircraft during a flight from Frankfurt Airport to London Heathrow Airport.

1. **Time** – The result of this stage will dictate how the rest of the process will be carried out.
   a. The captain and the first officer need to independently assess how much time is available to them before the aircraft needs to land.
   b. In likely order of severity, technical malfunctions that could severely limit the time available include:
      i. fire or indications of fire (in which case, it may be best to rely on System 1 rather than doing anything that may delay the aircraft from landing)
      ii. pressurization leak (checking this may alert the crew to go on oxygen)
      iii. icing, particularly if anti-icing systems are degraded:
         - worsening control problem
         - ice accumulation

      iv. fuel leak or fuel asymmetry

      v. hydraulic leak

      vi. electrics – running on battery power

      vii. worsening weather, either en route to or at the likely destination.

  c. Under many circumstances, the only limiting factor will be fuel. At this point, the pilots should elect what is the minimum amount of fuel they would be comfortable to land with (bearing in mind the possible need for a go-around). Clearly, the need for this step will be situation dependent and it probably would not be necessary for a technical failure shortly after take-off on a long-haul flight.

      i. Without a clear technical reason, the minimum fuel to land with should be the alternate fuel plus the final reserve fuel.

      ii. Once this figure is ascertained, the crew needs to consider how much fuel they have and calculate the fuel burn. This will give them the time they have remaining before they need to land.

      iii. Taking 5–10 minutes off this figure will give the time remaining before an approach needs to be commenced.

      iv. This information may have an impact on the rest of the decision-making process if it turns out that there is limited time and a suitable aerodrome still needs to be identified.

  d. The captain and the first officer need to compare their independent assessments of the time available to ensure that they are approximately the same. The most conservative time remaining until an approach needs to be started must be recorded in a prominent place or set as a countdown time on a flight deck timer.

  e. Below is an example of how an exchange between the captain and the first officer might go.

Captain: "How much time do you think we have before we need to land so that we have our alternate and final reserve fuel remaining?"

First Officer: "We have approximately 50 minutes before we need to land."

Captain: "I agree. In that case, we need to commence an approach no later than 45 minutes from now. This will be at time 20:05".

(This time should be recorded somewhere prominent.)

2. Diagnose – Both pilots will probably have a diagnosis in mind. The following procedure is meant to force the activation of System 2 and negate the effect of confirmation bias by having both pilots carry out independent assessments.

Captain: "What is your diagnosis of the problem?"

First Officer: "I think we've had an uncontained failure of the number 1 engine that has caused a complete loss of fluid from hydraulic system A".

Captain: "Why?"

First Officer (pointing to relevant instruments): "Engine 1 is reading no N1 or N2 rotation and the hydraulic fluid quantity of system A is reading zero".

Captain: "I agree. Can you detect any other system problems?"

First Officer [scans flight deck panels to assess whether any other systems have been affected, especially fuel, electrics and pressurization]: "No".

Captain [performs panel scan]: "I agree, but we will check with the cabin crew shortly to see if they can detect anything wrong".

3. Options – It is important to generate as many options as possible.

Captain: "What options are available to us and what do you think of each?"

There are normally at least three options: go back to the departure airport, divert or continue to the destination.

First Officer: "We could continue to London Heathrow Airport but there is a strong cross-wind forecast that could make landing with asymmetric thrust difficult. We could return to Frankfurt but we would need to commence an approach straightaway when we get there. We could divert to Amsterdam. We are nearby and there are several runways available to us that would limit the need for a crosswind landing. We could also divert to Brussels. It is nearer but there are fewer runways. The hydraulic loss checklist said we need to check performance as we'll have lost some spoilers but all of those runways should be easily long enough".

Captain: "The only other option I can see is that we could also continue to the UK and land at an airport other than Heathrow. This would be more convenient for the passengers but we can't guarantee the wind conditions for when we arrive".

4. **Decide** – At this point, the captain takes the lead in the process.
Captain: "Based on all the options, my decision is to divert to Amsterdam unless you think another option is safer".
First Officer: "I agree with that decision".

5. **Assign tasks** – The captain now needs to manage the diversion as well as informing the relevant people:
Captain: "I would like you to remain as the handling pilot and also to take the radios. Please tell ATC we'd like to divert to Amsterdam and we'd like a direct routing there now. Once en route, you could ask for the current weather and then ask for a routing to a holding fix nearest the approach for the runway that will give us the least crosswind. I'm going to talk to the cabin crew and get them to check all the engines. I'll call the cabin supervisor into the flight deck and do a NITS [Nature, Intentions, Time, Special instructions] briefing. Once that's done, I'll talk to the passengers as well. Is there anything else we need to do?"
First Officer: "Before you do all that, could we run the descent checklist and set up some guidance in the FMS just to provide some extra information? We'll also need to check our landing performance because we've lost some spoilers and our go-around performance is going to be degraded".
Captain: "Of course, I'll do that for you in a moment".

6. **Review** – Once both pilots begin carrying out their assigned tasks, the task-load is going to increase for both. This step needs to be carried out now and then repeated at every opportunity when the task-load is sufficiently low. It is designed to check that nothing has been missed in the decision-making process and that the situation hasn't changed or deteriorated to the extent that the plan needs to change.
Captain: "Let's do a review. We now have 30 minutes before we need to start an approach. We had an uncontained engine 1 failure leading to loss of fluid from hydraulic system A. All other systems are normal. The active problems are asymmetric thrust that we're compensating for and a loss of spoilers which means we need to check the landing performance. We're going to land on the most into-wind runway at Amsterdam but we can hold initially near the approach to finish our tasks. You're going to be handling pilot and I'm going to manage the other aspects of the flight. We'll do another review after the cabin crew have checked the engines. Can you think of anything at all that we've missed?"
First Officer: "No".

An effectively managed TDODAR can negate the effects of the representativeness and availability heuristics, confirmation and automation bias, and overconfidence. By asking direct questions to the first officer without revealing what his perception of the situation is, the captain can get "uncontaminated" analysis from the first officer, and by posing a specific question and asking him to justify his answers, there is a high chance that the first officer is going to have to engage System 2. It makes sense that the crew should not move from one stage to the next stage until they have reached an agreement. For example, there is no point discussing options if the crew cannot decide what the diagnosis is. At the decision stage, there is a chance that the captain will decide to do something that the first officer does not agree with. For example, in this scenario, he may have elected to continue to London Heathrow Airport. If the first officer is concerned by this, he needs to make this very clear to the captain. The reasons for this are covered in more detail in Chapter 3 (Error Management and Standard Operating Procedures for Pilots) and strategies for doing this are covered in Chapter 6 (Communication). However, the rule for making an effective decision using TDODAR is that if one pilot believes that a more conservative option is the better one, that should be the option that is selected. The reason for this is that the overconfidence heuristic may be having an impact on the captain's decision. For example, he may believe that a crosswind approach into London Heathrow Airport with asymmetric thrust is within his capabilities. The first officer may prefer not to test this. In short, when there is a disagreement about the best course of action, the more conservative course should be selected.

TDODAR is useful in any situation where there is any element of uncertainty. It would be of no use when there is absolutely no uncertainty or doubt about what to do next, for example, a cabin fire on the ground. The management of most abnormalities can only be improved by carrying out a well-timed, succinct TDODAR and this interactive process may bring to light some active problems that were not originally noted when the event occurred.

Even if TDODAR has been carried out, there is a chance that the situation might change. The risk now is that the crew experience plan continuation bias. If a change in situation has been noticed, the same discipline needs to be used to reconsider the problem from scratch. Without doing this, if the workload increases, both pilots may become susceptible to plan continuation bias. Workload management and using low-workload periods to review the situation gives the best chance of spotting a change in conditions that might warrant a second TDODAR or, at least, a review of the current plan.

2.7 Response

If we assume that our sensory information is giving us an accurate representation of our environment, our attention is in the right place, the mental model we have formed at the perception stage is accurate and our decision about what to do based on our perception is valid, there is only one remaining stage at which things can go wrong. If the plan of action we have decided on is correct, we may still end up with an undesired outcome if we fail to action it correctly. The response stage involves a coordinated motor response based on our plan of action. This could be flicking a switch, pressing a button or some other action. Performing a motor action requires coordination and feedback and when there is a problem with either of these, we may make a skill-based error known as a slip. If the plan we decided on requires several switches to be manipulated, we may simply forget to move one particular switch. This, again, is a skill-based error and is known as a lapse. Skill-based errors such as slips and lapses are covered in much greater detail in Chapter 3 but, briefly, they can be managed by confirming critical switch selections with the other pilot, making only one switch selection at a time and using checklists to identify if any lapses have occurred.

2.8   Fear-potentiated startle and freezing

Despite the limitations that we might experience at each stage of information process-ing, based on the strategies given in each section, we have a good chance of preserv-ing the integrity of our information processing and being able to come up with an appropriate response to a particular stimulus. There is, however, one event that can cause significant disruption to the entire information processing mechanism and lead to serious consequences – startle.

In an analysis of the effects of startle on pilots during critical events, it was noted that highly reliable automated systems and the ultra-high reliability of the aviation industry mean that pilots have become accustomed to the idea that they are highly unlikely to face a critical emergency in their career.[47] To quickly transition from this assumed level of safety to a flight condition that is confusing, sudden and potentially life threatening can induce fear-potentiated startle. Aside from the system that we use for normal information processing, there is a subsidiary system that processes sensory stimuli that have high emotional significance, such as potentially life-threatening stimuli. Sensory information is routed through the thalamus of the brain. Most of the time, this sensory information is conducted to different areas of the brain where it can affect attentional focus or be integrated using perceptual systems to form a mental model. However, there is also a connection between the thalamus and the emotional processing systems of the brain located in the limbic system, a system that is towards the center of the brain and is shared by many other animals. The amygdala, part of the limbic system, is responsible for some emotional processing, including the fear response. Fear-inducing stimuli can be recognized by the amygdala within 14 ms, significantly faster than the time it takes for the basal ganglia to activate a single production rule (50 ms). In fact, to formulate any sort of perceptual mental model takes proximately 500 ms and so the fear reaction is likely to be the first reaction that is perceived. The activation of the amygdala causes a variety of autonomic changes associated with the fight-or-flight reaction: increased heart rate and breathing rate, dilation of the pupils and production of stress hormones such as adrenalin and, most importantly, has been shown to inhibit the rest of the information processing system and limit the functioning of the perceptual, decision-making and response systems. This is because the amygdala is recruiting cognitive resources to mount a response to the perceived threat and the perceived fear, like any acute stress, is also occupying cognitive resources, as suggested by Attentional Control Theory.[33] This will lead to inhibition of the normal information processing channels, an observation that is sup-ported by the cases where humans have reacted in what would seem to be a highly inappropriate way after being startled (Box 2.11).

Box 2.11 Accident Case Study – Fear-Potentiated Startle

Air France Flight 447 – Airbus A330-200
Mid-Atlantic Ocean, 1 June 2009[46]

In the early hours of the morning, while flying at Flight Level 350 (35,000 feet or 10,000 m), the pitot probes of this Airbus A330-200 become blocked with ice crystals, leading to erroneous speed indications in the flight deck. This caused the autopilot and autothrottle to disconnect and the aircraft flight control law transitioned to alternate law, meaning that the normal Airbus stall prevention mechanisms were no longer active. At this time, the captain was not on the flight deck and there were two first officers occupying the seats in the flight deck. Shortly after the autopilot and autothrottle disconnected, the stall warning triggered briefly on two occasions. At this time, the airspeed indications were erratic. The handling pilot made rapid roll control inputs and then pitched the aircraft up to 11 degrees. The aircraft climbed towards 37,000 feet (~11,000 m). The monitoring pilots asked the handling pilot to descend and some inputs were made to reduce the pitch but the aircraft continued to climb. Shortly afterwards, the stall warning was triggered again in a continuous manner. The thrust-levers were advanced but the handling pilot made pitch-up inputs to the side-stick instead of pitch-down inputs as would be required in a stall situation. By the time the captain re-entered the flight deck, the altitude was 35,000 feet (10,000 m), the angle of attack was 40 degrees and they were descending at 10,000 feet (3000 m) per minute. The handling pilot continued commanding a pitch-up using the side-stick for nearly the entire duration of the event and it was deemed to have caused the excessive nose high attitude that led to the stall. Because the Airbus side-sticks are not linked, it is unlikely that the monitoring pilot was aware of this. The report notes that this unusual and persistent control input could have been caused by surprise. Unfortunately, the aircraft did not recover from the stall and it crashed into the sea, killing all 228 people on board.

For the handling pilot of Air France Flight 447, maintaining a persistent pitch-up command to the side-stick was probably a startle-mediated act caused by the sudden failure of multiple systems, the rapidly deteriorating situation and the understandable fear these events caused. Martin, Murray and Bates also identified this erroneous pitch-up tendency in two other crashes after the handling pilot was startled by the stick shaker: Colgan Air Flight 3407 and Pinnacle Airlines Flight 3701. To investigate this further, they conducted an experiment where 18 pilots were asked to fly a simulated instrument landing system approach in low-visibility conditions.[48] The cloud base was set at 100 feet (30 m) and the decision altitude was 200 feet (60 m), meaning that a go-around would have to be performed. At 240 feet (73 m), a startling stimulus was generated (a loud bang and the cargo fire warning bell activating). Five pilots performed optimally and went around and six pilots showed some delay in executing the go-around. Seven pilots performed significantly worse after being startled; two out of these seven initiated a go-around before reaching the decision altitude, three pilots continued to execute the go-around from below 100 feet and two landed. The startle caused an average delay of 36 feet (11 m) before the go-around was commenced and most of the participants noted that they felt a physiological response to the startle (an increase in heart rate or a perceived increase in adrenalin) and most reported a period of confusion or indecision after the startle. Other research has suggested that there are individual differences in how people react to a startle, with some being low reactors and some being categorized as "hyperstartlers".[49]

Given that fear-potentiated startle can inhibit information processing and delay responses (such as executing a go-around when there is no runway contact at decision altitude), there is another reaction that may be connected to this one: freezing. A sudden, unexpected and potentially life-threatening stimulus may be met not with an abnormal reaction, but with an absence of reaction. In a paper exploring this phenomenon, John Leach notes that freezing has a significant impact on an individuals' survival chances after a serious event.[50] He notes that several reports into evacuations from aircraft and ships stated that many people were unduly slow in evacuating and some even remained passive and stiff despite having the possibility of escape. After analyzing several of these cases, he suggests that response types can be divided into three groups:

- 10–15% of people will remain calm and their information processing system will be capable of assessing the situation, deciding on a plan and then acting on it
- 75% of people will be stunned and bewildered and their information processing system will be impaired; this effect is consistent with fear-potentiated startle
- 10–15% of people will demonstrate highly counterproductive behavior such as uncontrolled weeping, confusion, screaming and paralyzing anxiety.

Leach also suggests an interesting neural basis for why people might react in different ways. Translating his analysis into the language that we have used to talk about the neural basis of information processing, he suggests:

- If an appropriate response has been considered in advance and a production rule (or set of production rules) has been mentally primed and rehearsed in the procedural memory store, if the event were to actually happen, these production rules could be activated within 100 ms and allow the person to have immediate access to a course of action that can then be executed.
- If there is more than one rule to chose from, the decision-making mechanism will have to run some trial pattern matches, each taking about 50 ms. Given that it is likely that System 1 will be in use, the representativeness heuristic will settle on a rule that would work in a similar situation and then activate it. This process may take 1–2 seconds.
- If there are no production rules that are suitable for use, the individual has no choice but to try and construct a behavioral response from scratch. This could take 8–10 seconds, by which time the situation could have changed to the extent that the formulated plan of action is no longer valid. It is also possible that an intense fear reaction will completely inhibit this process and the person will be, for all intents and purposes, paralyzed. The report into the sinking of the MS Estonia ferry on 28 September 1994 that killed 852 people (significantly more than died in the largest aviation accident) describes passengers who did not respond at all to the events happening around them even when they were thrown lifejackets, people

who "… could find no options for rational action". Leach goes on to suggest that, by this point, they were probably incapable of coming up with any options for rational action.

Although this analysis was done with regard to passengers, there would seem to be a link to the fear-potentiated startle reactions that can occur in the flight deck. This may provide some clues about how best to manage the risk of startle and the associated risk of freezing.

2.8.1 Strategies for managing fear-potentiated startle and freezing

The evidence would suggest that startle and freezing have a similar neural basis. Unfortunately, any startle reaction will have a headstart on normal information processing by virtue of the rapid neural conduction that occurs between the thalamus and the amygdala and then from the amygdala to the stress response mechanisms. While it seems that there may be individual differences in how people respond to sudden, critical, life-threatening emergencies, there does seem to be one strategy that can limit the negative effects of startle and freezing: having a prepared plan of action. In an experiment conducted within an Australian airline, pilots were encouraged to discuss novel events with colleagues and explain what they would do in response, and why.[51] The usual starting point for such a discussion would be one pilot saying to the other: "What would you do if …?" The desired effect was to partly counteract the perception of ultra-reliability of aviation systems that can lead to such surprise on the rare occasions that they fail. Discussions would also lead to pilots having to formulate a plan to deal with such an emergency. Although the likelihood of needing such a plan remains highly unlikely, it is possible that this production rule will remain in procedural memory and, if it is mentally rehearsed occasionally, will be quickly accessible should the need arise.

We frequently train for critical emergencies but there are so many potential combinations of circumstances that it is not possible to train for all of them. Discussion and mental rehearsal of plans of action in hypothetical scenarios may allow a pilot to build up a store of production rules that could improve performance should a serious event occur.

2.9 A note about the models used in this chapter

Writing this chapter in a systematic way has proved quite a challenge. In attempting to present a cohesive model of information processing, I have had to bring together (and reconcile the differences between) several psychological and neuroscience-based theories. In particular, the section on decision making (Section 2.6) involved combining two major approaches to give as complete a picture as possible of what is thought to be going on in the brain when decisions are being made. The ACT-R model of how various brain regions interact to make decisions needed to be combined with our understanding of heuristics (the mental shortcuts that we are able to employ to make rough-and-ready decisions under high-workload conditions). Unfortunately, as well as the fact that research in both fields in ongoing, there is no Rosetta stone for translating the language of one approach into the language of the other and so some links that I have made between the two may be open to debate. However, as was stated and has often been repeated, human factors is a huge and growing science and I have had to be selective about which theories and approaches I include in this book. I have tried to choose those approaches that have a significant weight of research behind them and allow some practical guidance to be derived from them. The ACT-R model is, arguably, the most comprehensive model of human cognition ever devised. It can predict the behavior of the brain in a bottom-up fashion from an understanding of what is going on at the level of the neurons. Conversely, the work on the dual modes of human decision making takes a top-down approach and attempts to explain brain activity by observing and testing behavior. While I generally prefer to use models based on neuroscience, the findings of this psychological approach and the insights it gives into decision making, especially in uncertain situations, are too profound to be ignored.

And here you have the difficulty: one approach looks at the intricate mechanics of simple decision making and works upwards. The other looks at complex human behavior and works downwards. I have tried to fill the gap in the middle so that it all makes sense.

Chapter key points

1. Information processing is the primary subject upon which the rest of human factors is based.
2. Failures or inaccuracies of information processing are the cause of every type of error.
3. Different brain regions are responsible for different functions and each stage of information processing uses different areas of the brain.
4. The mechanisms of all the stages rely on how neurons interact with each other.
5. Neurons that are repeatedly activated at the same time will have more chance of being activated synchronously if any of them are activated again. This is the basis of learning.
6. The stages of information processing are sensation, attention, perception, decision making and response.
7. Sensation is limited by sensory thresholds, sensory habituation and sensory illusions. Many of these negative effects can be overcome by relying on flight deck instrumentation.
8. Attention may be selective, sustained or divided.
9. Selective attention has the benefit of allowing complex tasks to be carried out, but this can lead to failure to notice other salient changes in the environment due to the effects of change blindness.
10. Extra effort may be required to get someone's attention when they are selectively attending to a task.
11. Manual flight of the aircraft requires selective attention, and any stimuli or tasks that could disrupt this should be avoided.
12. Sustained attention is very difficult for humans to maintain for any meaningful period of time. A vigilance decrement occurs quickly.
13. Being well rested, handing over the task between pilots and sensible use of caffeine may delay the vigilance decrement, but it is preferable to avoid tasks that require sustained attention altogether.

14. Divided attention is possible but requires careful workload management, limiting the number of tasks and ensuring that tasks use different sensory and response modalities.

15. Perception allows us to build a mental model of the situation. This perception is based on integration of either a complete or an incomplete set of sense-derived information. When information is incomplete, the brain often tries to fill in gaps based on past experience. We are not aware of this happening.

16. Not having enough information to form a mental model leads to cognitive dissonance, a highly unpleasant feeling of not knowing what is going on.

17. Once a mental model has been formed, particularly if it has been difficult to form it, the brain is reluctant to revise it either partly or completely.

18. Perceptual illusions occur when our brains adjust our perception to align it with previously stored knowledge.

19. Despite the highly convincing nature of perceptual illusions, they can be managed by briefing their possibility based on current weather conditions, continuing to cross-check instrumentation during visual flight and avoiding night visual approaches wherever possible.

20. Decision making relies on complex interactions between aspects of the mental model, a knowledge of the goal and the memory stores of production rules and knowledge that may help to achieve the required goal.

21. The production module will attempt to match the relevant situation that requires a decision to be made with prestored production rules in order to come up with the best response. The number of matching attempts the production module can try is limited by workload.

22. Humans have two systems that allow them to make decisions: System 1 and System 2.

23. System 1 relies on preprogrammed heuristics and can make seemingly complex decisions under high-workload conditions.

24. Heuristics can be thought of as mental shortcuts that allow us to come up with approximately correct answers in a short period of time where it would take significantly longer to derive precise answers.

25. Although System 1 can be life-saving, the heuristics it uses are susceptible to systematic errors.

26. If time allows, the more rational System 2 can modify the output of System 1 and may correct these heuristic-induced errors.

27. System 1 outputs are more likely to be relied upon under conditions of high task-load and high time-pressure and where the problem is underspecified.

28. Heuristics are impossible to eliminate because they are hardwired into our brains through evolution.

29. Many heuristics gave our primitive ancestors certain advantages but can cause us significant problems because the conditions under which pilots have to operate are significantly different from those that our brains have evolved to cope with.

30. There are several heuristics that affect decision making, memory retrieval and social interactions that have a significant impact on flight operations and accident investigation.

31. The goal of a pilot when solving a problem should be to engage System 2 wherever possible and this can be done by managing task-load, time-pressure, problem underspecificity and the effects of System 1. This can be achieved using a structured decision-making tool like TDODAR.

32. Problems at the response stage are due to skill-based slips and lapses and can be addressed by confirming critical switch selections, making one switch selection at a time and backing up multiple selections using checklists.

33. Sudden, unexpected and critical emergencies may lead to fear-potentiated startle or freezing.

34. The information processing system can be severely disrupted by fear-potentiated startle. This may be so severe that the individual freezes.
35. Discussing and mentally practicing for emergency scenarios may minimize the effects of fear-potentiated startle and freezing.

Recommended reading

Kahneman D. *Thinking, fast and slow.* Macmillan; 2011. This book must be read by anyone who wants to understand decision making. It is an extraordinary read and is written by a Nobel Laureate who rewrote our understanding of how the human mind makes decisions. If you read only one thing recommended in this book, it should be *Thinking, fast and slow* [Difficulty: Easy/Intermediate].

Gilovich T, Griffin DW, Kahneman D, editors. *Heuristics and biases: The psychology of intuitive judgment.* Cambridge University Press; 2002. Contains many of the seminal papers that contribute to the dual system view of human cognition as well as papers on many of the heuristics that underlie System 1 [Difficulty: Advanced].

Dismukes RK, Berman BA, Loukopoulos LD. *The limits of expertise.* Ashgate; 2007. A fascinating reanalysis of 19 aviation accidents based on contemporary knowledge of applied psychology [Difficulty: Intermediate].

Gigerenzer G. *Gut feelings.* Penguin; 2007. A good introduction to the science of decision making [Difficulty: Easy].

Gigerenzer G. *Reckoning with risk.* Penguin; 2002. Covers how we assign risk values to different events [Difficulty: Easy].

Anderson J. *How can the human mind occur in the physical universe?* Oxford University Press; 2007. This incredible book lays out the ACT-R cognitive architecture theory for explaining how the mind works based on the actions of neurons. It is quite a complex read and assumes some knowledge of computational neuroscience [Difficulty: Very Advanced].

References

1. Kahneman D. *Thinking, fast and slow.* New York: Macmillan; 2011.
2. Hebb DO. *The organization of behavior: A neuropsychological approach.* New York: John Wiley & Sons; 1949.
3. Brown R, Galanter E, Hess EH, Mandler G. *New directions in psychology.* Oxford: Holt, Rinehart, & Winston; 1962.
4. Benson AJ. Sensory functions and limitations of the vestibular system. In: Warren R, Wertheim AH, editors. *Perception and control of self-motion.* Hillsdale, NJ: Lawrence Erlbaum Associates; 1990. pp. 145–70.
5. National Transportation Safety Committee. *Aircraft accident investigation report: Boeing 737–4Q8, PK–KKW, Makassar Strait, Sulawesi, Republic of Indonesia, 1 January 2007.* (KNKT/07.01/08.01.36) 2008.
6. Flight Safety Foundation Inflight spatial disorientation. *Human Factors and Aviation Medicine* 1992;**39**(1):1–6.
7. Zigmond MJ, Bloom FE. *Fundamental neuroscience.* San Diego, CA: Academic Press; 1999.

8. Aviation Safety Network. (n.d.). Retrieved March 16, 2014, from <http://aviation-safety.net/database/record.php?id=19780101-1>.

9. Koch K, McLean J, Segev R, Freed MA, Berry II MJ, Balasubramanian V, Sterling P. How much the eye tells the brain. *Current Biology* 2006;**16**(14):1428–34.

10. National Transportation Safety Board. (1973). *Aircraft accident report: Eastern Air Lines, Inc. L-1011, N310EA, Miami, Florida, December 29, 1972.* (NTSB-AAR-73-14).

11. Chabris CF, Simons DJ. *The invisible gorilla: And other ways our intuitions deceive us.* London: HarperCollins; 2011.

12. Levin DT, Simons DJ, Angelone BL, Chabris CF. Memory for centrally attended changing objects in an incidental real-world change detection paradigm. *British Journal of Psychology* 2002;**93**(3):289–302.

13. Sarter M, Givens B, Bruno JP. The cognitive neuroscience of sustained attention: Where top–down meets bottom–up. *Brain Research Review* 2001;**35**(2):146–60.

14. Warm JS, Parasuraman R, Matthews G. Vigilance requires hard mental work and is stressful. *Human Factors* 2008;**50**(3):433–41.

15. Helton WS, Dember WN, Warm JS, Matthews G. Optimism, pessimism, and false failure feedback: Effects on vigilance performance. *Current Psychology* 1999;**18**(4):311–25.

16. Warm JS, Parasuraman R. Cerebral hemodynamics and vigilance. In: Parasuraman R, Rizzo M, editors. *Neuroergonomics: The brain at work*, 2007. Oxford: Oxford University Press; 2007. pp. 146–58.

17. Lim J, Dinges DF. Sleep deprivation and vigilant attention. *Annals of the New York Academy of Science* 2008;**1129**(1):305–22.

18. Shaw TH, Matthews G, Warm JS, Finomore VS, Silverman L, Costa Jr. PT. Individual differences in vigilance: Personality, ability and states of stress. *Journal of Research in Personality* 2010;**44**(3):297–308.

19. Mackworth JF. Effect of amphetamine on the detectability of signals in a vigilance task. *Canadian Journal of Psychology* 1965;**19**(2):104–10.

20. Koelega HS. Stimulant drugs and vigilance performance: A review. *Psychopharmacology* 1993;**111**(1):1–16.

21. Salvucci DD, Taatgen NA. Threaded cognition: An integrated theory of concurrent multitasking. *Psychological Review* 2008;**115**(1):101–30.

22. Wickens CD. Multiple resources and performance prediction. *Theoretical Issues in Ergonomics Science* 2002;**3**(2):159–77.

23. Ellis HD, Lewis MB. Capgras delusion: A window on face recognition. *Trends in Cognitive Sciences* 2001;**5**(4):149–56.

24. Transportation Safety Board of Canada. *Aviation occurrence report: Tail strike on landing. Canadian Airlines Internations, Boeing 767-375 C-FOCA, Halifax, Nova Scotia, 08 March 1996.* (A96A0035) 1996.

25. Gibb RW. Visual spatial disorientation: Revisiting the black hole illusion. *Aviation, Space, and Environmental Medicine* 2007;**78**(8):801–8.

26. National Transportation Safety Board. *Aircraft accident report: Collision with trees on final approach, Federal Express flight 1478, Boeing 727-232, N497FE, Tallahassee, Florida, July 26, 2002.* (NTB/AAR-04/02 PB2004-910402) 2004.

27. Interstate Aviation Committee Air Accident Investigation Commission. *Boeing 737-505, VP-BKO, Aeroflot-Nord 821, September 14, 2008.* n.d.

28. Federal Department of the Environment, Transport, Energy and Communications. *Final report of the aircraft accident investigation bureau on the accident to the Saab 340B aircraft, registration HB-AKK of Crossair flight CRX 498 on 10 January 2000 near Nassenwil/ZH.* (No. 1781) 2002.

29. Flight Safety Foundation   Killers in aviation: FSF task force presents facts about approach-and-landing and controlled-flight-into-terrain accidents. *Flight Safety Digest* 1999;**17**(11–12), 18(1–2).

30. Kahneman D. Maps of bounded rationality: Psychology for behavioral economics. *American Economic Review* 2003;**93**(5):1449–75.

31. Anderson JR, Bothell D, Byrne MD, Douglass S, Lebiere C, Qin Y. An integrated theory of the mind. *Psychological Review* 2004;**111**(4):1036–60.

32. Anderson JR. *How can the human mind occur in the physical universe?* New York: Oxford University Press; 2007.

33. Eysenck MW, Derakshan N, Santos R, Calvo MG. Anxiety and cognitive performance: Attentional Control Theory. *Emotion* 2007;**7**(2):336–53.

34. Evans JSB. Dual-processing accounts of reasoning, judgment, and social cognition. *Annual Review of Psychology* 2008;**59**:255–78.

35. Klein GA. A recognition-primed decision (RPD) model of rapid decision making. In: Klein GA, Orasanu J, Calderwood R, Zsambok CE, editors. *Decision making in action: Models and methods.* Norwood, NJ: Ablex; 1993. pp. 138–47.

36. Gilovich T, Griffin D, Kahneman D, editors. *Heuristics and biases: The psychology of intuitive judgment.* Cambridge: Cambridge University Press; 2002.

37. Gigerenzer G. *Gut feelings: Short cuts to better decision making.* London: Penguin Books; 2007.

38. Air Accidents Investigation Branch. *Report on the accident to Boeing 737-400 G-OBME near Kegworth, Leicestershire on 8 January 1989.* (Aircraft accident report 4/90) 1990.

39. Dismukes K, Berman BA, Loukopoulos LD. *The limits of expertise: Rethinking pilot error and the causes of airline accidents.* Farnham, UK: Ashgate; 2007.

40. National Transportation Safety Board. *Aircraft accident brief: Southwest Airlines flight 1455.* (DCA00MA030) 2002.

41. Lichtenstein S, Fischhoff B, Phillips LD. Calibration of probabilities: The state of the art to 1980. In: Kahneman D, Slovic P, Tversky A, editors. *Judgment under uncertainty: Heuristics and biases.* Cambridge: Cambridge University Press; 1982. p. 306–34.

42. Lardner B, bin Lakim M. Animal communication: Tree-hole frogs exploit resonance effects. *Nature* 2002;**420**(6915):475.

43. Trivers R. The elements of a scientific theory of self-deception. *Annals of the New York Academy of Science* 2000;**907**(1):114–31.

44. Netherlands Aviation Safety Board. *Final report and comments of the Netherlands aviation safety board of the investigation into the accident with the collision of KLM flight 4805, Boeing 747-206B, PH-BUF and Pan American flight 1736, Boeing 747-121, N736PA at Tenerife airport, Spain on 27 March 1977.* (ICAO Circular 153-AN/56) n.d.

45. National Transportation Safety Board. *Aircraft accident report: Aborted takeoff shortly after liftoff, Trans World Airlines flight 843, Lockheed E-1011, N11002, John F. Kennedy International Airport, Jamaica, New York, July 30, 1992.* (NTSB/AAR-93/04 PB93-910404) 1993.

46. Bureau d'Enquêtes et d'Analyses. *Final report on the accident on 1st June 2009 to the Airbus A330-203 registered F-GZCP operated by Air France flight AF 447 Rio de Janeiro–Paris* 2012.

47. Martin W., Bates P.R., & Murray P.S. (2010). The effects of stress on pilot reactions to unexpected, novel, and emergency events. In Proceedings of the 9th International Symposium of the Australian Aviation Psychology Association, Sydney, Australia (pp. 263–6).

48. Martin W, Murray PS, & Bates PR. (2012). The effects of startle on pilots during critical events: A case study analysis. In Proceedings of the 30th Conference of the European Association for Aviation Psychology, Sardinia, Italy (pp. 388–94).
49. Simons RC. *Boo! Culture, experience, and the startle reflex.* New York: Oxford University Press; 1996.
50. Leach J. Why people freeze in an emergency: Temporal and cognitive constraints on survival responses. *Aviation, Space, and Environmental Medicine* 2004;**75**(6):539–42.
51. Martin WL, Murray PS, Bates PR. What would you do if …? Improving pilot performance during unexpected events through in-flight scenario discussions. *Aeronautica* 2011;**1**(1):8–22.

Error management and standard operating procedures for pilots

Chapter Contents

Introduction   77
3.1 Performance levels   78
3.2 Errors and violations at different performance levels   79
3.3 Detection of errors   82
3.4 The Swiss Cheese Model   84
3.5 Threat and Error Management 2   86
    3.5.1 Threat management   86
        3.5.1.1 Threat management opportunities   88
        3.5.1.2 Types of threat   88
        3.5.1.3 Threats associated with serious negative outcomes   90
        3.5.1.4 Threat identification framework   95
        3.5.1.5 Generic threat management strategies   97
        3.5.1.6 Threat management tool for briefings   97
    3.5.2 Unsafe act (error and violation) management   97
        3.5.2.1 Unsafe act prevention where no threat has been identified   98
        3.5.2.2 Unsafe act detection: checklists and crew communication   99
        3.5.2.3 Unsafe act management strategies   105
    3.5.3 Undesired aircraft state management   106
    3.5.4 Summary of TEM2   108
    3.5.5 History of Threat and Error Management and differences between original TEM and TEM2   108
3.6 TEM2 and unstabilized approaches   110
    3.6.1 Threat management for unstabilized approaches   111
    3.6.2 Unsafe act management for unstabilized approaches   113
    3.6.3 Undesired aircraft state management for unstabilized approaches   113
Chapter key points   115
Recommended reading   116
References   116

Introduction

Before you start reading this chapter, you should set aside your preconceptions of what the word "error" means. As you will see from what follows, error is a surprisingly tricky concept to get to grips with. At first glance, this is surprising because "error" is a word that we use regularly. It is rare to see a news report on an aircraft crash that does not mention the term "pilot error". Error is seemingly all around us and yet when you ask people to define what they mean by error, the responses can vary considerably. If error was just a concept like any other in human factors, this lack of clarity in the definition might not be so problematic. The problem arises because people have and will continue to lose their jobs on the basis of errors that others perceive they have made. This book's coverage of error is divided into two parts, each part being given a chapter of its own: the first chapter (this one) looks at error and error management as a flight crew might encounter it on a day-to-day basis. The other chapter (Chapter 4) takes an organizational perspective on error and how to manage it in the increasingly complex systems within which we work.

The aim of this chapter is to equip the pilot with some practical tools to reduce the risk of error and to improve the rates of error detection and correction. At its core is the concept of Threat and Error Management (TEM). For the purposes of this chapter, the original TEM model has been amended slightly so that it can be more fully integrated into day-to-day flight operations and includes some more contemporary research into human performance. To distinguish it from the original TEM model, it will be referred to as TEM2.

3.1   Performance levels

No discussion of human error is possible without mentioning Professor James Reason. His groundbreaking work, particularly during the 1990s, forms the basis of contemporary thinking in the field of industrial safety and marked a turning point in our understanding of the nature of error. His book *Human Error* is a masterpiece in psychological writing and without reading it any human factors practitioner is missing out on a veritable treasure trove of information that is still at the heart of error management more than 20 years later.[1] The early part of this chapter is largely based on his work, particularly *Human Error*, and although you will see that at the end of this chapter this book is listed as having an intermediate difficulty level, it is well worth the effort to gain an insight into how our understanding of error has changed over the years. It is also convenient that Reason dedicated his book to Jens Rasmussen, a highly respected researcher in the field of industrial safety, because it is one of his theories of human performance that forms the basis of the start of this chapter. Although some of these theories may be new to you, they have formed the foundation of today's approaches to industrial safety and, although they are not always referred to explicitly, the influence of this early thinking is clear when these new theories are considered. In fact, the first theory we are going to look at was not just groundbreaking in its day, but also prescient of one of the current theories of how information is processed and decisions are made in the human brain, a concept that you will have encountered in Chapter 2, namely the importance of production rules.

Back in 1983, Jens Rasmussen proposed that humans perform at three different levels.[2] The term "performance" is used to refer to some sort of interaction with an external element and, importantly, is often associated with attempting to solve some sort of problem. Performance levels are relevant because errors can occur at any level of performance and there are distinct characteristics in the types of error that occur, the rates at which they occur and, more importantly, the rates at which they are detected and adequately corrected. The three levels are as follows:

- Skill-based (SB) – Performance is preprogrammed and automatic so that the individual does not have to consciously engage with the problem in order to correctly react to it. For example, when taxiing an aircraft, if it diverges from the center of the taxiway, it is instinctive for the pilot to correct this without having to consciously think about it.

- Rule-based (RB) – Performance at this level is semi-automatic and requires more conscious engagement than the SB level. In order to perform at the RB level, as has been described in Chapter 2 (Information Processing), patterns are identified and these are associated with production rules stored in the form of "IF (this pattern), THEN (this response)". For example, once a pilot has been trained, he will know that a certain combination of warning lights means he should respond in a particular way. He will be consciously aware of the pattern of warning lights and, because of his training, will have a stored rule that says "IF this pattern is seen, THEN respond in this way". He can then activate this rule and so is able to respond quickly without having to consider the meaning of each warning light and work out their combined meaning when they are seen at the same time. It is worth noting that human performance at the SB level is actually based on rules as well, but these rules have become so well practiced and have been used so many times that they have become skills as they do not require conscious activation or monitoring. If you consider the example of learning to drive, it is initially difficult as rules regarding gear sequences and clutch use need to be consciously activated. Eventually, these rules become so well practiced that they can be activated without conscious engagement and we can now perform many driving tasks at the SB level.

- Knowledge-based (KB) – This level of performance occurs when the individual finds no suitable rule for determining what should be done in a particular situation. The only option is for the individual to examine his store of knowledge in an attempt to understand what is going on. This is a conscious, high-workload process and requires selective attention on the problem. If the problem is adequately solved using this KB process, it is highly likely that the original pattern of stimulation will be matched to the final response and the person will create a new production rule that will enable them to solve the same problem using the RB performance level should they encounter it again. This, in principle, is learning; through our experience we store more and more pattern–response production rules that allow us to solve problems more rapidly and accurately as our experience grows. In fact, James Reason defines "expertise" as never having to rely on KB problem solving; that is, as the person has more experiences and encounters more novel scenarios, many of which will require KB problem solving, this KB problem solving will result in the creation and storage of many more production rules. This means that the highly experienced person will be able to call on these production rules in future and will now be able to handle the vast majority of situations at the RB level; the need to resort to KB processing is increasingly unlikely unless a truly novel situation is encountered.

3.2   Errors and violations at different performance levels

Before we discuss how errors can occur at each of these levels of processing, we need to get into the details of what we actually mean by "error". Let's look at James Reason's definition of error: "Error will be taken as a generic term to encompass all those occasions in which a planned sequence of mental or physical activities fails to achieve its intended outcome, and when these failures cannot be attributed to the intervention of some chance agency".[1] To simplify this, we can say that error is defined as "the failure of planned actions to achieve their desired goal".

Having discussed the three levels of human performance, we can start to see a way of dividing this definition up further. If an error is the failure of planned actions to achieve their desired goal, there are two ways that this failure can happen; either the plan is wrong, in which case our actions are doomed to failure from the start, or our plan is correct but our execution of it fails for some reason. With this in mind, let's look at the types of error that occur at the SB, RB and KB performance levels:

- Errors at the SB level: slips and lapses – Our plan of action is correct and suitable but the error occurs during the execution of the plan. Inattention is a classic reason for this; for example, intending to drive to a location near your place of work and finding that you have driven to work instead. Another example would be forgetting to do something because of an interruption or a distraction. These are both examples of lapses. A slip is also an error at the SB level and occurs when an action is incorrectly performed, for example, switching another switch than was intended.
- Errors at the RB level – The production rule we have selected to deal with the current problem is not appropriate; that is, the plan is wrong. As you will see in the next paragraph, if we consciously recognize that we cannot solve a problem at the RB level, we should move up to the KB level and try and analyze the problem further and search our knowledge stores to be able to start solving it. If this is what should happen, why then do we not move to the KB level and, instead, stay at the RB level and try to use a rule that turns out to be unsuitable? The answer can be found in our earlier descriptions of RB and KB performance: KB performance takes a lot of conscious effort. It requires selective attention to be focused on the problem and an effortful re-evaluation of the problem parameters followed by a slow trawl through our knowledge stores to try and develop some sort of plan to respond to this novel situation. The reality of many of these novel situations is that we do not have the luxury of spending time and cognitive resources coming up with the best possible answer. There may be hundreds of other situational factors that are calling out for our attention, such as trying to land an aircraft that is on fire, with an incapacitated colleague and multiple system failures. We cannot always take each problem and devote the necessary time and resources to it in the hope of getting the best possible answer. High-workload situations make it much more likely that we will select a production rule to use for a particular problem that we think *might* be suitable or that we think will give an adequate solution rather than the perfect solution. In selecting these rough-and-ready but quickly accessible production rules, our brain is programmed to take several factors into account, the most powerful of which are[1]:
  - Similarity matching: Although the situation we are dealing with is novel and there is no production rule that we can use to come up with a response, is the pattern similar enough to another pattern that we have encountered where we know that we have a production rule that works? The reality is that we may overemphasize the similarity between two situations and so miss crucial differences that would mean that the production rule we are planning to use is not going to work.
  - Frequency gambling: A frequently used rule is more likely to be used than an infrequently used rule. Let us say that we use similarity matching to decide that a novel situation is similar to two situational patterns that we have encountered previously. We are then likely to use frequency gambling to decide which of the two situational patterns' corresponding production rule we use – the rule that we have used most often is the one that we are likely to pick. The reality is that we have engaged in a statistical act of risk taking. We are gambling that a rule that is used frequently will have a high chance of addressing the problem. Clearly, this is not always the case.
  - Overconfidence: For reasons that we described when we explored the decision-making aspect of information processing, studies have repeatedly shown that we have more

confidence in the accuracy of our knowledge and abilities than is justified. The ability to judge the accuracy of our knowledge is known as calibration, and humans are generally more confident in their judgment of the correctness and suitability of their actions than is justified, i.e. our calibration is poor.[3] Overconfidence reinforces the outputs of the similarity matching and frequency gambling programming that we have and increases the likelihood of making an RB error.

- Errors at the KB level: Provided the workload is low, we may have a conscious recognition that a novel problem cannot be solved at the RB level. Error can occur at the KB level when attempting to derive a solution by analyzing the problem more deeply followed by conscious interrogation of our knowledge stores in order to understand the situation and come up with a solution. The four main causes of KB errors are:

  ◦ Missing knowledge: A student taking a French translation examination may try to understand a phrase that he has not encountered before by looking at the context or trying to derive its meaning from similar phrases. However, if the required knowledge is not in his memory, there is a high risk of making a KB error. He simply does not know what that phrase means in English.

  ◦ Processing limitations: It may be that the situation is evolving so quickly that it is not possible for the individual to maintain an accurate mental model on which to base his decisions. This is not uncommon when trying to solve problems occurring in highly automated, highly interconnected technological systems like aircraft or nuclear power stations. As described in Chapter 1, the human operators at the Three Mile Island nuclear power plant were attempting to solve a serious problem. As the situation developed, various automatic interlocks and safety systems activated in rapid succession, significantly changing the operating parameters of the entire plant. The result was that while the operators were attempting to solve the initial problem, within 13 seconds, the system had reconfigured itself in a way that changed the parameters of the problem entirely.[4] They were still working on the original problem and could not keep up with the rapidly evolving situation.

  ◦ Oversimplification: A strategy that we sometimes employ in the face of highly dynamic situations is to try and replace a complex problem that is seemingly too difficult to solve with a simpler one. The risk is that in trying to eliminate complexity, we ignore some vital elements of the problem and so come up with an unsuitable plan.

  ◦ Overconfidence: In the same way as with RB errors, we generally have more confidence in the accuracy of our skills and knowledge than is justified and this will reinforce any KB solutions we come up with, even though they may be wrong.

Aside from error, there is one other type of unsafe act that we may have to deal with as pilots. This type of act is known as a violation and is where the individual intentionally does not follow a procedure or acts contrary to a procedure (compared with an error that is the failure of planned actions to achieve their desired goal). Although the term sounds entirely negative, violations are not always a sign of negligence. They can be divided into three types depending on the level of performance at which they occur[1]:

- Routine violations: occur at the SB level of performance – These occur unconsciously and could be interpreted as bad habits that have become so routine that the person does not even know he is doing them. Although the definition states a violation is intentional and given that our definition of SB performance says that it is automatic, there is a question about whether this should be called a violation at all. If the person is unaware of the procedure, then it is an error. If they are aware of the procedure but intentionally and repeatedly disregard it during RB performance, it will become a bad habit and transition from being an RB violation to an SB violation. For example, if someone repeatedly chooses to taxi quickly because it saves time, this is an RB violation as you will see from the next paragraph. However, if they become so accustomed to taxiing above the maximum allowed taxi speed that it now happens all the time without them making a conscious decision to deviate from the rules, this is now an SB violation.

- Situational violations: occur at the RB level of performance – These include departures from procedures that are usually well intentioned and are seen as necessary because of time pressure. This may be a response to overproceduralization; that is, doing everything "by the book" will take too long and we have a limited amount of time. This is a classic trade-off, known as an efficiency–thoroughness trade-off (ETTO),[5] and will be covered in more detail when we look at error management in organizations in Chapter 4.

- Exceptional violations: occur at the KB level of performance – Situations that require KB processing are normally quite complex given that they could not be solved at the RB level. An exceptional violation at this stage may be a final response to an insurmountable problem and, bizarrely, may be made with the best of intentions, in that the person can see no other way of solving the problem. The polar opposite of this is an exceptional violation that occurs after KB processing in a situation that does not demand it. This may be when the individual has made a conscious and calculated decision to violate procedures without any good reason, and may well constitute negligence or even sabotage.

For the purposes of the rest of this book, errors and violations, when considered together, will be referred to as "unsafe acts".

3.3 Detection of errors

Fortunately, we have internal error checking mechanisms that allow us to self-detect when we have made an error, whether it is an SB error, an RB error or a KB error. However, the success of this self-detection mechanism is variable and also depends on situational variables such as workload. A summary of the success rates and reasons for failure of our self-detection mechanisms are given below:

- Self-detection of SB errors (slips and lapses) – Slips have the highest rate of successful detection.[1] SB performance occurs when we are completing routine tasks that do not require conscious action selection under normal circumstances. It will normally be quickly apparent that an error has been made. For example, you may inadvertently select a switch that is located next to the one you intended. The action is executed and our self-detection mechanism may immediately realize that the error was made because the expected outcome was not obtained or the appearance of the switch panel does not fit the pattern that would have been expected. Motor skills are thought to include what is known as a recognition schema as a way of ensuring that our actions will lead to the desired result.[6] The motor skill of selecting a switch is associated with a stored pattern of what the outcome should look like: the switch being in a different position. If the resultant pattern does not fit the recognition schema, it activates the error detection system to tell us that we have made an error. Research tells us that SB errors are the most common type of error (about 70% of total errors, depending on the study used). They tend to be self-detected 75% of the time and adequately corrected 70% of the time,[7] although there will be considerable variability in these rates depending on the precise nature of the SB error. Lapses are more difficult to detect due to the fact that they represent an inaction rather than an action. It is understandably difficult to detect something that you have forgotten.

* Self-detection of RB errors – RB errors occur at a much lower frequency than SB errors, about 20% of all errors made.[7] Unfortunately, the rates of detection and correction are correspondingly low. To explain this in the context of human information processing, it is necessary to diverge from some of the earlier thinking on this topic and bring in some more recent research that seems to explain things more adequately and, more importantly, gives us a better grounding on which to discuss error management strategies. An RB error involves the use of a production rule that is not appropriate for the situation. If we have elected to try and solve a novel problem using RB problem solving rather than moving up to the KB level, it is likely to be because the situational workload is high and we cannot afford to take the time and resources needed to engage with the problem at the KB level. Similarity matching, frequency gambling and overconfidence may have led to us selecting a production rule that is not appropriate for the situation. A high-workload situation limits our internal self-detection mechanisms as we have probably been forced to move on to the next problem that needs our attention. The main factor that will make it less likely that we will detect an RB error is confirmation bias, as described in Chapter 2. Once a problem has been considered to have been solved, even if it actually has not been, our brains are reluctant to revisit the same problem again, particularly in high-workload situations. There are simply too many other things that it needs to be concentrating on rather than revisiting all its past decisions to reassess their efficacy. Instead, it uses confirmation bias to "close the file" on previous decisions; that is, it picks up on environmental cues that suggest that the chosen course of action has been correct. The result is that in concluding that the chosen production rule has successfully solved the problem, this implied that the initial assessment of the problem must have been correct as well and so this "fact" is reinforced in the mental model of the situation. The case study of British Midland Flight 92 (Box 2.7 in Chapter 2) is a clear example of this phenomenon.
* Self-detection of KB errors – KB errors occur least frequently of all but, as you can imagine with a problem that is so novel that it requires KB processing to solve it, the risk of error at this level is high simply because engaging KB processing is usually the last resort. Self-detection and self-correction rates are also very low because we often simply do not know what the outcome will be or what to do if it is not as expected. The novel situation has led to a novel solution with no guarantee of success. Failure at the KB level requires another complete cycle of KB processing to derive another potential solution.

Our self-detection mechanisms are good at detecting SB errors (slips and lapses) but bad at detecting RB and KB errors, for the reasons we have just covered. Does this mean we are doomed to live with the outcomes of failing to detect these errors in our day-to-day operations? Fortunately not, because we have three other useful resources that can detect our errors for us if we utilize them correctly:

* Detection by other liveware (other pilot or crew member) – We do not spot our own error but a colleague does and (hopefully) brings it to our attention.
* Detection by software (standard operating procedures-SOPs) – A checklist or procedure brings an error to our attention, normally something that we have missed because of an SB error, especially if it was a lapse.
* Detection by hardware (the aircraft systems) – The system itself detects the error and brings it to our attention.

The resources available for error detection in the event that we have failed to self-detect it are very useful, and when we talk about TEM2 we will see how we can make best use of these resources. As an aside and as a primer to one of the topics in Chapter 4 about error management in the organization, bear in mind these escalating steps of error detection:

- liveware detection (self) – best for detection of SB error
- liveware detection (others) – good for detection of RB and KB errors
- SOP detection – good for detection of SB errors, particularly lapses
- hardware detection – can potentially detect all forms of error.

When we consider implementing error detection methodologies, it becomes progressively more difficult and expensive as you work through the list, but hardware detection methods have the best chance of trapping errors.[8] It was expensive to design and test the hardware (and computer programs) that guarantee flight envelope protection in the Airbus design philosophy but it is now almost impossible for a pilot to exceed the flight envelope in error. A cheaper, liveware solution would have been simply to train pilots not to exceed the flight envelope but this would not guarantee safety to the same extent as the hardware solution.

Before we consider the practical implications of SB, RB and KB performance, the sorts of unsafe acts (errors and violations) that can occur at each level and how these are best detected and corrected, we need to look at a model of error that will form the foundation for the practical tools that will be introduced later in the chapter. This model is colloquially known as Reason's Swiss Cheese Model.[9]

3.4   The Swiss Cheese Model

The wonderful thing about this model is that it was the first time that people had considered that errors do not necessarily occur in isolation. Several versions of the model occur and may contain anything from four to six layers. The version that best suits the rest of this chapter and gives us a framework on which to build an error management strategy is partly based on the Human Factors Analysis and Classification System developed by Wiegmann and Shappell as a way of expanding the Swiss Cheese Model so that it can be used in accident investigation.[10] Figure 3.1 shows this model.

The layers in the adapted version of the Swiss Cheese Model are as follows:

1. organizational influences – pre-existing pressures from higher up in the organization that may affect the operation
2. unsafe supervision/monitoring – deficiencies in the immediate supervision or monitoring of an individual that make it more likely that an unsafe act could be committed
3. error-promoting conditions (threats) – environmental or personal conditions that make it more likely that an individual will perform an unsafe act
4. unsafe acts – these could be errors or violations
5. inadequate defenses – high-risk industries often have defensive systems that are designed to prevent unsafe acts from having an adverse effect on safety; in some cases, these defenses are absent or inadequate
6. negative outcome – a near-miss, incident or an accident.

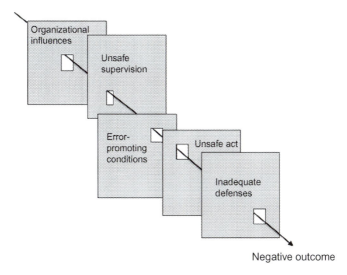

Negative outcome

Figure 3.1 James Reason's Swiss Cheese Model.

If we rearrange some of the layers and rename them, we start to get closer to a model of error management that will be expanded on later. The new "slices" would be:

1. error-promoting conditions (threats)
2. unsafe acts (errors and violations)
3. unsafe supervision/monitoring (unsafe act detection failure)
4. inadequate defenses (failure to mitigate effects of unsafe act)
5. negative outcome.

Readers familiar with the original model of TEM may be beginning to recognize some of these concepts. The reality is that although James Reason's Swiss Cheese Model and the model of TEM were being developed at roughly the same time, they were not developed by the same groups of people. However, there is a lot of similarity between them and it is possible to map the principles of one on to the principles of the other. As the two models were developed independently, it suggests that both approaches have hit upon a really valuable idea and have just expanded this idea in slightly different ways. This concept can described as follows and is a summary of the principles of both:

> *The complex, dynamic nature of real-world operations means that we do not always work under optimal conditions. We frequently encounter threats: error-promoting conditions that increase the risk of an unsafe act occurring (an error or a violation). If we consider this possibility in advance, we may be able to prevent the unsafe act before it happens through planning. In the event that an unsafe act occurs, we may be able to detect it ourselves, others may detect it or the procedures or hardware may detect it. Once it is detected, we may be able to correct it. If the effects of the unsafe act are not corrected, we may find ourselves in a situation*

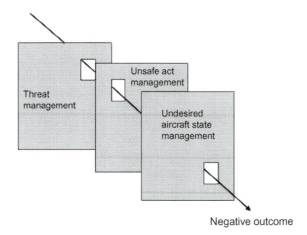

Figure 3.2 Swiss Cheese Model adapted for Threat and Error Management.

where our safety margin is seriously eroded. We now have a limited opportunity to use any remaining defenses in order to correct this abnormal situation. If it is not corrected and the abnormal situation persists, the outcome is likely to be negative – a near miss, incident or accident.

This is TEM in a nutshell. We are now going to put this into an operational setting for pilots and describe how we can use this principle to our advantage to keep us safe. The pilot-specific model is called TEM2.

3.5 Threat and Error Management 2

Figure 3.2 shows a summary of the adapted stages from the Swiss Cheese Model that form the basis of TEM. The unsafe act detection failures and the failure to mitigate the effects of the unsafe acts are represented as the holes in the second and third layers.

For practical purposes, threat management can occur at any time but will mainly happen during the departure and the arrival briefings. Again, for practical purposes, the departure briefing will be assumed to include discussion of preflight preparation, starting, taxi-out, take-off, climb and cruise. The arrival briefing will be assumed to include discussion of the descent, approach, landing, taxi-in, parking and shutdown. While threat management will tend to occur during briefings, unsafe act management and undesired aircraft state (UAS) management may occur at any time during the flight.

3.5.1 Threat management

The principles laid out in this section can be adapted to suit the briefing format favored by your own organization. The format given here is generic and is organized according to some of the principles of effective communication that will be covered in more detail in Chapter 6. General guidance for an effective briefing that includes TEM2 is as follows and is adapted from information in Chapter 6:

- The purpose of a briefing is to ensure that the operations being covered by the briefing can be carried out as safely as possible and that both pilots have a shared mental model that includes a shared understanding of how the aircraft is going to be maneuvered on the ground and in the air, what threats exist, how those threats are going to be managed, and what each pilot's actions are going to be under normal and (select) abnormal conditions. To do this, the following things need to have been achieved by the end of the brief:
  - a shared mental picture of how the flight is planned to be conducted
  - identification of all threats that could impact the safe operation of the flight
  - agreement of threat management strategies to mitigate the risks associated with the threats identified.
- An effective briefing assumes that both pilots will be active in safely operating the flight and implementing TEM2.
- Given that both pilots are going to be active in creating safety, both pilots should actively participate in the briefing.
- As well as having an interactive briefing, the following principles can help to ensure that information that is communicated is retained by the other person:
  - Do not use a standard "script". It is very easy for the other person to become inattentive if they think they know exactly what is going to be said.
  - Asking is better than telling as it actively engages the other person and forces them to both process the meaning of the question and generate a response.
  - Use eye contact to maintain engagement.
  - Avoid distractions.
- Discuss the plan and any threats and strategies in a logical and chronological way. For example, the plan and any threats that might affect taxiing should appear early in the departure briefing but later in the arrival briefing. A method for doing this will be shown when we classify threat types.
- Recap the threats and the management strategies again at the end of the brief.

A significant barrier to effective threat management at the planning stage is how to ensure that all relevant threats are identified. Different companies use different aides-mémoire to try and remind their pilots to consider particular threats. There are several drawbacks to this system that limit its effectiveness and can, ironically, introduce additional risks:

- There is a risk that pilots will only consider the threats included in the aide-mémoire.
- There are far too many potential threats to include in any aide-mémoire.
- On days where no threats exist, the pilot may continue to use their system to repeat the fact that no threats exist in a rote manner that wastes time and results in their colleague no longer paying attention.

A sample form used for line-oriented safety audits (LOSAs) included codes for 30 different threats (including code 99, "Other Threats").[11] It would be unrealistic to expect pilots to run a checklist to ensure that they have identified every potential threat as there are so many (far more than the 30 listed on the LOSA audit form) that it would take far too long. To try and resolve this difficulty, TEM2 attempts to provide a tool for pilots to use that will facilitate identification of threats so that they can be managed. To develop TEM2 threat management, the following steps were carried out:

1. Identify at which points during flight operations threat management is carried out.
2. Identify all the threats that pilots could potentially pick up at each stage.
3. Highlight the threats that, if missed, could lead to the most serious negative outcomes.
4. Define a generic management system that could be applied to each threat.
5. Develop a tool that could assist pilots with threat management.

3.5.1.1 Threat management opportunities

Although threat management can occur spontaneously if a threat is identified at any point during the day, there are some periods during the course of a flight duty where threat management will naturally occur:

- crew briefing before going to the aircraft
- departure briefing
- arrival briefing.

Aside from the timing and location of these briefings, the most salient difference between them is the information that the crew has at hand to be able to identify any relevant threats. We will be returning to this idea of information availability later on as it forms the basis of the threat management tool that is part of TEM2. The information available to the crew at each stage will differ depending on how the particular operation is organized. For example, in some companies, the crew may know about any technical problems with the aircraft long before they get on board. In other companies, the pilots may only get a chance to find this out once they are on board. The crew briefing before going to the aircraft can be very useful in exchanging threat management information between the pilots and the cabin crew, for example the pilots warning the cabin crew of the possibility of turbulence en route or the cabin crew warning the pilots about recent changes to their procedures that will affect the flight. However, because crew briefings before going to the aircraft may or may not occur and, if they do, may rely on information that is incomplete, this section will focus only on the departure and arrival briefings that are carried out by the pilots. It is assumed that the cabin crew have informed the pilots about any threats they have identified that they think the pilots need to know about.

3.5.1.2 Types of threat

The next stage in developing a useful threat management tool is to consider some of the common threats that pilots may encounter, as shown in Table 3.1. There are four main threat categories, each with further subcategories. You will notice that this framework is based on the adapted SHEL model (software, hardware, environment and liveware) that was presented in Chapter 1. The range of threats listed here is not exhaustive and your operation may encounter additional threats.

It should be noted that when addressing threats from liveware external to the flight deck, "known problems" imply a prior knowledge of shortcomings in the performance of people outside the flight deck that may have an impact on flight safety. An example

Table 3.1 Main threats to consider during threat management

LIVEWARE THREATS	ENVIRONMENTAL THREATS	
	METEOROLOGICAL	NON-METEOROLOGICAL
Liveware in Flight Deck Fatigue Unfamiliarity: airport, airspace, local terrain/weather Degraded manual flying skills Susceptibility to sensory/perceptual illusions Time/operational pressure Other acute stress **Other Liveware** Known problems with ATC quality Bilingual ATC leading to impaired mental models Known problems with cabin crew Distractions from cabin crew Known problems with dispatcher Frequent errors with paperwork/load sheets Known problems with ground staff Potential problems with passengers **HARDWARE THREATS** Technical problems: GPWS, TCAS, weather radar, etc Automation problems Unusual configuration required Possibility of Rejected Take-Off Dangerous goods or cargo **SOP THREATS** Complex performance calculations required Potential performance limitations Alternative procedures needed for some reason	**METAR/ATIS/TAF** Wind: tailwind, crosswinds, gusts, windshear Visibility: difficulty tracking centreline, illusion risk Cloud: low cloud base, cumulonimbus, etc Precipitation: (freezing) rain, snow, thunderstorms Temperature: icing, hardware/liveware issues Dewpoint: chance of decreasing visibility QNH: low QNH – altitude busts, transitions Additional ATIS/METAR threats: e.g. runway contamination, birds, etc **Significant Weather Charts** Cumulonimbus and thunderstorms Icing Jetstreams, particularly when close to high terrain Turbulence Other, more rare, weather phenomena, e.g. tropical storms, volcanic activity **Synoptic Weather Charts** Low pressure systems Weather fronts **Wind Charts** Headwinds and tailwinds en route Temperature differences from ISA: performance **SIGMETS** Areas of storms Areas of turbulence Any other severe weather	**Airport Layout** Complex taxi routing Crossing active runways Poor or defective signage/lighting Runway incursion hotspots Multiple runways: parallel arrivals/departures, threat of lining up on wrong runway **Terrain/Obstacles** High terrain/obstacles in vicinity of flight path **Airspace** Danger/restricted areas Airspace congestions VFR traffic Other traffic: parachutists, gliders, etc **Procedural (SIDs, STARs, IAPs and runways)** Complexity: departure, approach and go-around Require unusual navigational aid setup e.g. holding a DME or alteration of FMS routing Altitude restrictions Small difference between cleared departure Flight level and transition altitude Speed restrictions Approach slope: e.g. more than 3 degrees Runway: narrow, short, sloping, etc Approach guidance, e.g. non-precision Visual approach guidance e.g. no PAPIs Runway lighting in low visibility Any unusual notes on the procedure

ATC: air traffic control; GPWS: ground proximity warning system; TCAS: traffic collision avoidance system; SOP: standard operating procedure; METAR: meteorological terminal air report; ATIS: automatic terminal information service; TAF: terminal area forecast; QNH: altimeter subscale setting to obtain elevation when on the ground; ISA: international standard atmosphere; SIGMET: significant meteorological information; VFR: visual flight rules; SID: standard instrument departure; STAR: standard terminal arrival; IAP: instrument approach procedure; DME: distance measuring equipment; FMS: flight management system; PAPI: precision approach path indicator.

could be knowing that fuelers at a particular airport have encountered difficulties with the aircraft refueling panel, the threat being that it is not set correctly. If there is no evidence of problems or deficiencies, there is little need to address non-existent threats.

3.5.1.3 Threats associated with serious negative outcomes

Threats need to identified and managed in a manner that is appropriate to the risk they pose. Given that time is often limited during flight operations, although consideration needs to be given to all threats, those that pose the greatest risk should have more time devoted to them. The next step in developing a threat management tool is identifying which threats pose the biggest risk to safety. Since 2011, the UK Civil Aviation Authority (CAA) has published several reports into what it calls "the Significant Seven", the seven most significant safety issues based on global fatal accidents and high-risk occurrences involving large commercial aircraft from the UK.[12–14] The seven factors identified were:

- loss of control
- runway excursions
- controlled flight into terrain (CFIT)
- runway incursions
- airborne conflict
- ground handling
- fire.

The purpose of the CAA research into these areas was to improve safety using a variety of measures: training, procedural, investigatory, and so on. In this section, we are going to use these seven safety issues and work backwards to identify salient threats that could lead to each negative outcome. You will notice that more emphasis is given to the undesired aircraft states (UASs) and the threats rather than any associated unsafe acts. The reason for this is that there are so many potential unsafe acts that could contribute to a threat leading to a UAS that it is impossible to summarize them in any meaningful way. However, any unsafe act would still have to be either an error or a violation at the SB, RB or KB level and should be managed accordingly. Analysis of each of the Significant Seven using TEM2 will look at how these negative outcomes could occur in flight. Normally, there will be similar threats that are relevant for the departure phase [covering start, push, taxi, take-off (or rejected take-off), climb and cruise] and the arrival phase [covering descent, approach, landing (or go-around), taxi, park and shutdown] and so the threats listed are relevant for both departure and arrival briefings. Threats that are relevant to only one phase (e.g. complex go-around procedure) will be obvious to the reader.

Loss of control
- Negative outcomes – Stall, uncontrolled flight path (e.g. spiral dive, inversion).
- UASs – Low airspeed, unusual attitude, incorrect aircraft configuration (especially flaps and slats), incorrect weight/balance, unnecessary weather penetration (including severe clear air turbulence), aircraft experiencing windshear.
- Unsafe acts – Many possible unsafe acts could lead to each UAS but SB errors during execution of a go-around may lead to a UAS.

- Threats – The most significant threats and threat categories are *italicized*:
  - *Liveware threats*
    - fatigue
    - *degraded manual flying skills*
    - *susceptibility to sensory/perceptual illusions*
    - frequent errors with load sheets (made by others)
    - known problems with ground staff (incorrect loading)
  - Hardware threats
    - *automation problems*
    - unusual configuration required
  - SOP threats
    - complex performance calculations required
    - *potential performance limitations*
  - *Environmental threats: meteorological*
    - *wind: windshear*
    - *precipitation: ice/snow, freezing rain*
    - *visibility: low visibility, no visual references*
    - *cumulonimbus and thunderstorms*
    - *jet streams, particularly when close to high terrain: these can lead to severe turbulence*
    - *icing*
    - any other severe weather
  - Environmental threats: non-meteorological
    - *procedural complexity: go-around.*
- Additional notes – The CAA concluded that since 2000, loss of control occurred for the following reasons:
  - 68% non-technical reasons (13% of these due to inappropriate use of automation).
  - 32% technical faults (55% of these due to autopilot malfunction or disengagement).

Runway excursions
- Negative outcomes – Veering off runway, overrunning end of runway.
- UASs – Unstabilized approach (high energy), not tracking runway centerline, high-speed rejected take-off, incorrect aircraft configuration (flaps, slats, reversers, spoilers), long/floated landing, incorrect performance (e.g. wrong thrust settings or incorrect weight/balance).
- Unsafe acts – Many possible unsafe acts could lead to each UAS, although RB errors have been strongly implicated in the continuation of unstable approaches.
- Threats – The most significant threats are *italicized*:
  - *Liveware threats*
    - fatigue
    - *unfamiliarity with airport*
    - *degraded manual flying skills*
    - *time/operational pressure*
    - susceptibility to sensory/perceptual illusions
    - *known problems with air traffic control (ATC) quality*
  - Hardware threats
    - unusual configuration required
    - possibility of rejected take-off

- ◦ SOP threats
  - − complex performance calculations required
  - − *potential performance limitations*
- ◦ *Environmental threats: meteorological*
  - − *wind: tailwind, crosswinds, gusts, windshear*
  - − *visibility: difficulty tracking centerline*
  - − *precipitation: rain (leading to wet runway)*
  - − *additional METAR/ATIS threats: runway contamination*
- ◦ *Environmental threats: non-meteorological*
  - − approach slope
  - − *runway: narrow, short, sloping*
  - − runway lighting (lack of centerline lights in low visibility).
- • Additional notes − The CAA research discovered the following about runway excursions:
  - ◦ 89% occurred on landing (42% as a result of unstabilized approaches).
  - ◦ 70% were veer-offs and 30% were overruns.
  - ◦ 24% of all excursions occurred on wet runways.
  - ◦ Runway excursions are the most commonly reported accidents, with two happening every week globally.

Controlled flight into terrain

- • Negative outcome − Collision with terrain.
- • UASs − Loss of separation between aircraft and terrain, unstabilized approach (low energy).
- • Unsafe acts − CFIT involves a mismatch between where a pilot thinks the aircraft is and where the aircraft actually is. Although it is conceivable that there are technical failures that could lead to the pilot being given incorrect information from his instrumentation and so lead to an incorrect mental model being formed, the pilot may cause either of these UASs through a variety of unsafe acts. For example, making an SB lapse in resetting QNH (the altimeter subscale setting to obtain elevation when on the ground) in the descent may lead to the aircraft being lower than the pilot's mental model suggests.
- • Threats − The most significant threats and threat categories are *italicized*:
  - ◦ *Liveware threats*
    - − fatigue
    - − *unfamiliarity with local terrain*
    - − *susceptibility to sensory/perceptual illusions*
    - − *known problems with ATC quality*
  - ◦ Hardware threats
    - − *technical problem, e.g. ground proximity warning system (GPWS)*
  - ◦ *Environmental threats: meteorological*
    - − *visibility: difficulty seeing terrain*
    - − *cloud: low cloud base*
    - − *QNH: low pressure can lead to aircraft being lower than indicated if QNE (standard pressure altimeter subscale setting used above transition altitude to give a flight level) not reset to QNH*
  - ◦ *Environmental threats: non-meteorological*
    - − *procedure complexity: departure, approach and go-around*
    - − *high terrain/obstacles in the vicinity of flight path*
    - − *procedure requires unusual navigational aid set-up, e.g. holding distance measuring equipment (DME) or alteration of flight management system (FMS) routing.*

- Additional notes – The CAA research discovered the following about CFIT:
  - In one year there were 24 reported GPWS warnings on UK-registered/operated aircraft.
  - 16% of those warnings were hard warnings such as "Pull Up!"
  - Only one was a nuisance warning.

Runway incursions

- Negative outcome – Collision with another aircraft or a vehicle.
- UASs – Proceeding towards wrong taxiway or runway, being on the wrong taxiway, runway incursion, loss of separation between aircraft and another aircraft/vehicle on the ground.
- Unsafe acts – Although a variety of unsafe acts could lead to any of these UASs, RB errors are most likely to lead to runway incursions. The pilot may have an incorrect mental model of the layout of the airport, particularly in low visibility, and may see the turning/entry that he thinks is appropriate and action a rule (turn right at third entry) only to find that the rule has been used inappropriately. The overconfidence heuristic and confirmation bias make it more likely that he will proceed even if he is not entirely sure of his position.
- Threats – The most significant threats are *italicized*:
  - *Liveware threats*
    - *fatigue*
    - *unfamiliarity with airport*
    - *known problems with ATC quality*
  - Environmental threats: meteorological
    - *visibility: unable to clearly see taxiways/markings*
  - *Environmental threats: non-meteorological*
    - *complex taxi routing*
    - *crossing active runway*
    - *poor or defective signage/lighting*
    - *runway incursion hotspots*
    - *multiple runways.*
- Additional notes – The CAA research discovered the following about runway incursions:
  - In the UK, in one year, there were 190 runway incursions.
  - 71% involved aircraft and 25% involved vehicles.
  - 33% of runway incursions involved overrunning a stop bar or indicator.

Airborne conflict

- Negative outcome – Collision between two aircraft.
- UAS – Loss of separation between two aircraft in the air, vertical or lateral deviations from desired flight path, incorrect systems configuration (altimeter), incorrect automation use (incorrect altitude set).
- Unsafe acts – The act of incorrectly setting or failing to set aircraft systems such as altimeters or mode control panel altitude selections is the most likely to cause airborne conflict. This may be associated with inappropriate ATC clearances or mishearing/misinterpreting an appropriate ATC clearance.
- Threats – The most significant threats are *italicized*:
  - *Liveware threats*
    - *fatigue*
    - *unfamiliarity with airspace*
    - *degraded manual flying skills*
    - *bilingual ATC leading to impaired mental models*
    - *known problems with ATC quality*

* Hardware threats
  - *technical problems: traffic collision avoidance system (TCAS) and automation, especially altitude capture and altitude alerting systems*
* *Environmental threats: meteorological*
  - *wind: windshear*
  - *visibility: aircraft unable to see each other*
  - *cloud: cumulonimbus requiring many aircraft to seek weather avoidance vectors or early climbs/descents*
  - *QNH: not transitioning from QNH to QNE and vice versa*
* *Environmental threats: non-meteorological*
  - *multiple runways: parallel departures and arrival*
  - *visual flight rules traffic*
  - *other traffic: parachutists, gliders, etc.*
  - *procedural complexity: go-around; may put aircraft into conflict with other aircraft if not flown correctly*
  - *altitude restrictions on departure/arrival procedures*
  - *small difference between cleared departure flight level and transition altitude.*
* Additional notes – The CAA research discovered the following about airborne conflicts:
  * There is an increasing trend towards high-severity events.
  * There are approximately six losses of separation in the UK every week.
  * In one year, there were six high-risk losses of separation involving UK-registered/operated aircraft. Five of these involved loss of separation between a commercial aircraft and a light aircraft.
  * The countries where UK-registered/operated aircraft encountered loss of separation most frequently were the USA, France and Spain.

Ground handling

The negative outcomes that come under the heading of "ground handling" are not really amenable to deconstruction within the TEM2 framework simply because there are so many of them, and most of them are for reasons beyond the control of the pilot. The most significant outcome of ground handling errors has been found to be incorrect loading of aircraft (62% of ground handling occurrences in the UK). Eleven percent of ground handling occurrences happened during pushback. Threat management relies on prior knowledge of the ground handling services that could be expected at a particular airport. If that airport or ground handling services are known to have quality issues, the pilots may make additional efforts to double-check any safety-critical tasks that the ground staff have carried out. Beyond that, it may be difficult for pilots to successfully mitigate unsafe acts made by ground handling staff.

Fire

Fortunately, fire is an extremely unlikely event to occur on board an aircraft. From 2000 to 2010 there were five in-flight fires. However, all five resulted in fatal accidents. Two were on passenger flights and three were on board cargo flights. Owing to the rarity of this event, there is no meaningful way of carrying out threat management. If smoke or fire were to occur on board, the aircraft would immediately be in an undesired state and the recommended course of action would be to land without delay.[15]

Summary of main threats

Given that ground handling and fire issues are largely outside of the pilot's control we will focus on the five situations where the pilot can mitigate the risks. Based on these five negative outcomes out of the CAA's Significant Seven (loss of control, runway excursion, CFIT, runway incursion and airborne conflict), there are several threats that were strongly related to more than one negative outcome:

*   Low visibility – all five
*   Known problems with ATC quality – four out of five: runway excursion, CFIT, runway incursion and airborne conflict
*   Unfamiliarity: airport, airspace, local terrain/weather – four out of five: runway excursion, CFIT, runway incursion and airborne conflict
*   Complex procedures (taxi routes, departures, go-arounds, etc.) – four out of five: loss of control, CFIT, runway incursion and airborne conflict
*   Degraded manual flying skills – three out of five: loss of control, runway excursion and airborne conflict
*   Windshear – three out of five: loss of control, runway excursion and airborne conflict
*   Performance – two out of five: loss of control and runway excursion
*   Altimetry – two out of five: CFIT and airborne conflict
*   Automation problems – two out of five: loss of control and airborne conflict.

3.5.1.4   Threat identification framework

Now that we have identified potential threats, matched them to the most severe negative outcomes and identified a subset that occurs most frequently, this information can be used to construct a framework to help pilots to identify threats that they might otherwise miss. Although low visibility is evidently a significant threat, it is usually one that pilots are specially trained to manage. Most operators will lay down guidance for crews in the event that they are operating in low visibility and these skills are often tested in the simulator. Given that low visibility is also something that the crew will be very aware of during departure or arrival briefings, its unlikely that this will be missed. The next three most significant threats are interrelated: unfamiliarity with airport, airspace and local terrain/weather; (known) problems with ATC quality; and procedural complexity. Operating out of or into an airport that is unfamiliar to one or both pilots requires a much more detailed briefing than operating out of or into the pilot's home base, especially if the quality of ATC may be in doubt or the procedures required for guiding the aircraft on the ground or in the air are complex. Having to employ manual flying skills when the pilot has not practiced them for some time is also a significant threat and one that is related to automation problems. The most significant weather-related threat after low visibility is windshear. Problems in either calculating aircraft performance or operating close to performance limits in terms of climb gradients or runway distances are threats that could lead to loss of control (especially in icing or windshear conditions) and runway excursions both on take-off and on landing. Problems related to altimetry (forgetting to set QNH/QNE and having a low QNH and a transition altitude that is close to the cleared procedural flight level) could also lead to negative outcomes.

A threat identification framework needs to ensure that the pilot has identified these very important threats but has also considered other threats that may be present. Rather than trying to develop an aide-mémoire that addresses every potential threat, the threat identification framework proposed here is an example of a generic one that could be used. As this is only an example, there may be a more appropriate aide-mémoire that would be more relevant to the particular threats more frequently encountered in your own organization. Guidance for employing this generic threat management tool is given below:

- Threat identification should be part of an interactive briefing.
- Consider the following threats (or sources of information about threats) and ask the following questions either to your colleague or to yourself (the aide-mémoire for this threat identification framework is L-M-N-O-P-Q-R-S-T-U):
  - Low visibility: Is there or will there be decreased visibility?
  - Meteorology: What threats are present in the METARs/terminal area forecasts (TAFs), SIGMETs, significant weather charts, synoptic charts and wind charts? What does the ATIS say? Is there a risk of windshear? What does it look like out of the window (e.g. storms moving in)?
  - Notice to Airmen (NOTAM): Are there any NOTAMs that will affect operations?
  - Other traffic: Is there an increased threat of loss of separation with traffic on the ground or in the air?
  - Performance: Are we going to be performance limited with regard to runway distances or climb gradients? What are the runway conditions?
  - QNH: What is the current barometric pressure? Is it falling? If there is a small difference between the maximum flight level of the departure procedure and the transition altitude, could this cause a level bust?
  - Radiotelephony: What is known about the quality of ATC?
  - Systems: Are there any technical defects with the aircraft?
  - Terrain: Is there any terrain that we should be aware of?
  - Unfamiliarity: How familiar is the airport?

One of the main considerations that would inform your decision about how many of these threats you would consider is your familiarity with the airports being reviewed. Operations into and out of your home base would be unlikely to require a full reconsideration of every threat on this list on every occasion unless there had been some sort of change. However, on occasions when the crew is operating into or out of an unfamiliar airport, perhaps one that has never been visited before or one that may not have been visited for some time, a review of all the threat categories may bring to light some relevant threats. Another consideration is if one pilot is not familiar but the other pilot is. What kind of briefing should be carried out then? Given that the purpose of the briefing is to share a mental picture, identify and plan for threats, the briefing should be conducted as if neither pilot is familiar. This has the advantage of minimizing the chances of missing anything and gives the pilot who is familiar an opportunity to share his experience with the other pilot.

3.5.1.5 Generic threat management strategies

Once a threat has been identified, the process is not complete until a suitable threat management strategy has been identified, discussed and agreed on. There are three types of threat management strategy, the acronym for which is ABC:

1. Avoid – Can the threat be avoided completely? For example, if the threat is the storm moving through the airspace around the airfield, can we avoid this threat completely by delaying the flight until it has passed?
2. Buffer – Can we utilize more resources to buffer the effects of the threat should we encounter it? If there are multiple storm cells moving through the airspace around the airfield but they are not severe enough to delay the flight, what can we do to minimize the potential effect they may have on our flight?
   a. Liveware (in flight deck) – Brief the monitoring pilot to be ready to request weather avoidance vectors as required.
   b. SOPs – Change performance so that maximum thrust is used.
   c. Hardware – Ensure the weather radar is set correctly, use anti-icing systems and continuous ignition, use automation to decrease workload and allow more resources for planning a path around the storms.
   d. Liveware (outside flight deck) – Request alternative routing with tower/departure controller before take-off to avoid storms, and brief the cabin crew about the possibility of turbulence.
3. Contingency plan – If we do, for some reason, commit an unsafe act and/or we end up with a UAS due to this threat, what do we do? If the proximity of a storm on departure leads to a windshear encounter shortly after take-off, we are now in a UAS. Our contingency plan would be to carry out a windshear recovery maneuver and this should be rehearsed during the briefing.

3.5.1.6 Threat management tool for briefings

Table 3.2 summarizes threat identification and management. The brace indicates that any of the strategies can be used to mitigate any of the threats.

3.5.2 Unsafe act (error and violation) management

We may encounter unsafe acts at any time. They may be a result of a mismanaged threat or they may occur spontaneously. Note that for the purposes of TEM2, "error"

Table 3.2 **Threat management summary**

Threats		Strategies
Low visibility	QNH	Avoid
Meteorological	Radiotelephony	
NOTAMs	Systems	Buffer
Other traffic	Terrain	
Performance	Unfamiliarity	Contingency plan

NOTAM: notice to airmen; QNH: altimeter subscale setting to obtain elevation when on the ground.

includes errors and violations; that is, any unsafe act. Although the original TEM model refers to error management, TEM2 (although keeping "E" in the name to maintain the link with the original TEM model) will refer to unsafe act management as this includes the management of violations as well as the management of errors. The original TEM model gives three categories of error[16]:

*   aircraft handling errors
*   procedural errors
*   communication errors.

This categorization was useful when it came to noting errors during audits, and recording forms have vast numbers of possible errors that could occur in the flight deck. However, when we try to develop ways of managing unsafe states in our day-to-day operation, not much is gained by listing every possible unsafe act that could occur in the flight deck; there are simply too many. What will dictate the best management strategy is not necessarily the precise technical nature of the unsafe act but rather what type of unsafe act it is; that is, what performance level was being used when the unsafe act occurred. We have covered SB, RB and KB performance, when they are used and how they are prone to errors and violations, and this is will form the basis of the unsafe act management strategies used in TEM2.

Although unsafe act prevention is the goal of the threat management stage, some unsafe acts can occur spontaneously without a threat being present. Therefore, before we move on to detection and management of unsafe acts, it is worth looking at some strategies for preventing unsafe acts that could occur spontaneously.

3.5.2.1 Unsafe act prevention where no threat has been identified

SB errors can occur spontaneously. When we give our threat management briefing, it would be unusual to talk about the threat of retracting the flaps when we meant to retract the gear or the threat of switching off the generators in flight when we actually meant to switch off the anti-icing system. SB errors are very common but tend to occur without warning (although fatigue will increase their likelihood). Here is some general guidance for preventing SB errors:

*   Confirm critical switch/system selections before making them, particularly if the workload is high – this will minimize the risk of making an SB slip.
*   If switching multiple switches, switch them one at a time – making an SB slip can be serious if it involves one switch, but may be far more serious when a slip is made while switching several switches simultaneously.

RB errors tend to occur in high-workload situations. We cannot always predict when technical issues or other problems are going to occur that will increase the situational workload and so we cannot always manage this threat in advance. Below is some general guidance and questions to consider when managing workload:

*   Reduce workload as much as possible before solving a complex problem:
    *   Can some workload be delegated?
    *   Can some workload be delayed?

- Can resolving the current problem be delayed until a lower workload phase of flight?
- Will slowing down allow more time to solve the problem?

Preventing KB errors that can occur when solving complex, novel problems relies on good decision-making strategies and these are covered in Chapter 2 (Information Processing), specifically TDODAR (Time available, Diagnose, Options, Decide, Assign tasks, Review).

Prevention of SB, RB and KB violations relies on one very important principle: does the individual know what the procedures are? If the procedures are not known or are too complex or extensive to be effectively used, violations are more likely to occur. General guidance for preventing violations is relatively straightforward in the first instance: know the procedures and follow them wherever possible. If the procedures are too complex or extensive, violation prevention is in the hands of the organization as it will be necessary to rewrite these procedures so that they are more usable. The role of the organization in managing error is covered in Chapter 4.

3.5.2.2 Unsafe act detection: checklists and crew communication

If a mismanaged threat has led to an unsafe act or we have not been able to prevent a spontaneous unsafe act from occurring, we must then manage this. The first step in unsafe act management is detection. If an unsafe act is not detected, it cannot be corrected. Based on the error theory we have covered so far, Table 3.3 gives all the different types of error and violation, the likelihood that they will be self-detected and an example. Remember, SB unsafe acts can occur during routine and abnormal operations. RB and KB unsafe acts can only occur if some sort of decision is being made or some sort of problem being solved. For this reason, they are more likely to occur when there is any deviation from routine operations, such as a minor or major technical problem, deciding whether to deice the aircraft, operating into or out of a new airport, etc.

Table 3.3 **Chance of self-detection for each type of unsafe act, with examples**

Type of unsafe act	Chance of self-detection	Example
SB error – slips and lapses	High for slips, low for lapses	Unintentionally hitting a switch
SB violation (routine)	Low	Habitually taxiing too fast
RB error	Low (due to workload)	Misinterpreting system warning
RB violation (situational)	High	Shortening brief to make a slot
KB error	Low (lack of knowledge)	Mishandling a new, complex problem
KB violation (exceptional)	High	Violating procedures because there is no other choice (exceptional). Ignoring procedures for no reason (negligent)

SB: skill-based; RB: rule-based; KB: knowledge-based.

Detection of skill-based errors: checklists

Some SB errors, especially slips, may be self-detected as soon as they are made. If you accidentally switch something that you did not mean to, you will probably detect and correct it immediately. However, other SB errors, especially lapses, may not be self-detected and they are usually difficult to detect by the other pilot. A lapse is missing a certain SB action that should have been carried out, usually due to inattention or distraction. The reason that a lapse is difficult to self-detect and is very hard for someone else to detect is that it is due to an *inaction*, rather than an action. One useful strategy against lapses is the proper use of checklists. Configuring systems for different phases of flight usually requires a preprogrammed flow of actions that require little conscious awareness. Flaps are set at a particular time, engine air supply and air conditioning systems are set in a routine fashion, and most of us carry out these action flows automatically. Occasionally, though, we miss something, so there is a lapse in the normal flow. Fortunately, checklists are there to ensure that essential actions are carried out during various phases of flight, but checklist discipline is essential for them to work effectively. Because these checklists are performed after the relevant actions have been completed, they are known as do–confirm checklists and work on the principle of challenge–response; thus, the pilots carry out their action flows, one pilot asks for a checklist and then the other pilot reads out each item while the first pilot confirms that the item is set correctly and responds accordingly. Abnormal checklists are normally read–do–confirm and may involve just one pilot or both pilots. In these sorts of checklist, no actions are taken before starting the checklist (except, occasionally, some memory items) and then each item in the checklist is read out, actioned and then confirmed. In the case of one pilot having to carry out an abnormal checklist while the other pilot is occupied flying the aircraft, he may read out the item "select ENGINE ANTI-ICE on", switch the engine anti-icing systems on and then confirm this: "ENGINE ANTI-ICE on". For critical selections, such as shutting down an engine, the checklist may call for the pilot to confirm that the correct switch, button or lever is being touched before moving it. Even if the checklist does not call for it, the pilot may still ask his colleague to confirm this; for example, "select THRUST LEVER to fuel off", the pilot then places his finger on the relevant thrust lever and asks his colleague, "confirm THRUST LEVER number 1". His colleague will then reply, "THRUST LEVER number 1 confirmed". Once he has confirmation, the first pilot will then select the thrust lever to the fuel-off position and confirm this before moving to the next step in the checklist by stating, "THRUST LEVER number 1 is at fuel off".

No matter which type of checklist is being used, there are some strategies for good checklist discipline:

- If a normal checklist is triggered by a particular event (e.g. the after take-off checklist being triggered by complete flap retraction), be rigorous in calling for the checklist at that point or, if the workload is high, making the checklist clearly visible so that it can be actioned as soon as possible after the normal trigger point.
- Checklists should only be carried out when both pilots are ready. Do not start a checklist until the other pilot is in a position to participate. If he has not called for a checklist and you think he should have, suggest it rather than start it.

- Announce the title of the checklist before starting so that your colleague can confirm it is the correct one.
- Always use the checklist. Do not be tempted to carry it out from memory.
- Use standard phraseology every time.
- To mitigate the risk of pilots just responding to checklists from memory, both pilots should ensure that they direct their attention to the relevant switch before continuing, so that the pilot reading the checklist will also visually check the switch position.
- Use whatever means necessary to keep your place on the checklist, for example moving your finger down the list step by step. This will minimize the risk of losing your place when you look away to confirm switch positions.
- Do not interrupt checklists.
- Do not let other people interrupt your checklists. If someone enters the flight deck while you are carrying out a checklist on the ground, it is preferable to let them stand there for a few more seconds while you continue. They will soon realize that what you are doing is important and will be less likely to interrupt you in the future.
- If a checklist is interrupted, start it again.
- If a checklist cannot be completed because there are one or two items outstanding, the checklist should be left in an obvious place and should be completed as soon as possible and before moving on to a subsequent checklist.
- If there is an item outstanding, allow the pilot who normally gives the response to say this rather than saying it yourself and then moving on to the next item. As you will see in Chapter 6 (Communication), this makes it more likely that they will remember that this item is outstanding as they have had to think about it more deeply than had you just made the response for them.
- Once a checklist is completed, announce that it is completed.

It can be very difficult to use good checklist discipline if checklists are poorly designed, (e.g. no clear trigger point, too long or ambiguous). However, this is the responsibility of the organization and will be covered in Chapter 4.

Detection of skill-based violations, and rule- and knowledge-based errors: crew interaction

As you can see from Table 3.3, there are several unsafe acts that will probably be detected only if the other pilot notices them (SB violations, RB and KB errors). Given that RB and KB errors occur during problem solving, we reach a crucial concept and one that justifies having two pilots in the flight deck. Historically, the first planes built could fly with only one pilot. As planes became more complex, a second crew member was needed to help operate the plane. This was in the days before ergonomics and human factors; therefore, flight decks in early planes had no systematic arrangement of instruments and controls and two people were needed to successfully operate them. The legacy of having two pilots in commercial aircraft continued until the aircraft became even more complex with the advent of jet engines, radio navigation and hydraulic control. At this point, a third crew member, the flight engineer, needed to be added to successfully operate the aircraft. Some aircraft needed even more crew. Then came the age of computers. Complexity did not decrease, but computers were able to manage a lot of the functions that a flight engineer used to manage and so we went back down to two crew on the flight deck.

In 2010, the aircraft manufacturer Embraer said that by 2020, commercial aircraft could be operated with just one pilot.[17] Although there has been a lot of debate about whether the public would be willing to fly on an aircraft with just one pilot, arguments for a single-pilot commercial airliner include mechanisms for remote control in the case of incapacitation and an increased reliance on advanced air traffic management systems. It seem that the industry is really considering the possibility of single-pilot operation for commercial airliners. Pilot reaction is, understandably, skeptical. Most arguments against such a scheme rely on the second pilot providing some level of redundancy. Based on the information covered so far in this book, there is a more compelling argument about why it is safer to have two pilots in the flight deck. Human beings are exceptionally good at finding novel solutions to complex problems. Although the crash of United Airlines Flight 232 in Sioux City in 1989 claimed the lives of 111 people, the insight and skill of the pilots in using differential thrust to control the aircraft after an uncontained engine failure disabled all three hydraulic systems saved 185 lives.[18] Our 86 billion neurons and 500 trillion synapses are capable of abstract thinking, innovation and inspiration in the face of seemingly insurmountable problems. However, because our brains evolved to operate in environments far less dynamic and complex than modern airliners, they are prone to systematic errors, usually as a result of using System 1 decision making (see Chapter 2), errors that we are not always good at detecting. The absolutely fundamental point about error management is recognizing that error is inevitable in humans. No matter what strategies we use to reduce our error rates, we have to accept that we will continue to make them although, hopefully, less frequently. However, a pilot could still make an error of a critical nature that could jeopardize flight safety and be unable to detect it himself. What should give us hope and what should allow us to do our jobs as well and as safely as we can is the knowledge that we have another set of 86 billion neurons and 500 trillion synapses located about 3 feet (90 cm) away from us: the other pilot. Because humans are notoriously bad at detecting errors they make at the RB and KB levels of performance, levels of performance that are used when things are going wrong, we should be relieved and reassured by the fact that we have a living, breathing, awesomely well-designed error-detecting machine sitting next to us. It is far easier for someone else to detect our RB and KB errors than it is for us to detect them ourselves. From a human factors perspective, this is the purpose of the two-crew flight deck and it goes back to the saying, "two heads are better than one". That saying is true, but only if you use the other head effectively.

Let's start with the errors that are difficult to self-detect. RB and KB errors occur when decisions are being made, usually in response to a problem. As we have stated, RB errors occur in high-workload conditions where the brain takes a shortcut in problem solving by using a solution that is known to work for a similar problem, rather than spending time and cognitive effort trying to derive a better solution, i.e. used System 1 rather than System 2 decision making. KB errors occur because knowledge is missing or the problem is too complex or dynamic to be solved in the time available. The best way of picking up on these errors is to take a fresh perspective on the problem. In Chapter 2, the TDODAR decision-making framework that was recommended is designed to work on the basis of getting a fresh, "uncontaminated" perspective on the situation from the other pilot. The hope is that this will prevent RB and KB errors from happening at all by highlighting any mismatches between the mental models of the two pilots. In cases where action has been taken by one pilot that has resulted in an RB or a KB error, it may only be the non-acting pilot who spots the problem. If the first pilot has acted without consulting the second pilot, the second pilot has not been through the same thought process and so will be unbiased by it; that is, he will not have used System 1 yet and so will not be susceptible to errors as a result of the heuristics used. He may easily spot the error as he is "uncontaminated" by the assumptions of the other pilot. To complete the process, we need to ensure that he voices his concerns.

Unsafe act management is based on cross-checking and crew communication. When time allows, actions taken in abnormal situations should be agreed by both pilots. If the abnormality is complex enough, the TDODAR system can be used to elicit a fresh diagnosis from one of the pilots and can highlight potential RB and KB errors before they happen. Failing that, if an RB or a KB error has been made, the only way this can be highlighted is through communication between crew members. Highlighting someone else's error can be potentially difficult, depending on the relationship between the pilots. In Chapter 6 we will cover different modes of communication, different channels that can be used and assertiveness strategies. For now, it is sufficient to say that there needs to be the right tone in the flight deck so that either pilot can highlight anything they think is an error without any hesitation or reservation. Some practical considerations and strategies for doing this are as follows:

- Errors are inevitable. Even the most experienced, well-trained pilot will make errors.
- If you spot your own error, admit it. This is probably the most positive way that you can encourage an effective error-detecting environment. If you have made an error, highlighting it demonstrates that errors do actually occur. Also, if it is an error that the other person could have detected, this will raise their alertness level to detect any further errors. It will also make it more likely that the other person will admit to their errors.
- It is easier for the captain than for the first officer to highlight an error.
- No matter how senior the captain or how junior the first officer, highlighting a potential error should never be criticized, even if it turns out to be incorrect.
- How sure are you that any crew member would be completely comfortable telling you that they think you have made an error?
- Studies have repeatedly shown that the heuristic of overconfidence gives us more assurance in the correctness of our knowledge and actions than is justified. You may be completely confident in your action but must set this surety aside and reassess your view of the problem if someone challenges you on it.
- Avoid point-scoring at all costs. Highlighting an error is not highlighting failure. We have established that error is inevitable and will continue to be a part of everyday life. A lot of work has been done looking at interpersonal dynamics between crew members in the flight deck, but two pilots trying to score points through a tit-for-tat tallying of each other's perceived errors is pathological and highly problematic and runs counter to effective TEM.
- Someone pointing out an error for you is much better than having it go unnoticed. While there may be a sense of embarrassment at having made an error, it vindicates the purpose of a two-crew operation and proves that the system works.

- Thanking or congratulating someone for pointing out an error is an excellent way to keep that highly capable, living, breathing error-detecting machine sitting next to you as motivated and alert as possible to detect future errors.
- If, as a captain, you have to point out a first officer's error, particularly in training, a little self-deprecation can help to ensure that the first officer's confidence is not affected. At the risk of repeating this too many times, it is worth restating that error is inevitable. You will seldom see an error that someone else has made in the flight deck that you have not made yourself at some point.
- Have a look at Chapter 6 (Communication) for more guidance about the best use of language to achieve an open, safe working environment, particularly if cultural barriers are present.

SB violations, like RB and KB errors, are also difficult to self-detect. If a pilot has developed bad habits that mean that his SB operation of the aircraft is not in accordance with procedures, he may not even know that he is doing it. He needs to rely on his colleague to bring the violation to his attention. The guidance for ensuring open channels of communication in this regard are the same as for highlighting RB and KB errors. If an SB violation has become habitual, it may take repeated reminders from the other pilot before the violation stops.

Detection of rule- and knowledge-based violations

Because RB and KB violations require a conscious choice on the part of the individual, there is a high chance that they will be aware that they are acting contrary to procedures; that is, they self-detect the violation. In the case of an RB violation where a shortcut is taken to complete a task when time is limited, the violation is situational and may be an accepted norm of the operation. In this case, the "violation" takes the form of an efficiency-thoroughness trade-off (ETTO); these are common in complex systems where there are often competing goals and the human must make trade-offs to satisfy as many goals as possible.[5] When an RB violation occurs that is not a norm of the system, the other pilot may have to raise the issue if he feels that the violation is not justified by the situation. In this case, the RB violation occurs as a response to a problem that does not appear to have a procedurally correct solution. When the other pilot notices this, the violation can be highlighted but, unless he can offer an alternative, more procedurally acceptable solution, the violation may still occur. However, if the pilot who committed the violation has done so when an alternative option exists, one that is compliant with procedures, this alternative course of action should be highlighted at the same time as the violation is noted.

KB violations normally occur under extreme circumstances. In some cases, there may be no way to solve a particularly complex problem other than by committing a KB violation. In 1983, an Air Canada 767 ran out of fuel and had to glide to a landing at a former air force base in Gimli, Manitoba. On the final approach, the aircraft was too high to land on the runway but too low to execute an orbit to lose the excess height. To solve this problem, the pilot intentionally carried out a maneuver that would not normally be permitted on a commercial airliner: a sideslip.[19] This was a KB violation that was used to solve a novel, complex problem in a highly unusual situation. The aircraft landed safely with no loss of life. In contrast, if a pilot carried out a sideslip because he felt he was high on a normal instrument landing system (ILS) approach

Table 3.4 **Unsafe act detection strategies**

Type of unsafe act	Example	Detection strategy
SB error – slips and lapses	Unintentionally hitting a switch	Self-detection for slips, checklists for lapses
SB violation (routine)	Habitually taxiing too fast	Highlight violation until the bad habit stops
RB error	Misinterpreting system warning	Offer alternative interpretation/ solution
RB violation (situational)	Shortening briefing to make a take-off slot (ETTO)	Only highlight violation if there is an alternative
KB error	Mishandling a new, complex problem	Offer alternative interpretation/ solution
KB violation (exceptional)	Violating procedures because there is no other choice (exceptional) Violating procedures for no reason (negligent)	If there is no choice, only highlight violation if there is an alternative. If negligent and unsafe, prevent violation by any means

SB: skill-based; RB: rule-based; KB: knowledge-based; ETTO: efficiency–thoroughness trade-off.

rather than discontinuing the approach, this would still be a KB violation but would constitute negligence because the pilot is knowingly carrying out a potentially risky, unapproved and highly abnormal maneuver with no situational justification. The guidance regarding the detection of KB violations is similar to that for RB violations, in that the person often knows that they are committing a violation but unless you've got a better option, it is probably not going to change anything. The only exception is the KB violation where there is no justification. In this case, emergency action such as taking over control of the aircraft may be necessary to ensure the safety of the flight. This is explored in more detail in Chapter 6, specifically Section 6.5 on assertiveness.

Summary of unsafe act detection strategies
Table 3.4 shows a summary of error and violation detection strategies.

3.5.2.3 Unsafe act management strategies

Once the unsafe act has been detected, it must be managed. There are no concrete rules for unsafe act management because of the huge number of potential unsafe acts that could occur. Research has shown that mismanagement of unsafe acts can lead to further problems and may result directly in a UAS.[16] Correcting the problem in an appropriate way is vitally important. A couple of general strategies appear below:

• Do not rush your correction of an unsafe state or you will be prone to making another unsafe act, especially an RB error (more likely in time-pressured situations).
• Treat unsafe state correction like any other problem on the flight deck. Get your colleague's perspective on the problem or do a TDODAR if necessary.

- If in doubt, can you return the system to its original configuration and then reconsider the problem?
- This is the golden rule of TEM2 – some unsafe acts are not correctable. You may find yourself in a UAS and you must stop any unsafe act management and immediately start managing the UAS. The priority is to recover from the UAS rather than continuing to try and manage an unsafe act that has already led to a serious problem.

3.5.3   Undesired aircraft state management

A UAS is a state that, if it persists, will lead to a negative outcome such as an accident, an incident or a near-miss. UASs have been adapted from the original TEM model and are categorized in the TEM2 model as follows:

1. Aircraft handling UAS
   a. vertical or lateral deviations from desired flight path (e.g. during a go-around)
   b. speed deviation
   c. unusual attitude
   d. unstable approach (high or low energy)
   e. continued landing after unstable approach
   f. long, floated, firm or off-centerline landing
   g. not tracking runway centerline during take-off or landing run
   h. high speed rejected take-off
   i. unnecessary weather penetration (including severe clear air turbulence)
   j. aircraft experiencing windshear.
2. Ground navigation UAS
   a. proceeding towards wrong taxiway or runway (risk of incursion)
   b. being on wrong taxiway
   c. runway incursion
   d. taxiing too fast.
3. Aircraft configuration UAS
   a. incorrect aircraft configuration (e.g. flaps, slats, gear, reversers, spoilers)
   b. incorrect systems configuration (e.g. altimeters, pressurization)
   c. incorrect automation use (e.g. using an inappropriate mode)
   d. incorrect performance (e.g. wrong thrust settings or incorrect weight/balance).
4. Loss of separation UAS (may not be caused by pilot)
   a. loss of separation between aircraft and another aircraft/vehicle on the ground
   b. loss of separation between two aircraft in the air
   c. loss of separation between aircraft and terrain.

UASs will not necessarily result in a negative outcome if they are managed correctly. As was stressed when the TEM model was first developed and is equally true for TEM2, the biggest risk in a situation deteriorating towards a negative outcome is that the pilots are still a stage behind in their management; that is, they are still trying to manage the unsafe act rather than actively trying to manage the UAS.

As with unsafe acts, there is a basic framework for managing UAS:

1. Prevention of spontaneous UAS – sometimes the aircraft could be at risk of being in a UAS without the pilots committing an unsafe act. For example, if an air traffic controller issues an incorrect clearance that puts the aircraft into conflict with another aircraft, this will lead to a UAS.
2. Detection of UAS.
3. Correction of UAS.

Preventing a spontaneous UAS occurring for reasons beyond our control is difficult and requires vigilance and suspicion. Loss of separation between aircraft or between an aircraft and terrain can occur for reasons other than unsafe acts on the part of the pilots. These UASs can be prevented by maintaining awareness of other clearances being given and being willing to question clearances when they do not seem to make sense. More than in threat management or unsafe act management, UAS management relies more on the hardware in the flight deck; for example, prevention of a spontaneous UAS may be possible if a clearance is given by ATC that, if followed, could put the aircraft into conflict with another aircraft shown on the TCAS display. General strategies for preventing spontaneous UASs are given below:

- Remember that unsafe acts are not down to pilots alone. Any of the other humans in the system may commit an unsafe act that could lead the aircraft to be in a UAS.
- The only way to prevent a UAS may be to question someone else's actions, for example questioning a clearance that could lead to conflict with other aircraft.
- Preventing a UAS is paramount in staying safe, and any instruction or clearance that threatens this should not be followed until the situation has been clarified.

If we do find ourselves in a UAS, the first step is detection. The source of this detection can be one of the familiar four categories mentioned in Section 3.3 on detecting unsafe acts:

- liveware (self)
- liveware (others)
- SOPs
- hardware.

Some strategies for recovering from a UAS are given below:

- If we personally detect the UAS or if our colleague does, the first thing to do is to communicate this fact. Everyone in the flight deck should be aware that the aircraft is in a UAS and that the priority is to now to recover from this. Ensuring that your colleague feels comfortable enough to communicate this fact if they are the first to notice it relies on the same guidance for ensuring open channels of communication as covered in Section 3.5.2 on unsafe act management (see above).
- SOPs may highlight when an aircraft configuration is incorrect or that an approach is unstable, but these procedures are only useful if they are known and if they are used.
- Once everyone is aware that a UAS has been detected, the UAS must immediately be recovered from. The strong temptation that will arise from the confusion and disorientation of having to readjust your mental model to the new reality is to ask the question: "How did this happen?" This question will lead the pilot back to unsafe act management and that is not the priority. The first priority once a UAS is recognized is to immediately recover to a safe condition and to make the aircraft stable. Analysis of what went wrong is best done once on the ground.
- It may be that the only way of reliably detecting a UAS is with the hardware of the aircraft. TCAS and GPWS can detect critical UASs when all other liveware and SOP detection methods have failed. If hardware in the flight deck issues warnings, these should immediately alert the pilot to a UAS and they should be acted upon rather than attempting unsafe act management and trying to work out what has gone wrong.
- UAS management takes priority over all other tasks because flight safety is in jeopardy until the UAS is recovered from.

Table 3.5 **Summary of amended Threat and Error Management model (TEM2)**

Threats		Strategies
Low visibility Meteorological NOTAMs Other traffic Performance	QNH Radiotelephony Systems Terrain Unfamiliarity	Avoid Buffer Contingency plan

⬇

Unsafe acts	Strategies
SB error (slips and lapses)	Self-detection for slips, checklists for lapses
SB violation (routine)	Highlight violation until the bad habit stops
RB error	Offer alternative interpretation/solution
RB violation (situational)	Only highlight violation if there is an alternative
KB error	Offer alternative interpretation/solution
KB violation (exceptional)	If there is no choice, only highlight violation if there is an alternative. If negligent and unsafe, prevent violation by any means

⬇

UAS	Strategies
Aircraft handling Ground navigation Aircraft configuration Loss of separation	Communicate Rely on hardware systems Recover immediately Do not revert to unsafe act management

NOTAM: notice to airmen; QNH: altimeter subscale setting to obtain elevation when on the ground; SB: skill-based; RB: rule-based; KB: knowledge-based; UAS: undesired aircraft state.

3.5.4   Summary of TEM2

Table 3.5 summarizes the key points of TEM2 in the flight deck. For threats and UASs, the addition of a brace between the threat or UAS and the list of strategies means that any strategy may be applicable to any of the items to the left of the brace. For unsafe acts, specific strategies are required for specific unsafe acts.

3.5.5   History of Threat and Error Management and differences between original TEM and TEM2

TEM was originally designed as an audit tool and so a lot of work was put into classifying and coding various different threats, errors and UASs.[16] It was initially developed as a collaboration between the University of Texas and Delta Airlines in 1994 and later went on to include 27 other airlines from all around the world. Thousands of line-oriented safety audits (LOSAs) were carried out on scheduled flights with an observer on the jumpseat who recorded threats, errors and UAS and how the crews managed these. The data were collected anonymously as the aim was to obtain a picture of normal operations across the airline rather than judging the performance of individual pilots. The data were then analyzed and TEM training was carried out for all pilots. The audit was then redone and there was a significant improvement in standards, thus validating the usefulness of the TEM model. Since the 1990s, TEM has been adapted for a variety of purposes such as accident/incident investigation and for use in air traffic management. TEM2 is a further development and is a way of applying the original TEM principles to day-to-day pilot operations. To do this as comprehensively as possible, it has been necessary to diverge from some of the original definitions and approaches used in the first TEM model. Readers are encouraged to look at the original research in this field, particularly if planning to implement a TEM-based system of LOSAs to assess the operational safety in their own airline. TEM2 is meant to be a practical tool that pilots can use to improve the safety of their operation, and to best achieve this, some additions, deletions and modifications have been made to the original model:

- Threats were originally defined as events beyond the influence of the flight crew but TEM2 includes threats that are not beyond the influence of the crew (e.g. fatigue).
- Organizational threats are not directly manageable by the crew on the day and so this category has been omitted.
- The threat categories used in TEM2 (flight deck liveware, other liveware hardware, SOPs and environmental) are unique to TEM2 and are based on the generic model of human factors adapted for aviation.
- The ABC system of managing threats was not present in the original model.
- Errors were originally categorized as aircraft handling, procedural or communication, but this system has been replaced with one based on the performance level being used (SB, RB and KB).
- The original TEM model does not refer to SB, RB or KB unsafe acts, or to the different management strategies that are required.
- Violations are not covered in the original model.
- TEM2 uses the term "unsafe acts" in preference to "error" as unsafe acts also include violations.
- Strategies for prevention of spontaneous unsafe acts are unique to TEM2.
- UASs were originally classified as having to be flight-crew induced, but for practical purposes TEM2 recognizes that UASs can occur for reasons beyond the pilot's control.
- Strategies for prevention of spontaneous UASs are unique to TEM2.

None of these alterations should detract from the outstanding achievement of the original TEM model and its encapsulation of a principle that seems obvious to us now but had not been fully appreciated before TEM was developed, namely that very often threats lead to unsafe acts that lead to UASs that lead to negative outcomes. TEM2 is about arming the pilots with the most up-to-date tools based on the best research available.

3.6   TEM2 and unstabilized approaches

An unstabilized approach is a serious UAS that can have a disastrous outcome. Sixty-eight percent of aircraft accidents occur during the landing phase and 66% of these accidents result from an unstabilized approach, making it one of the most significant safety risks in aviation today.[20] Because it is also a phenomenon that is completely avoidable through careful planning and good crew interaction (the core concepts of TEM2), it is being covered in this section as an example of how TEM2 can be implemented to improve safety.

Low-energy unstabilized approaches tend to result in CFIT while the more common high-energy unstabilized approaches tend to result in runway overruns. Based on the available statistics, unstabilized approaches represent the most significant risk to commercial aviation today. For an approach to be stable, the following criteria must be met by 1000 feet (300 m) above runway threshold elevation (ARTE)[21]:

1. The aircraft is on the correct flight path.
2. Only small changes in heading/pitch are required to maintain the correct flight path.
3. The aircraft speed is not more than V_{REF} + 20 knots and the indicated airspeed is not less than V_{REF}.
4. The aircraft is in the correct landing configuration.
5. Sink rate is no greater than 1000 feet (300 m) per minute; if an approach requires a sink rate greater than 1000 feet per minute, a special briefing should be conducted.
6. Power setting is appropriate for the aircraft configuration and is not below the minimum power for approach as defined by the aircraft operating manual.
7. All briefings and checklists have been conducted.
8. Specific types of approaches are stabilized if they also fulfill the following: ILS approaches must be flown within one dot of the glideslope and localizer; a Category II or III ILS approach must be flown within the expanded localizer band; and during a circling approach, wings should be level on final approach when the aircraft reaches 300 feet (90 m) above airport elevation.
9. Unique approach procedures or abnormal conditions requiring a deviation from the above elements of a stabilized approach require a special briefing.

If any of these criteria are not met, or are lost after passing the stabilization height, the approach is deemed to be unstable and a go-around must be flown.

While there is considerable variation among operators in the rates of unstable approaches and the rates of go-arounds following unstable approaches, the aggregated statistics for the industry paint a bleak picture. Approximately 3% of approaches are unstable but this is not a problem in itself.[22] The truly concerning figure is that less than 5% of unstable approaches result in a go-around, suggesting that more than 95% of pilots will continue an unstable approach to a landing.[23] Reducing this figure is a high priority for aviation safety managers and understanding why over 95% of unstable approaches continue to a landing is no easy task. The reasons range from high workload leading to impaired decision making, lack of awareness, lack of assertiveness, overconfidence and an aversion to flying go-arounds (see Box 3.1). The aim of this section is not to examine all the human factors contributing to unstable approaches and their continuation to landing, but rather demonstrate how TEM2 can

---

Box 3.1 Accident Case Study – Unstabilized Approach

Garuda Indonesia Flight 200 – Boeing 737-400
Adisucipto International Airport, Indonesia, 7 March 2007[24]

At 10.1 nautical miles from the runway, this Boeing 737-400 aircraft was approximately 1500 feet (450 m) higher than would have been normal for an instrument landing system approach and was traveling at 283 knots. The captain attempted to capture the glideslope by descending steeply, an action that caused the speed to increase. Although the captain called for flap deployment, the first officer elected not to do this because they were in excess of the flap operation speed limit. The ground proximity warning system sounded 15 times and the first officer called for a go-around. The captain elected to continue and the aircraft touched down at a speed of 232 knots and subsequently overran the runway, killing 22 people.

---

be used by pilots to reduce the risk of an unstable approach and increase their chance of going-around should one occur.

3.6.1   Threat management for unstabilized approaches

Using the TEM2 threat identification framework (as shown in Table 3.2), a pilot will be able to identify some key threats that could potentially lead to an unstabilized approach. Because non-precision approaches involve very specific threats of their own, the assumption is that this approach will be performed using an ILS. Below is a summary and explanation of the main threats that a pilot could identify during an approach briefing:

- Low visibility – (will not act as a significant threat for an unstabilized approach).
- Meteorological – What threats are present in the METARs/TAFs and SIGMETs? Is the weather forecast to be bad? Will we need vectors around storm cells that could potentially shorten the approach? What does the ATIS say? Is there a risk of windshear? What does it look like out of the window (e.g. storms moving in)? Tailwinds may make it more difficult to decelerate. The presence of cumulonimbus cells in the area could require vectoring upwind of cells and may result in a shortened approach, especially given that the runway in use normally has a headwind and so vectoring upwind would bring the aircraft closer to the runway. If the temperature is low, having aircraft anti-icing systems on may make it more difficult to decelerate.
- NOTAMs – Is the ILS functional or will we need to carry out a non-precision approach?
- Other traffic – Will there be increased pressure on ATC to minimize separation and request higher speeds in order to maximize runway usage?
- Performance – Are we going to be performance limited with regard to runway distances? Is the runway wet, contaminated or sloping? Are there any steps we can take to shorten our landing distance or decrease our approach speed?
- QNH – What is the current barometric pressure? If the QNH is very high, when transitioning from QNE to QNH, the aircraft may suddenly "gain" 1000 feet (300 m) and require a steeper descent to lose this height.

- Radiotelephony – Is ATC aware of the stabilization requirements? Does ATC usually keep aircraft high and fast? Are there any known problems with ATC?
- Systems – Are there any technical defects with the aircraft? For example, are any drag-increasing systems inoperative that will make it more difficult to lose speed?
- Terrain – Will terrain in the approach area mean that we are kept high?
- Unfamiliarity – How can we manage unanticipated situations that we may encounter during the approach that could lead to an unstabilized approach?

The threats listed above refer to normal ILS approaches but the risks are increased with visual or non-precision approaches. Visual approaches in particular have a high risk of becoming unstable as they are often commenced before being visual with any Precision Approach Path Indicators (PAPIs), meaning that there is a period before turning on to the final approach track where the pilot may have no guidance about whether the aircraft is above or below the normal 3 degree approach path. Unless visual approaches are being carried out frequently, it is very difficult to have a reliable mental picture of what a normal approach should look like, given that commercial pilots are generally more accustomed to flying ILS approaches.

Once the threats have been identified, strategies can be put into place that will mitigate the risk that these threats will lead to an unstable approach. Possible threat management strategies are:

- Avoid – Can the threat be avoided completely? We can avoid the threat of ATC leading us into an unstable approach by agreeing to decline any clearance that could threaten stability. We may be able to avoid the increased risk of an unstable approach following a visual approach by electing to do an ILS approach to a different runway.
- Buffer – Can we utilize more resources to buffer the effects of the threat should we encounter it?
  - Liveware (in flight deck): Brief the monitoring pilot on the latest points at which you are planning to make any aircraft configuration changes in order to be stable by 1000 feet (300 m). If this plan is not being followed for any reason, he is to make you aware that you are not configuring the aircraft as briefed. The monitoring pilot should also be briefed that they will probably have a lower workload than the handling pilot during the approach and will be more capable of assessing whether an approach is becoming unstable.
  - SOPs: Ensure that an approach is permitted given the weather/operational conditions. If a visual approach is being carried out, consider how you will judge whether or not your approach is too high or too low before being able to see the PAPIs next to the runway. This may involve writing out distances and expected heights with reference to threshold elevation to use as guidance before turning on to the final approach track. This can be cross-checked with FMS information giving distance to the runway.
  - Hardware: Decelerate/configure early if using anti-icing systems or when any drag-inducing devices are inoperative.
  - Liveware (outside flight deck): Decline any unacceptable ATC request or ask for more track miles in order to lose height and decelerate.
- Contingency plan – Should an unstable approach occur for any reason, both crew should agree that a go-around will be flown and review the steps and calls required to carry out this maneuver safely.

Table 3.6 **Unsafe acts that could lead to an unstabilized approach and their management strategies**

Type of unsafe act	Example	Management strategy
SB error	Forgetting to slow down at the agreed point	Monitoring pilot warns of deviation from agreed plan
SB violation (routine)	Consistently flying approaches too quickly	Highlight violation until the bad habit stops
RB error	Accepting an ATC clearance that will cause instability	Monitoring pilot warns of risk of instability
RB violation (situational)	Shortening approach to land sooner	Monitoring pilot warns of risk of instability
KB error	Keeping too high a speed because of severe icing	Monitoring pilot offers alternative, e.g. "icing is too severe here, let's abandon the approach"
KB violation (exceptional)	Disregarding stabilized approach criteria altogether and intentionally flying an unstable approach	Monitoring pilot calls for go-around, tells ATC that the aircraft is going-around or takes control

SB: skill-based; RB: rule-based; KB: knowledge-based; ATC: air traffic control.

3.6.2 Unsafe act management for unstabilized approaches

Despite the pilot's best efforts at the planning stage, it is possible that an unsafe act may be committed that increases the risk of making an unstabilized approach. Using the TEM2 categorization of unsafe states, Table 3.6 gives some examples of how different types of unsafe act could lead to an unstabilized approach and how these unsafe acts can be managed.

All of these unsafe act management strategies rely on an engaged and assertive monitoring pilot and this is sometimes difficult if the captain is the handling pilot and the first officer is the monitoring pilot. Setting an appropriate atmosphere in the flight deck has been covered earlier in this chapter and we will look at monitoring pilot assertiveness in more detail when we consider UAS management.

3.6.3 Undesired aircraft state management for unstabilized approaches

If an unstabilized approach does occur for any reason, the aircraft is, by definition, in a UAS, and UAS management must immediately be used to recover to a safe and stable flying condition as quickly as possible. If an unstabilized approach does occur, the first step is detection. Detection may come from one of four sources:

- Liveware (self) – The handling pilot recognizes that the approach is unstable.
- Liveware (others) – The monitoring pilot recognizes that the approach is unstable.

- SOPs – The landing checklist highlights something that shows that the approach is unstable.
- Hardware – The GPWS may alert us that we are not in the correct configuration or that we are descending too quickly.

As we saw in Chapter 2, in high-workload conditions as would occur during a rushed approach, it is possible that we will know that we are unstable but become mentally "locked" into continuing the approach because of plan continuation bias and so continue the landing anyway. However, if this is not the case and we either detect the unstabilized approach ourselves or the SOPs or hardware detect it, we can then execute a go-around. The difficulty arises when it is only the monitoring pilot who detects the unstabilized approach and may be reluctant to communicate this, particularly if he is a first officer and the captain is the handling pilot. If there is a particularly high or a particularly low flight deck gradient (see Chapter 5: Personality, Leadership and Teamwork), the monitoring pilot, particularly if he is the first officer, may be reluctant to call for a go-around as it may open him up to criticism from the captain. Conversely, if the first officer and the captain are friends, the first officer may be reluctant to highlight the unstabilized approach as pointing this out implies that his friend has made a mistake. The absolute bottom line is that the monitoring pilot is primarily responsible for monitoring the actions of the handling pilot. In the vast majority of cases, it should be the monitoring pilot who calls for a go-around. A safe approach is the responsibility of both pilots equally and captains should also consider this when discussing the matter with first officers. Calling for a go-around is a difficult call to make and first officers should not be criticized for this as it may impede their confidence in calling for a go-around in the future.

There are many other reasons why over 95% of unstable approaches are not managed properly and continue to a landing:

- The unstabilized approach is not recognized – Sometimes unstable approaches are missed. They may be subtle but it is primarily the role of the monitoring pilot to ensure that stabilization criteria are met. The other reason they may not be recognized is that the workload in the flight deck is too high and monitoring tasks are usually the first to be shed. Avoiding high-workload induced plan continuation requires the approach to be discontinued as soon as the possibility of instability is considered. Continuing a rushed, potentially unstable approach down to the minimum stabilization height increases the chance of getting locked into plan continuation and landing from the unstabilized approach.
- It is assumed that the approach will be stable just after stabilization height – Unfortunately, there is no way to guarantee that it will and so the stabilization height is the absolute and final limit for stabilization.
- Fear of going-around – Go-arounds are uncommon and are often badly executed. Briefing the possibility of a go-around before commencing the approach may go some way to addressing this, but this is primarily a problem for training departments.
- The monitoring pilot assumes that it is the handling pilot's responsibility to call for a go-around – As stated above, the monitoring pilot should be the one to call for a go-around in the majority of cases (although, of course, the handling pilot may also call for a go-around).
- A first officer thinks it is the captain's responsibility – Both pilots are responsible for carrying out a safe approach and the monitoring pilot, whether it is the first officer or the captain, should normally be the one to call for a go-around.

Unstabilized approaches rarely occur spontaneously. Like any other UAS, they normally start with a mismanaged threat followed by a mismanaged unsafe act. While ATC often has some role in causing an approach to become unstable, the final responsibility rests with the pilots and careful application of the principles of TEM2 can prevent this from happening.

Chapter key points

1. Human performance can be skill-based (SB), rule-based (RB) or knowledge-based (KB).
2. SB performance is preprogrammed and automatic.
3. RB performance is semi-automatic and requires some conscious engagement.
4. KB performance is effortful and only occurs when no suitable rules for a particular situation can be activated.
5. Error is defined as the failure of planned actions to achieve their desired goal.
6. SB errors occur when the plan is good but there is a problem with execution. This is either getting an action wrong or missing out an action.
7. RB errors occur during decision making and problem solving and are often related to the heuristics used by System 1 during decision making (see Chapter 2).
8. KB errors occur owing to lack of knowledge about a problem or its solution, processing limitations when trying to solve highly dynamic problems and oversimplification as a means of trying to solve complex problems. The overconfidence heuristic also contributes to KB errors.
9. Unlike errors, violations are where the individual intentionally does not follow a procedure or acts contrary to a procedure.
10. Violations can occur at all three performance levels. SB violations are bad habits, RB violations are often an attempt to satisfy competing goals and KB violations occur as a result of completely novel, abnormal situations or as a willful act of negligence.
11. Unsafe acts include all types of error and violation.
12. Different unsafe acts are most likely to be detected in different ways and several resources can aid in this: liveware (self), liveware (others), SOPs and hardware.
13. The Swiss Cheese Model suggests that many negative outcomes are a result of a sequence of actions, inactions and missed opportunities.
14. The slices of the Swiss Cheese Model can be adjusted to give a model that will allow the pilot to prevent these negative outcomes: Threat and Error Management 2 (TEM2).
15. TEM and TEM2 propose that threats may lead to unsafe acts and that these may lead to undesired aircraft states (UAS), which will lead to negative outcomes if they are not managed appropriately. However, unsafe acts and UAS may also occur spontaneously.
16. Threat management normally occurs during briefing.
17. Pilots need to make an effort to identify threats, and the following framework is proposed to allow them to do this during briefings: low visibility, meteorological, NOTAMs, other traffic, performance, QNH, radiotelephony, systems, terrain and unfamiliarity (L-M-N-O-P-Q-R-S-T-U). A different threat identification framework may be more suitable for the threats that are relevant to your operation.
18. Threats can be managed by avoiding them, buffering their effects or making contingency plans (avoid, buffer, contingency plan – ABC) in the event that they affect the flight.
19. Unsafe act management depends on the type of unsafe act that has occurred and may rely on checklists or crew communication.

20. UAS management requires the cessation of unsafe act management and immediate recovery actions to avoid a negative outcome.

21. TEM2 can be used to minimize the risk of unstabilized approaches occurring and to minimize the risk of an unstabilized approach continuing to a landing.

Recommended reading

Reason J. *Human error*. Cambridge University Press; 1990. An excellent summary of all the research done on human error up to 1990 [Difficulty: Intermediate].

Reason J. *A life in error*. Ashgate; 2013. An updated and simplified summary of Professor Reason's work in the field of error [Difficulty: Easy/intermediate].

Hollnagel E. *The ETTO principle: Efficiency–thoroughness trade-off*. Ashgate; 2012. A very interesting introduction to the dilemma often faced by pilots: do things by the book or do things in the time available [Difficulty: Easy/intermediate].

Gawande A. *The checklist manifesto*. Profile Books; 2011. Written by a surgeon, this book explores how checklist philosophy developed in aviation is being introduced into healthcare. It also gives some of the theory regarding checklist construction and use [Difficulty: Easy].

References

1. Reason J. *Human error*. Cambridge: Cambridge University Press; 1990.
2. Rasmussen J. Skills, rules, and knowledge; Signals, signs, and symbols, and other distinctions in human performance models. *IEEE Transactions on Systems, Man and Cybernetics* 1983(3):257–66.
3. Lichtenstein S, Fischhoff B, Phillips LD. Calibration of probabilities: The state of the art to 1980. In: Kahneman D, Slovic P, Tversky A, editors. *Judgment under uncertainty: Heuristics and biases*. Cambridge: Cambridge University Press; 1982. pp. 306–34.
4. Perrow C. *Normal accidents: Living with high risk technologies*. Princeton, NJ: Princeton University Press; 1999.
5. Hollnagel E. *The ETTO principle: Efficiency–thoroughness trade-off: Why things that go right sometimes go wrong*. Farnham, UK: Ashgate; 2012.
6. Schmidt RA. A schema theory of discrete motor skill learning. *Psychological Review* 1975;**82**(4):225–60.
7. Allwood CM. Error detection processes in statistical problem solving. *Cognitive Science* 1984;**8**(4):413–37.
8. Jarvis, S. Putting human factors on the bus. *9th International Symposium of the Australian Aviation Psychology Association*, Sydney, April 2010. Canberra: Australian Aviation Psychology Association; 2010.
9. Reason JT, Reason JT. *Managing the risks of organizational accidents*. Farnham, UK: Ashgate; 1997.
10. Wiegmann DA, Shappell SA. *A human error approach to aviation accident analysis: The human factors analysis and classification system*. Farnham, UK: Ashgate; 2012.
11. International Civil Aviation Organization. (2002). *Line operations safety audit (LOSA)*. (Doc. 9803 AN/761). Montreal: ICAO.

12. Civil Aviation Authority. (n.d.). Safety plan 2011–2013. West Sussex: CAA.
13. Civil Aviation Authority. (2011). *CAA "significant seven" task force reports. (CAA paper 2011/03).* West Sussex: CAA.
14. Civil Aviation Authority. (n.d.). Significant seven statistics. Retrieved from http://www.caa .co.uk/docs/2445/Significant%20Seven%20Statistics%20Posters.pdf. Accessed on March 17, 2014.
15. Federal Aviation Administration. *Advisory circular: In-flight fires.* Washington, DC: FAA; 2004 (pp. 120–180).
16. Merritt AC, Klinect JR. *Defensive flying for pilots: An introduction to threat and error management.* University of Texas Human Factors Research Project, The LOSA Collaborative; 2006.
17. Flightglobal. (2010). *Embraer reveal vision for single-pilot airliners.* Retrieved from http://www.flightglobal.com/news/articles/embraer-reveals-vision-for-single-pilot-airliners-343348/.
18. National Transportation Safety Board. *Aircraft accident report: United Airlines flight 232, McDonnell Douglas DC-10-10, Sioux Gateway Airport, Sioux City, Iowa, July 19, 1989.* (NTSB/AAR-90/06 PB90-910406) 1990.
19. Aviation Safety Network. (n.d.). Retrieved from http://aviation-safety.net/database/record .php?id=19830723-0. Accessed on March 17, 2014.
20. Flight Safety Foundation. Killers in aviation: FSF task force presents facts about approach-and-landing and controlled-flight-into-terrain accidents. *Flight Safety Digest* 1999;**17**:11–12. *18*(1–2).
21. Flight Safety Foundation. *ALAR Briefing Note 7.1 – Stabilized approach.* Virginia: Flight Safety Foundation; 2000.
22. Direction Générale de l'Aviation Civile. *Stabilised approaches – Good practice guide.* Paris: Direction Générale de l'Aviation Civile; 2009.
23. Flight Safety Foundation. *Go-around safety forum, 18 June 2013, Brussels: Findings and conclusions.* Virginia: FAA; 2013. Retrieved from http://www.skybrary.aero/bookshelf/ books/2325.pdf.
24. National Transportation Safety Committee. *Aircraft Accident Investigation Report: Boeing 737-497 PK-GZC Adisucipto Airport, Yogyakarta, Indonesia*; 2007.

Error management and standard operating procedures for organizations

Chapter Contents

Introduction   119
4.1 Beyond human error   120
4.2 Systems thinking   120
    4.2.1 Normal Accident Theory   121
    4.2.2 The Old View versus the New View of human error   122
4.3 Resilience engineering   124
4.4 Safety culture   127
    4.4.1 Just culture   127
    4.4.2 Moving from Safety I to Safety II   129
4.5 Principles of managing organizational resilience   130
Chapter key points   131
Recommended Reading   131
References   132

Introduction

During an interview in 2008, the renowned psychologist and industrial safety expert Erik Hollnagel suggested something very interesting: in recognizing that error means different things to different people, it might be best for this poorly defined and misunderstood term to be abandoned altogether.[1] This seems like a radical move, given that so much of what we do in human factors is based on eliminating or reducing error, but the idea becomes more attractive when we consider the drawbacks of the term. When accidents happen, the media are always poised to see whether the event was down to human error or not; that is, whether there is someone to blame. It is almost as if "human error" is an acceptable explanation for an accident or incident when, in fact, it explains nothing and is often misused as a pretext for making someone a scapegoat rather than addressing the more serious, underlying organizational or industrial problems. Are we now at a place in our understanding of human factors where the concept of error could be dispensed with entirely? If we do move our focus away from error, how can we possibly manage safety in our organizations? This chapter explores some of the potential strategies for improving safety management in our organizations by introducing a set of concepts that moves beyond human error. For the purposes of this chapter, error can mean any unsafe act: errors or violations.

Practical Human Factors for Pilots. DOI: http://dx.doi.org/10.1016/B978-0-12-420244-3.00004-2

4.1   Beyond human error

We spent all of Chapter 3 talking about error and other unsafe acts from the perspective of front-line crew trying to improve safety. We now need to take a step back and look at how error relates to our organizations. Before we do so, let's look again at Professor James Reason's definition of error: "Error will be taken as a generic term to encompass all those occasions in which a planned sequence of mental or physical activities fails to achieve its intended outcome, and when these failures cannot be attributed to the intervention of some chance agency". To simplify this, we can say that error is defined as "the failure of planned actions to achieve their desired goal". That definition served us well when we were looking at error management and standard operating procedures (SOPs) for pilots in Chapter 3. However, when we take an organizational view, this definition brings up some difficulties. If we want to manage error and prevent it from happening, this definition can limit our efforts. Based on this definition, we only know that we have made an error if we have failed to achieve our intended outcome and so we can only know that we have made an error *after* it has happened. While this may seem like a small point, this observation has created a new branch of industrial safety that we will explore later. It is possible to be working in an incredibly dangerous environment without error but where a single event could lead to catastrophe. This chapter considers the organizational environment within which error can occur and considers how the management of that environment can have as big an impact on the occurrence and outcomes of errors especially with regards to the actions of the people directly involved with them. The organizational environment was one of the slices of James Reason's original Swiss Cheese Model, as mentioned in Chapter 3. In his seminal book entitled *Human Error*, he referred to these organizational conditions as "latent errors", and in a later book he expanded this idea and called them "organizational factors".[2] Before we look into these organizational factors, we need to define something that has been briefly mentioned in Chapter 3 but will become a lot more relevant now: the system.

4.2   Systems thinking

If our expanding understanding of human error has shown us anything, it is that we can no longer think about errors as occurring in isolation. James Reason's work was the first to show explicitly that aside from the act itself there are normally multiple contributory, often organizational factors involved. A further refinement of this is to consider the nature of the organizations within which we work. Consider an airline. An airline is a collection of humans, machines and procedures that come together to achieve a purpose: the safe and efficient transport of people and/or cargo from one point to another. In this way, we can consider an airline to be a system. To complicate matters slightly, the airline system is composed of smaller subsystems, each of which can be considered as a system in its own right. For example, two pilots operating a flight constitute a system. The cabin crew operating that flight also constitute a system. These two systems come together to form another system: the crew. For an airline with several aircraft, there may be several of these crew systems operating at the same time. These flights form part of the airline's network, itself a system, and the airline's network is part of the global transport system. Like Russian dolls, large systems often contain multiple smaller systems.

So here we have a system: a collection of interconnected components working as a whole. In the case of an airline, it is a collection of people, machines and procedures that work together to achieve a specific objective: the safe and efficient transport of people and/or cargo from one point to another. A large airline may have thousands of people, hundreds of aircraft and millions of lines of procedures, laws and computer code that are required to come together to achieve the desired objective. How, in this highly complex and interconnected system, can we possibly predict where error is going to occur? The simple answer is that we cannot. Let us imagine a system that is composed of 1000 components (people, procedures or machines). Let us further imagine that each of these components has only two states: functioning and non-functioning. If we were to try to predict all the different states that that system could be in, there are 2^{1000} possible configuration. That number is 10 novemnonagintillion (10 with 300 zeros after it). Even the most proactive safety department might have some trouble doing a risk analysis on this system! Unfortunately, the situation is made more complicated by the fact that each of the components has far more than two possible states. Consider yourself: when you are at work you may be performing well, above average, average, below average or terribly. There are, of course, lots but states in between these but it is just meant to illustrate that most components of the system operate on a spectrum that starts at being non-functional and goes all the way up to maximum functionality. When we consider this, we can see that accurately predicting how such a system could fail is truly impossible as there are simply too many variables. This observation leads us to consider another concept that marked a turning point in how we understand and manage error: Normal Accident Theory.

4.2.1 Normal Accident Theory

Charles Perrow wrote a wonderful book in 1984 called *Normal Accidents: Living with High Risk Technologies*. I have classified the difficulty level of this book as "Easy" in the Recommended Reading list and it is a fascinating and accessible analysis of accidents that have happened in complex systems such as commercial aviation, spaceflight and the nuclear industry. If you are looking for an introduction to the management of safety in complex systems, this book is the place to start. Nuclear safety was very controversial at the time Perrow was writing his book because of the Three Mile Island nuclear disaster in 1979. Charles Perrow's description of the events at Three Mile Island formed the basis of the opening case study in Chapter 1 (Introduction to Human Factors) and, as stated, it was this accident that led to the realization that human factors play a significant role in industrial safety. Perrow was the first person to appreciate that because of the complex, interconnected nature of the systems within which we work, it was becoming impossible to predict how failures would evolve and spread throughout the system and how the performance of other components could potentially cancel out or, conversely, amplify these effects.[3] In essence, failure and success are two sides of the same coin and should be considered as normal, emergent properties of complex systems. The term "emergent" is very popular in the field of industrial safety because it implies that the behavior of a system cannot be predicted based on our knowledge of the component of the system. There are so many variables that we are essentially dealing with chaos theory.

For those of you not familiar with chaos theory, it suggests that very small disturbances in a system can either disappear or spread throughout the system and cause surprisingly large consequences in a way that is not predictable. The classic example attributed to one of the founders of chaos theory (Edward Lorenz who, incidentally, was a meteorologist) is that a single beat of a butterfly's wing in Brazil could lead to a hurricane in Texas seven weeks later.[4] A single, seemingly inconsequential event can determine whether a catastrophe happens or not. In the same way, a single, seemingly inconsequential event could be the tipping point for a system leading to its collapse or could prevent a catastrophe from happening in a way that we might be completely unaware of. This brings us on to the Old View versus the New View of human error.

4.2.2 The Old View versus the New View of human error

Based on Sydney Dekker's description of the Old View and the New View of human error in one of his scientific papers, we can describe them as follows.[5] The Old View says:

- Human error is the cause of most accidents.
- Systems are designed to be safe. It is the unreliable humans that lead to failures.
- Systems can be made safe by protecting them from unreliable humans by getting rid of those that are found to be unreliable and by enforcing procedures to minimize the risk of errors.

The New View of human error takes into account systems thinking and Normal Accident Theory and says:

- Error is a symptom of a problem, not necessarily the cause of a problem in the system.
- Systems are not inherently safe. They may contain many contradictions and competing goals that humans have to deal with.
- To understand safety and risk in a system, you have to try and understand how it actually works, not how you think it works.
- Humans generally try to do their best.
- Humans are the best tools for creating safety through their capacity to adapt.

The idea that systems are not inherently safe may seem slightly daunting. Surely if people just did things by the book, nothing would ever go wrong? Unfortunately, for the reasons we mentioned when we covered Normal Accident Theory, systems are just too complex to be designed to be perfectly safe. The humans working within the system are often faced with situations where they must deal with multiple goals in a limited period and so have to make trade-offs to complete work in as safe and as efficient a way as possible. This classic trade-off is described by Erik Hollnagel as an efficiency–thoroughness trade-off (ETTO).[6] An ETTO occurs when an individual has to make a trade-off between completing a task "by the book" and completing the task in the time available.

Consider this fundamental paradox described by Professor Hollnagel. In whichever industry you care to name, when things go wrong, employees are often criticized for not having followed procedures. This is classic Old-View thinking: the system is designed to work perfectly and it is the humans that cause it to fail. However, what do groups of employees do when they want to protest to their management without actually going on strike? They "work to rule". This form of industrial action involves following the "rules" of the workplace precisely, taking no shortcuts and not carrying out any additional tasks beyond those specified in the written procedures. While this might appear to be any manager's dream, the result is an inevitable slowdown in operations, possibly to the extent that the system does not function at all. Essentially, workers stop themselves making trade-offs and concentrate only on thoroughness at the expense of efficiency. This form of industrial action has been used by medical staff, police, firefighters, air traffic controllers, flight crew and cabin crew, as well as in many other industries. It is generally seen as a safe form of industrial action because it is difficult to fire someone for following the rules precisely. The lawyers, however, have come up with one of my favorite phrases to cover this type of action: malicious compliance (complying with the rules even though you know it will harm the operation).

And there we have the fundamental paradox of working in a complex system: when the system fails, the humans are blamed for not following the rules, but when the humans follow all the rules, the system fails. The reality is that systems often do not work *unless* the humans take shortcuts and make trade-offs. Hollnagel concludes this argument by making an excellent point, one that should be watermarked into every page of procedures that is ever published in any industry: complex systems containing human components are successful because the humans have adaptive capacity.[1]

The systems we work in are so complex that they are unpredictable, and both success and failure are emergent properties of the complex interactions between components. An action that an individual takes on one day may have absolutely no negative outcome but the same action made by another individual under the same conditions on a different day may result in catastrophe because some other, unobservable part of the system has changed. Both individuals acted in exactly the same way but because only one event had a catastrophic outcome (for reasons that may be well outside the person's control), they are the one who will be retrained, fired or even sent to prison. The Old View puts far more emphasis on outcomes than on the actions that led to them.

Have a look at the case study in Box 4.1. The ironic thing about this non-event is that you have probably never heard about it. But what if the Federal Express plane had been heavier? Or, even worse, what if it had been a fully loaded Boeing 747? In both cases, the departing aircraft would have needed a longer take-off run and would have collided with United 1448 at high speed. If one variable at the other end of the runway from where the "human error" had occurred had been different, we would be talking about Providence in the same way as when we talk about the worst air disaster in history, the collision of two Boeing 747s at Los Rodeos Airport in Tenerife in 1977, which killed 583 people. In the current climate, if the pilots of United 1448 had

---

Box 4.1 Incident Case Study – Errors and Outcomes

United Airlines Flight 1448 – Boeing 757
Theodore Francis Green State Airport, Rhode Island, USA, 6 December 1999[7]

After landing at night-time in low-visibility conditions, the crew of this Boeing 757 taxied off runway 05R and were given instructions to taxi to their gate via runway 05L, a runway that was not in use at the time. Unfortunately, the crew took a wrong turn in the thick fog and ended up re-entering runway 05R. A Federal Express Boeing 727 aircraft taking off from runway 05R narrowly missed them as it rotated. On the cockpit voice recording taken from United 1448, the sound of the engines of the departing Boeing 727 can be clearly heard in very close proximity. Both pilots were unsure which runway they were on and this confusion filtered into the control tower. The controller became convinced that they were on the unused runway 05L. Despite the confusion, the controller gave another aircraft clearance to take off from runway 05R, a clearance that the pilot of that aircraft wisely chose to decline given the confusion. A second take-off clearance was given and this was also declined. Eventually, it became clear that United 1448 had strayed onto the active runway and all departures were stopped until the situation was resolved.

---

survived the hypothetical crash, do you think that they or the air traffic controller in the tower would still be employed? Lawsuits would be brought, prosecution sought and careers ended. But, because of blind luck, the errors did not lead to any negative outcome and all the humans involved lived to fight another day.

Fortunately, the Old View is beginning to die out in the aviation industry. It still persists, mainly because it is an easy way to deal with problems. An accident or incident happens, an investigation is carried out, the person who made the error is identified and then they are dealt with. Action is taken and confidence in the system is restored. It is far more difficult to take a New-View perspective as it involves looking for a problem in the system, not just in the workforce. Later on in this chapter, we will cover some of the practical aspects of encouraging New-View thinking when it comes to investigating problems in your organization. There is one more step we can take, though, and that is to put the relentless crusade to prevent error to one side and talk about another way of improving the safety of our system: engineering it to be resilient.

4.3 Resilience engineering

Resilience engineering came about when a large group of researchers in the field of industrial safety realized that accepting the New View of human error created a new opportunity to build safer organizations. Traditional methods of safety management focused on things that went wrong. Resilience engineering focuses on things that go right. If an accident or incident occurs during one out of every 100,000 flights, as an industry we spend a great deal of time and money attempting to understand what went wrong during that one flight. But what about the other 99,999 flights? Did all of these flights occur on perfectly clear days with no bad weather, no high terrain, no technical difficulties and nothing else out of the ordinary? Of course not. The only difference is that in the vast majority of flights the crews managed to deal with these external disturbances successfully, preserve system function and have a good outcome. Might there not be something to learn from how these crews managed to do this?

While the goal of reducing the number of errors that occur is an understandable one, these scientists realized that there was another opportunity, namely to design the systems themselves to be resilient. Normal Accident Theory and the New View of human error make it clear that error can never be entirely eliminated in a complex system. The challenge now becomes: how can we design and manage these systems so that the risk of error is minimized and, should an error occur (perhaps one that could propagate throughout the system and lead to its collapse), how can we make the system resilient? Rather than expecting that every situation that a crew may encounter can be dealt with by following procedures exactly, a resilient perspective would suggest that there are situations where crews have to use their adaptive abilities to solve complex problems. These solutions, trade-offs and workarounds may well be the things that preserve the safety of the organization and, hence, what makes it resilient.

In one of his early works in the field of resilient engineering, David Woods said that there were four properties of systems that could be used to assess their resilience[8]:

- Buffering – the size or kinds of disruption that the system can absorb without failing
- Flexibility versus stiffness – how the system restructures itself in the face of continued external pressure
- Margin – how close the system is operating to its performance boundary, the boundary beyond which the system will stop functioning
- Tolerance – how the system behaves as the margin to the performance boundary decreases: it may continue to function normally until it collapses or it may degrade first.

These properties are useful if you are assessing the resilience of an organization from outside it, but what are the key properties that people running an organization should encourage to make it as resilient as possible? To answer this question, David Woods and Erik Hollnagel identified four cornerstones that can lead to maximum resilience in an organization[9]:

- the ability to anticipate system requirements
- the ability to monitor the system
- the ability to respond to what happens
- the ability to learn from past experiences.

We can translate these four cornerstones into an aviation setting by considering the first slice of James Reason's Swiss Cheese Model: organizational influences. One of Reason's most useful insights is that organizations often have latent threats within them that make it more likely that unsafe acts will be committed and that negative outcomes will occur. Resilience engineering considers how these organizational influences can be managed to promote resilience elsewhere in the system. To illustrate this, let us consider the organizational procedures in place to run an airline, specifically the part of the airline that involves the flights themselves. A resilience engineer would ask the following questions:

- Is the system designed to satisfy the requirements of the front-line crew when it comes to carrying out a safe and efficient flight? Are the procedures understandable, easy to use and appropriate to be carried out in the time available to the crew?
- What markers can we use to monitor the system and assess whether it is maintaining an adequate safety margin or whether it is beginning to have to use more buffers to protect itself from unexpected disturbances? Is the system operating with a sufficient safety margin or are the components of the system constantly having to make trade-offs to satisfy different demands that we are imposing? The classic example of this is the "Faster, Better, Cheaper" approach that the National Aeronautics and Space Administration (NASA) implemented to improve efficiency. These are three potentially competing demands that can put excessive pressure on components within a system unless the trade-offs that are required to achieve all three are carefully managed. Cybernetics is the science of control and communication in biological, mechanical and organizational systems, and it has a useful principle that addresses this question of how to monitor a system to detect warning signs that indicate that things are going wrong. An algedonic signal is one that is triggered by a problem or failure at the component level but is communicated to a much higher level because it is an important marker of impending failure.[10] How easy is it for your organization to detect algedonic signals from people working in an environment that may be very far removed from senior management?
- What events or disturbances can we envisage as an organization and how are we going to respond to them? Should we try to write a procedure for every eventuality that we imagine, at the risk of overproceduralizing the entire system, or rely on the training and adaptive ability of our staff to handle novel situations as best they can? In highly abnormal situations, is it better to try and centralize command and control, even though a situation may be evolving too rapidly for communication channels to keep up with developments, or do we decentralize command and control and, again, rely on the training and adaptive ability of our staff to handle novel situations as best they can?
- If things have gone wrong, what can we learn? Should we always make some sort of change to our system in response to a failure or do we just accept that this failure was so rare it is unlikely to happen again and that by changing something, we just introduce more uncertainty into the system? If something does need to be changed in order to learn from events, which components in the system should be adapted? Remember from Chapter 3 that there are three main classes of components in a system and that there is an inverse relationship between how easy/cheap it is to make changes in these components and how effective that change is, i.e. an easy fix is unlikely to be very effective. The changes that are most likely to give the greatest improvements are often the ones that are the most difficult or expensive to implement:
  - Liveware: Once we have learned what happened, we can change the system by retraining the humans involved. This is relatively cheap but may not be the best way to improve resilience. Remember, unsafe acts are often a symptom of an underlying problem with the system and by simply changing the liveware, the underlying system problem will still remain.
  - SOPs: Can we adjust the procedures to mitigate the risks based on what we have learned?
  - Hardware: Can we modify the hardware based on what we have learned? This is likely to be the most difficult/expensive approach but often gives the best chance of improving resilience.

4.4 Safety culture

Organizational safety culture has been defined in a variety of ways, but is often summarized as "how people in the company behave when no one is looking". There are several "subcultures" that contribute to the overall safety culture of an organization and these are related to the cornerstones of resilience as mentioned in Section 4.3[2]:

- Flexible culture – Like the ability to anticipate system requirements, having a flexible culture means being able to reconfigure the system to deal with the pressures that are acting on it at the time.
- Reporting culture – Like the ability to monitor, people working in the organization feel completely comfortable to report threats, errors or undesirable conditions as well as any near-misses. These potential warning signals (or algedonic signals, in the language of cybernetics) can provide managers with vital information about the levels of resilience in the system and where resilience may be eroded.
- Informed culture – Like the ability to respond, the components of the system know what they need to do and what is expected of them as part of the larger system.
- Learning culture – Like the ability to learn, the system can reconfigure itself based on what it has learned from past experience.
- Just culture – People working in the system trust that they will be treated fairly in the event of unsafe acts occurring. The key to a successful just culture is the acceptance of the New View of error. This is such an important contributor to the overall safety culture that it will be covered in more detail in Section 4.4.1, below.

Patrick Hudson identifies five levels of safety culture that an organization could have[11]:

- Pathological – It does not matter what we do, as long as we do not get caught.
- Reactive – We do a big safety drive after things go wrong, and then we stop.
- Calculative – We have systems that can manage all hazards.
- Proactive – We continue to work on problems that we identify.
- Generative – We are constantly looking for new areas of risk and we do not take past success as a guarantee against future failure.

An organization can evolve up through these stages (or, indeed, devolve back down through them), and one of the key determinants of the safety culture is the extent to which a just culture exists.

4.4.1 Just culture

Having a just culture means having a systematic way of investigating unsafe acts, incidents and accidents that is open, consistent and fair. It is worth pointing out that this is not the same as having a no-blame culture. It is possible that someone might be blamed after an incident if culpability can be clearly demonstrated. This involves following a series of investigative steps, the most crucial of which is the substitution test (i.e. given the same set of circumstances, is it possible that another person would have done the same thing?). The difference between having a robust just culture in an organization and having a poor just culture is like the difference between living in a modern country with a fair and independent judicial system and living in a country

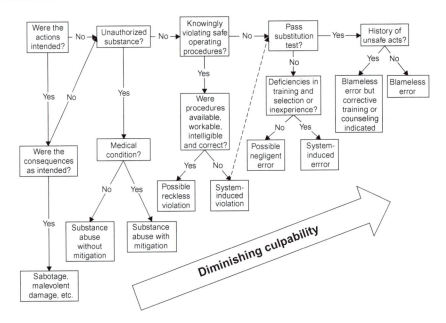

Figure 4.1 Culpability flowchart.

where the judicial system is inconsistent, corrupt and controlled by the government. In the former case, you can be sure that your treatment will be fair and sensible. In the latter case, not even your basic rights are guaranteed and you could find yourself being made a scapegoat to draw attention away from a more severe and significant problem higher up in the organization. Developing a just culture in your organization can be the springboard that improves safety culture as people will be more likely to report problems if they know they are not going to be criticized in a prejudicial manner for the role they played. Figure 4.1 illustrates one procedure for determining culpability during an investigation. There are several versions of this type of decision tree but this one demonstrates the framework on which the others are built.[2]

You will notice that the flowchart differentiates errors and various types of violation, and is based on the error/violation theory we covered in Chapter 3. Whatever system you use, the challenge is to make it as consistent and fair as possible. Consider your just culture procedures to be similar to those of a courtroom. In order for the people working in your organization to have faith that the system works, investigations and deliberations should not be prejudiced by facts beyond those relevant to the matter at hand. Ideally, events should be judged by trained individuals who do not know the identities of the people involved. In a small organization, this can prevent prejudices clouding the investigation. The main risks in accident/incident investigations are that the investigator succumbs to hindsight bias and fundamental attribution error as described in Chapter 2 (Information Processing), specifically Section 2.6.3.4 on social heuristics. It is the challenge for the investigator to put himself in the place of the people involved and try to imagine the situation as it would have been unfolding around them at the time, with no knowledge of the final outcome. If this is done successfully, it may become apparent that the crew had no way of knowing that their actions were going to result in a negative outcome and that other crews would probably behave in the same way.

4.4.2   Moving from Safety I to Safety II

The current industry drive to improve safety culture builds on the Old View and the New View of error to bring about change for the better. The transition from old-style safety management to a more resilient kind is described as the move from Safety I to Safety II. The characteristics of Safety I and Safety II are summarized in a paper written for Eurocontrol Air Traffic Management in 2013.[12] These differences are given in Table 4.1.

The authors note that some incidents and accidents do have a single-point cause that can be easily identified using Safety I techniques. Safety II is important because the increasing complexity of the systems within which we work means that many events need a Safety II perspective on them because of the range of system components involved. As well as using Safety II when things go wrong, taking a Safety II view of normal performance (particularly when humans have to vary their performance and make trade-offs to satisfy multiple competing goals) can give operators useful information that they can use to refine their procedures and the system as a whole.

Table 4.1 Differences between Safety I and Safety II

	Safety I	Safety II
Definition of safety	That as few things as possible go wrong.	That as many things as possible go right.
Safety management principle	Reactive: respond when something happens or is categorized as an unacceptable risk.	Proactive: continuously trying to anticipate developments and events.
View of the human factor in safety management	Humans are predominantly seen as a liability or hazard.	Humans are seen as a resource necessary for system flexibility and resilience.
Accident investigation	Accidents are caused by failures and malfunctions. The purpose of an investigation is to identify the causes.	Things basically happen in the same way, regardless of the outcome. The purpose of an investigation is to understand how things usually go right as a basis for explaining how things occasionally go wrong.
Risk assessment	Accidents are caused by failures and malfunctions. The purpose of an investigation is to identify causes and contributory factors.	To understand the conditions where performance variability can become difficult or impossible to monitor and control.

4.5   Principles of managing organizational resilience

Normal Accident Theory, chaos theory and resilience engineering are three very large topics and are research disciplines in their own rights. Because they are so large, it has only been possible to touch on each of them in describing the role they play in understanding the organizational aspects of human error and human performance. In trying to present a summary of this chapter, what has been covered so far will be presented as a set of principles that an organization needs to understand if it is going to move from a Safety I to a Safety II oriented operation. While some of them may not seem to be directly related to the information that has been presented so far, they can be inferred from the relationships between the various theories in both this chapter and Chapter 3:

- Even the safest system can fail unexpectedly. All you can do is manage the probability of this happening.
- If your system has not failed yet, this does not imply safety. It could fail tomorrow, next week or next year.
- Complex systems are not intrinsically safe and humans are not always to blame when things go wrong.
- Some events do not have a root cause and sometimes there is no one to blame.
- Forget about the outcome. Focus on the act when trying to understand the importance or severity of an event. Try to avoid fundamental attribution error by imagining yourself in the place of the people involved as the situation was unfolding around them. They had no idea of the outcome at that time and it is all too easy to go back and say what they "should" have done, now that you have the benefit of hindsight. Doing this may feel like a satisfactory step but it ignores the more important question: why did the people involved act in the way that they did?
- The vast majority of people working in your organization try to do their best.
- Bad things can happen to anyone.
- Not following procedures is not necessarily a bad thing. It can be a sign that the procedures are unworkable or too time-consuming, or that there is a better way of doing something. If people are consistently taking shortcuts, consider that they are probably trying to reconcile two opposing goals and you should adjust the system so that this does not have to happen.
- The gap between work-as-imagined (procedures as written down in manuals) and work-as-done (the shortcuts people take to reconcile multiple goals) should be narrowed as much as possible by incorporating the successful strategies that people are using to manage trade-offs between multiple goals.
- If there is a big gap between work-as-imagined and work-as-done, you probably have no idea what is happening lower down in your organization.
- It is not always necessary to immediately change things in the face of failure.
- If you are going to change something, consider what you are going to change and how those changes may introduce new disturbances into your system. Is it worth the risk?
- The easiest thing to change (the liveware) is not always the best thing to change. If lots of people are failing to follow a procedure, changing the procedure may be more successful than retraining everyone.
- Learn from success as well as failure.
- Identify what warning signs you should look out for that may indicate problems in your organization.

- Being proactive in developing your organization's safety culture can be the single best way of improving overall safety.
- Ensure you have systems in place to guarantee a robust, trustworthy just culture. It will only take one case where someone is dealt with unfairly or inconsistently for your safety culture to take a step backwards.
- The aim should be to move from Safety I to Safety II.

Chapter key points

1. There are many organizational factors that can lead to unsafe acts.
2. A system is a collection of interconnected components (humans, machines and procedures) working together as a whole to achieve a specific objective.
3. According to Normal Accident Theory, complex systems are at risk of system accidents because their many possible states cannot be predicted.
4. The Old View of human error states that systems are safe, and that when things go wrong it is usually the fault of the humans involved.
5. The New View of human error states that systems are not inherently safe, particularly if they are complex, and that humans usually try to do their best.
6. Resilience engineering builds on the New View and tries to enhance the things that go right in a system rather than focusing entirely on trying to eliminate the things that go wrong.
7. Enhancing a company's safety culture can help to achieve this, especially by instituting a just culture and making changes to transition from Safety I to Safety II.
8. There are several principles that an organization needs to understand if this change is going to be successful.

Recommended reading

Reason J. *Human error*. Cambridge University Press; 1990. An excellent summary of all the research done on human error up to 1990 [Difficulty: Intermediate].

Reason J. *Managing the risks of organizational accidents*. Ashgate; 1997. Expands the idea of latent errors into an organizational framework [Difficulty: Intermediate].

Perrow C. *Normal accidents*. Princeton University Press; 1984. This is an excellent introduction to Normal Accident Theory and formed one of the cornerstones of resilience engineering. It would definitely be of interest to those involved with safety management systems or designing procedures. It was written as a response to the Three Mile Island nuclear accident as that was a turning point for human factors when its importance in that accident was fully appreciated [Difficulty: Easy].

Hollnagel E. *The ETTO principle: Efficiency–thoroughness trade-off*. Ashgate; 2009. A very interesting introduction to the dilemma often faced by pilots: do things by the book or do things in the time available [Difficulty: Easy/Intermediate].

Hollnagel E, Woods DD, Leveson N. *Resilience engineering: Concepts and precepts*. Ashgate; 2006. This was the first book that laid the groundwork for the discipline of resilience engineering and proves a useful starting point in linking the historical view of error with the current and future work in the field [Difficulty: Intermediate].

Dekker S. *Ten questions about human error*. Lawrence Erlbaum; 2005. Introduces the New View of human error [Difficulty: Easy/Intermediate].

Dekker S. *Just culture*. Ashgate; 2007. Introduces many of the important concepts upon which a just culture needs to be built [Difficulty: Easy/Intermediate].

Beer S. *The brain of the firm*. John Wiley; 1994. This recommendation is included despite the fact that the subject it covers, organizational cybernetics, is somewhat out of fashion in academic circles at the moment. Cybernetics is the science of control and communication and this book introduces a fascinating model based on the human nervous system (the Viable System Model) that can be applied to organizations to understand how they succeed and how they fail [Difficulty: Intermediate].

References

1. Vandermuelen J. (Interviewer). (December 2008). Erik Hollnagel – To err is human: The ETTO principle [Audio podcast]. Retrieved from: http://www.namahn.com/we-share/interviews/erik-hollnagel-err-human-etto-principle
2. Reason JT. *Managing the risks of organizational accidents*. Farnham, UK: Ashgate; 1997.
3. Perrow C. *Normal accidents: Living with high risk technologies*. Princeton, NJ: Princeton University Press; 1999.
4. Lorenz E. *Predictability: Does the flap of a butterfly's wings in Brazil set off a tornado in Texas?* Washington, DC: American Association for the Advancement of Science; 1972.
5. Dekker SW. Reconstructing human contributions to accidents: The new view on error and performance. *J Saf Res* 2002;**33**(3):371–85.
6. Hollnagel E. *The ETTO principle: Efficiency–thoroughness trade-off: Why things that go right sometimes go wrong*. Farnham, UK: Ashgate; 2012.
7. National Transportation Safety Board. (2000). Safety recommendation [runway incursions]. Retrieved from: http://www.ntsb.gov/doclib/recletters/2000/a00_66_71.pdf
8. Woods DD. Essential characteristics of resilience. In: Hollnagel E, Woods DD, Leveson N, editors. *Resilience engineering: Concepts and precepts*. Farnham, UK: Ashgate; 2006, pp. 21–34.
9. Hollnagel E. The four cornerstones of resilience engineering. In: Nemeth C, Hollnagel E, Dekker SWA, editors. *Resilience engineering perspectives: Preparation and restoration*, vol. 2. Farnham, UK: Ashgate; 2008, pp. 117–34.
10. Beer S. *Brain of the firm: The managerial cybernetics of organization*. New York: John Wiley; 1994.
11. Hudson PTW. Safety management and safety culture: The long, hard and winding road. In: *Proceedings of the First National Conference on Occupational Health and Safety Management Systems*, Sydney, Australia (pp. 3–32). Melbourne: Crown Content, 2001.
12. Eurocontrol. (2013). From Safety-I to Safety-II: A white paper. Retrieved from: http://www.eurocontrol.int/sites/default/files/content/documents/nm/safety/safety_whitepaper_sept_2013-web.pdf

Personality, leadership and teamwork

Chapter Contents

Introduction   133
5.1 Personality   134
    5.1.1 Personality structure   134
    5.1.2 Personality and behavior   138
    5.1.3 Personality management strategies   139
5.2 Leadership and command   140
    5.2.1 Leadership and personality   141
    5.2.2 Non-technical skills for leadership   143
    5.2.3 Leadership, personality and flight safety   144
        5.2.3.1 The toxic captain   146
        5.2.3.2 The toxic first officer   150
5.3 Flight deck gradient   152
    5.3.1 Culture and flight deck gradient   152
5.4 Cooperation and conflict solving   156
    5.4.1 Individual differences and conflict-solving strategies   157
    5.4.2 Situational variables and conflict-solving strategies   158
Chapter key points   159
References   160

Introduction

There are few concepts in human factors that cause more difficulty or are surrounded by more controversy than personality. Over the past 100 years, many different approaches have been taken to try and dissect the complex nature of human personality in order to uncover a systematic way of categorizing and understanding it. Personality is an important topic to discuss in human factors because it is one of the primary determinants of behavior. It is also one of the features that differentiates us from other people because the subtle interactions between an individual's personality type and other factors that determine behavior all contribute to our individualism. Understanding how personality is structured and how it can influence us can give us insights not only into our own behavior but also into the behavior of others, and can equip us with strategies to allow us to make best use of our own abilities and those of other people. Our natural leadership capabilities are encoded in our personality structure and certain traits have been associated not only with the emergence of leadership but also with the effectiveness of leadership. Culture plays a large role in dictating how we employ and how we respond to leadership, and an understanding

Practical Human Factors for Pilots. DOI: http://dx.doi.org/10.1016/B978-0-12-420244-3.00005-4

of cultural dynamics can help to ensure that we get the best out of ourselves and our colleagues. Owing to the complex, wide-ranging nature of the research into personality, it has been necessary to simplify some of the concepts and measures described in this chapter and so it should serve only as an introduction to the topic. Any readers interested in knowing more, possibly with a view to introducing psychological screening processes in their organization, are advised to obtain expert advice. Although the title of this chapter includes "teamwork", there is no section that is devoted solely to this subject as much of the content of the chapter relates to teamwork in some way.

5.1 Personality

5.1.1 Personality structure

In choosing a particular model of personality to use as the basis for this chapter, a feature that seems to be characteristic of many human factors topics once again becomes an issue: there is a tendency in academic literature to approach the same phenomenon from multiple different angles, each with its own labeling system for mechanisms that are fundamentally the same. This can be seen in studies regarding decision making, communication skills and, especially, personality. Despite the multiple approaches taken to measure and classify personality traits, as with any other part of psychology, there are structural components in the brain that give rise to what we call personality. Consider the stages of information processing, the basic mechanism that determines how we interact with the world: sensation, attention, perception, decision making and response. As we will see later on, different personality types have a tendency to act in a particular way. These personality-induced tendencies can only affect our behavior because of how they affect some of the stages of information processing, primarily the attention, perception and decision-making stages. Two major approaches to categorizing personality, the Sixteen Personality Factor Questionnaire (16PF)[1] and the Five-Factor Model[2] (FFM), share many similarities but also have some methodological differences when categorizing results.

In this chapter, we will be using the FFM as the basis for our exploration of personality. There are five domains that form the FFM and these represent the five main dimensions of personality. Within each domain there are six facets that can be assessed. Various tools that are used to assess personality using the FFM rate each of these facets from 1 to 5 (1 being maladaptively low, 2 being low but normal, 3 being neutral, 4 being high but normal and 5 being maladaptively high). The term "maladaptive" is used in preference to "abnormal" as it is meant to convey that while scoring extremely high or low on one of the facets does not in itself imply any particular pathological personality variant, the behavioral tendencies that this facet variant may generate can result in this person exhibiting behavior that is mismatched (i.e. maladapted) to a particular situation. Certain combinations of maladaptively high or low facet variants can be associated with personality disorders but these are relatively rare compared with non-pathological variants. The five domains, their component facets and examples of maladaptively low and maladaptively high variants are given below:

1. Extraversion:
   a. warmth (from cold and distant to forming intense attachments)
   b. gregariousness (from socially withdrawn to attention seeking)
   c. assertiveness (from resigned and uninfluential to dominant and pushy)
   d. activity (from lethargic to frantic)
   e. excitement seeking (from dull and listless to reckless and foolhardy)
   f. positive emotions (from grim to manic and melodramatic).
2. Agreeableness:
   a. trust (from cynical and suspicious to gullible)
   b. straightforwardness (from deceptive to guileless)
   c. altruism (from greedy and self-centered to self-sacrificial)
   d. compliance (from combative to subservient)
   e. modesty (from boastful to self-effacing)
   f. tender-mindedness (from callous and merciless to soft-hearted).
3. Conscientiousness:
   a. competence (from disinclined and lax to perfectionist)
   b. order (from careless, sloppy and haphazard to preoccupied with organization)
   c. dutifulness (from irresponsible and undependable to rigidly principled)
   d. achievement (from aimless and shiftless to workaholic)
   e. self-discipline (from negligent to single-minded doggedness)
   f. deliberation (from rash and hasty to ruminative and indecisive).
4. Neuroticism (emotional instability):
   a. anxiousness (from being oblivious to signs of threat to being fearful and anxious)
   b. anger hostility (from not protesting exploitation to being rageful)
   c. depressiveness (from unrealistic and overly optimistic to depressed)
   d. self-consciousness (from glib and shameless to uncertain of self and ashamed)
   e. impulsivity (from overly restrained to unable to resist impulses)
   f. vulnerability (from fearless and feeling invincible to helpless and overwhelmed).
5. Openness:
   a. fantasy (from concrete to unrealistic)
   b. aesthetics (from disinterested to bizarre interests)
   c. feelings (from unable to connect with emotions to intense feeling of emotions)
   d. actions (from being stuck in routine to eccentric)
   e. ideas (from close minded to peculiar)
   f. values (from dogmatic and intolerant to radical).

The first two domains of extraversion and agreeableness have the biggest impact on our ability to relate to others. The domain of conscientiousness has a large impact on how we relate to work. The facets in the domain of neuroticism have an impact on our relationships both inside and outside work as emotional stability is generally regarded as a useful quality. The last domain of openness is less stable than the others and is sometimes described as unconventionality versus conventionality.

Although any system designed to assess personality will never be able to perfectly categorize and classify every facet of an individual's character with 100% accuracy, the FFM has proved remarkably successful as a classification system. Although we like to think of our personalities as being nebulous, intricate and deeply complex, the

FFM does seem to be able to highlight those facets that have a large impact on our behavior. Even restricting itself to a five-point rating scale, there are still 5^{30} variant combinations possible (9.3×10^{20}, i.e. 93 with 19 zeros after it), more than enough to cover all the possible personality types out there. Most people would be expected to score 2, 3 or 4 on each of the 60 facets, this equating to exhibiting a low but normal, neutral, or high but normal tendency towards that facet. For example, when considering the facet of warmth in the extraversion domain, a low but normal variant is being formal and reserved. A high normal variant is being affectionate and warm. Neither of these variants is indicative of anything maladaptive and we would probably be able to classify most of our friends and family members as one or the other. A neutral score suggests that the person demonstrates no dominance towards either end of the spectrum and this is also normal. Even if we exclude the maladaptive variants and restrict ourselves to a three-point rating scale, there are still 3^{30} variant combinations possible (2 million billion). Although scoring one or five for a particular facet may suggest a personality variant that may be problematic either for interpersonal relationships or for the individual's ability to manage daily life, it does not necessarily suggest a personality disorder. Some people are callous, sloppy and haphazard, and although this may cause them personal problems and limit their success in the workplace (i.e. it is maladaptive), it does not mean that there is any sort of pathological process at work. However, if you are recruiting surgical trainees, it would be useful to know about such a personality tendency in advance as it may have a significant impact on the quality of their work and the well-being of their patients.

Some combinations of variants in specific personality facets are associated with the group of mental illnesses called personality disorders. While having maladaptively high or maladaptively low variants with regard to a particular facet is not necessarily abnormal, certain combinations of high and low scores are associated with personality types that can be considered pathological.[3] Unlike certain mental illnesses where there may be specific biochemical or physical changes in the brain, personality disorders result from abnormalities in the basic structure of the individual's personality to the extent that it causes them harm or predisposes them to harm others. While some of these disorders are associated with some physical changes in the brain,[4] these changes are not as apparent as in some other mental illnesses. Two example of personality disorders are the histrionic and antisocial types. The histrionic personality disorder is characterized by excessive emotions and constant attention seeking.[5] Antisocial personality disorder (encompassing psychopathy and sociopathy) is characterized by deception, aggression[6] and abnormally low scores in the domain of agreeableness.[2] Personality disorders can be very difficult to treat because they are the result of an abnormality in the key determinant of behavior: the individual's personality. There is also considerable evidence of the genetic role in determining personality traits and it appears that inherited genes plus environmental influences will determine personality structure, whether normal or pathological.[7] The structure of an individual's personality is generally fixed by adulthood and is unlikely to change. It may, however, be possible to adjust behavior even though the underlying personality structure remains constant.

A lot of work has been done to uncover the genetic basis of personality and characterize how it affects human behavior by hypothesizing the impact of personality on information processing. For example, consider the FFM facet of deliberation. Depending on where an individual scores on this facet of the FFM, they may be indecisive or hasty. This may relate to the basic activity level of their pattern-matching system located in the basal ganglia of their brain. Similarly, goal setting is a key part of information processing when it comes to problem solving. An individual's psychological traits could influence the basic neural programming of how goals are set. There are several approaches being taken at the moment to look at superimposing a personality structure on to the ACT-R model of information processing as described in Chapter 2.[8] As well as being able to model some of these personality traits, there is evidence regarding the genetic nature of personality and how genes affect brain structure and function in this regard. Although it is unlikely that single genes will dictate individual facets of personality, it is likely that combinations of genes among the 10,000 or so that are expressed in the brain will be found to have an impact on personality structure.[7] In cases of pathological personality structure, the search is easier because analysis can be done on people who have been diagnosed with specific personality disorders. Alterations in certain genes that affect one of the systems used for neurotransmission, specifically, the serotonin system, have been associated with the borderline personality disorder, a disorder characterized by anxiety, depression and impulsivity. As with many other disorders, evidence for a genetic component is given by the 65–75% heritability risk; that is, if one twin is diagnosed with the disorder, there is a 65–75% chance of the other twin having it.[9] To put it simply, we have a system that allows us to process information based on matching stimulus to response. Personality structure sets the default positions for many of the components of the information processing system, although it is possible to override this.

The relatively constant nature of personality is what gives us the enduring characteristics that make us who we are. If personality was easy to change, we may be self-disciplined and flexible one day and negligent and intolerant the next. Given that personality gives us a tendency to behave in a particular way, changes in our personality would alter how we interact with the world and how we interact with others, and essentially make us into a different person who may be unrecognizable to the people who know us. While the rigidity of personality structure is unfortunate for people who have personality disorders, it is essential for those of us that do not, so that we know who we are from day to day. As well as determining our relationships with others, our personality type has a significant impact on our relationship with work. For this reason, employers tend to look for people with personality types that will be suited to the job. For example, someone who is deemed to be social and outgoing in their personality facet of gregariousness may be well suitable for a customer contact role. On the other hand, someone who is deemed to be quick at making decisions in their personality facet of deliberation may not be suited to handling investments. As you can probably see from these examples, it is possible for employers to draw up a personality profile of the type of person that they believe would be best suited to a particular job.

Although we are focusing on the FFM, there is another personality classification system known as the Myers–Briggs Type Indicator (MBTI).[10] Rather than having 5^{30}

potential different classifications, the MBTI classifies people's tendencies towards one or the other end of four spectra covering different facets of personality:

- introversion versus extraversion
- sensing versus intuition (information gathering – preferring concrete, factual information or abstract, intuitive information)
- thinking versus feeling (decision making – preferring to make decisions based only on data available or preferring a more holistic approach)
- judging versus perception (the tendency to show the information gathering or decision-making facets to the world).

Because there are only two options for each facet, the MBTI has only 16 possible outcomes. The system is, in fact, a lot more complex than shown here as there is considerable interrelation between the four different dimensions and the result of one facet can have an impact on the others. The reason for including the MBTI is to demonstrate how particular personality types are associated with particular jobs. One of the 16 Myers–Briggs types is ESTJ – extravert, sensing, thinking, judging. Here is a description of this personality type:[11]

> *Practical, realistic, matter-of-fact. Decisive, quickly move to implement decisions.*
> *Organize projects and people to get things done, focus on getting results in the*
> *most efficient way possible. Take care of routine details. Have a clear set of*
> *logical standards, systematically follow them and want others to also. Forceful in*
> *implementing their plans.*

For any pilots reading this chapter, this description will probably sound familiar as it describes many of the people we work with and may even be a good description of you. It could be that people with this personality type are drawn to becoming pilots or it could be the selection process that means that someone with this personality type has a high chance of being employed as a pilot. This is one example of the association between personality variants and suitability for particular jobs.

5.1.2   *Personality and behavior*

If personality was a purely internal phenomenon, it would have very little relevance to the subject of human factors. It seems, however, that our personality forms the basis of how we perceive ourselves and how others perceive us because it has a major impact on our behavior. The relationship between personality and behavior is complex. Just because someone has been assessed as having a certain personality type, that does not guarantee that in every situation they will behave as their personality type dictates. A personality classification is a reflection of the *tendency* of that individual to act in a particular way in a given situation. Someone who scores low in the facet of self-discipline may not be negligent in every aspect of their life. However, they will have a greater *tendency* to be negligent than someone who does not score so low. When questionnaires are used to assess people's personality, they normally ask questions regarding how someone would behave in particular situations. A commonly used tool for assessing someone's personality according to the FFM is the Revised

NEO Personality Inventory (NEO-PI-R).[12] The NEO-PI-R uses 240 questions to assess the 30 facets that comprise the FFM. These questions tend to propose a behavior or attitude and then ask the participant whether they agree or disagree that they would behave in that manner or have that attitude. Tests designed to assess personality are open to considerable criticism and have been accused of giving an uncharacteristic snapshot of an individual. For this reason, they should only be administered by trained professionals and the results require careful interpretation. It is also possible for candidates to answer according to what they believe to be desirable traits in an attempt to appear to be more suitable for the job than they really are. Some questionnaires even include questions designed to detect this type of deception. However, when these tests are administered and interpreted carefully, the results can help in the selection process. Training up a prospective employee who does not have a personality type suited to the role may result in a waste of time, money and effort when it turns out that the person is unable to fulfill their duties. The debate about the usefulness of personality questionnaires is likely to continue but in aviation, it appears that they are here to stay.

As we have said, behavior is based partly on personality. In a particular situation, the relevant personality facets act as filters on the perception of a situation and will affect any decision making. While someone with a personality type may not react in the same way every time, they will probably react in a manner that is in keeping with their personality-based tendencies the majority of the time. For example, someone who has a tendency to being overly trusting or even gullible may not always react that way, but will have a tendency towards that particular characteristic. Given that the structure of our personality is relatively fixed, it is only by adjusting our behavior that we can overcome any perceived deficiencies in our personality. For example, someone who is generally irresponsible and undependable can modify their behavior through constant reinforcement. Although the personality trait remains, a new behavioral response can be trained. This idea underlies the cognitive–behavioral therapy that is used to treat some personality disorders[13] and also in self-help workshops that are aimed at helping participants to overcome some feature of their personality (e.g. social anxiety) that they feel is holding them back.[14]

5.1.3 Personality management strategies

Although some changes can occur in adulthood, we generally have our personality structure programmed in a certain way by genetics and the influence of our environment when we are growing up, and this programming remains relatively constant throughout our lives. This personality program has a big impact on how we react to situations, although it is possible to modify our behavior if necessary. In this way, we can say that personality will predict how we will *tend* to act but not how we will *always* act. Some strategies for managing your own personality and the personalities of people who you work with are given below:

- Is your personality causing you difficulty? – Have a look at the examples of very high and very low scoring variants according to the FFM. Although the variants given were from the extreme ends of the spectrum, there may be one or two that you feel may describe you and that you think may be causing difficulty in your work or your personal life. If you think people would consider you cold and distant and you would like to change this, perhaps it is time to consider getting some help. Although it is difficult to change someone's underlying personality trait, it may be possible to change their behavior and so negate its effect.

- Recognize and be patient with the personality variance of others – Because our view of the world is framed by our own personality, it is easy to think that others should think and act in the same way as us. For someone who is consistently practical, realistic and pragmatic, it may be difficult to relate to someone who is imaginative, creative and curious. We may believe that our perspective on the world is more valid simply because we know that it is well suited to the environment that we operate in. However, imagination, creativity and curiosity most certainly have a purpose in the world, one that is no more or less valid than practicality, realism and pragmatism. Insisting that a colleague, friend or family member change their personality to more closely match that which you perceive as normal is unrealistic and potentially damaging.
- Take advantage of the personality variance of others – Given that employers take advantage of their knowledge of our personality, we can, to a certain extent, do the same to get the best out of our colleagues. For example, a crew member who is sociable, outgoing and personable may be better suited to talking to passengers than someone who is formal and reserved.

5.2   Leadership and command

At first glance, leadership and command appear to be two sides of the same coin. However, when we analyze these concepts more deeply, the picture becomes slightly more confusing. Legally, there are certain licensing requirements in order for someone to be designated as pilot-in-command. However, simply meeting these requirements does not imply the *right* to be pilot-in-command. It is the operator who can designate a pilot as being the pilot-in-command, and their method for determining who should be designated as such is left to their discretion.[15] At the discretion of the operator, a pilot meeting the minimum requirements could be given the minimum amount of command training and then be designated pilot-in-command of a commercial flight. Fortunately, most operators have their own standards when determining the transition process from first officer to captain. Command upgrades are often given according to experience, the implication being that having spent a sufficient amount of time as a first officer equips an individual with the necessary skills to become a captain. Although many airlines have a process for selecting candidates for command based on other factors, there is still a seniority-based philosophy behind many command upgrade protocols. Logic would suggest that for the most part, an experienced candidate will be a more effective commander than an inexperienced one; however, there are several other variables, many of which are not experience related, that will have an impact on an individual's suitability (or lack of suitability) for command. The question we must consider now is this: if simply having operational experience is not sufficient, what additional skills are needed to be an effective commander?

Command of an aircraft is a position of authority. Command authority is extensive and is enshrined in international law. It is perfectly feasible for a commander to be able to conduct a flight, including giving instructions to his crew, using only his command authority. However, relying on authority alone has its problems, as we will see later. There is another way other than constantly using his authority, that the commander can ensure that the crew will operate together effectively to complete the flight as safely and efficiently as possible: he can use leadership.

Before we talk about the qualities of an effective leader, let's consider a definition of leadership: "leadership is a process of social influence in which one person is able to enlist the aid and support of others in the accomplishment of a common task".[16] This definition does not refer to command or authority. From a legal perspective, the pilot-in-command of an aircraft does not need to use social influence to enlist the aid and support of others. Crew members are legally obliged the follow any instruction given by the commander in the interest of safety and it is perfectly legal for a captain to discharge his duties by telling crew members exactly what to do. Why then should the commander feel the need to use social influence to accomplish the same task? If he uses his command authority, there does not need to be any discussion, nor does he require the input of other crew members when making decisions. As you can imagine, this would lead to quite an authoritarian atmosphere and research has shown that an authoritarian flight deck environment poses a significant threat to safety because there is ineffective use of available resources. In Chapter 2 (Information Processing), we discussed the systematic errors that can occur when decision making is performed in dynamic, time-limited situations using System 1. Successful decision making, like effective Threat and Error Management, relies on open communication channels between crew members and it is usually the captain who can best ensure that these channels are open. It is here that the phenomenon of "social influence" becomes relevant and proves to be a more effective way of commanding an aircraft than relying on authority. The commander's authority is still there and can be used when necessary, but it is preferable to use leadership in the first instance. Given the title of this chapter, it is worth considering the personality factors that are associated with effective leadership.

5.2.1 Leadership and personality

A study that reviewed all the research regarding personality traits and leadership qualities found that correlations in all five domains of the FFM were associated with high leadership effectiveness.[17] People who demonstrated effective leadership generally had high scores in the domains of extraversion, openness, agreeableness and conscientiousness, and low scores in the domain of neuroticism. A more recent study again identified positive scores in the domains of extraversion, openness, agreeableness and conscientiousness as predictive factors for leadership effectiveness in a military training environment, and highlighted conscientiousness as a key factor in maintaining high leadership effectiveness over time.[18] This study also concluded that there are distinct profiles that relate to both an individual's natural leadership ability and the potential for developing an individual's leadership ability during a period of training. These profiles are given below:

- Consistently high leadership effectiveness – started training with high scores for extraversion, openness, agreeableness and conscientiousness, and a low score for neuroticism (the same profile as identified by the previous study[17]).

- Improving leadership effectiveness during training – started training with higher than average scores for neuroticism, extraversion and openness. Also had average scores for agreeableness and low scores for conscientiousness. This suggests that the higher than average neuroticism and lower than average conscientiousness can be overcome through behavioral training and improved confidence; for example, these individuals will become less neurotic about making decisions as they gain confidence in their abilities and can also be trained to take a more conscientious approach to their duties.
- Consistently low leadership effectiveness – started with low scores for extraversion, openness, agreeableness and conscientiousness, and a higher score for neuroticism.

This study is interesting because it suggests that personality profiling can *predict* whether leadership effectiveness will improve through training. Although some personality types do not seem to benefit from leadership training, there does appear to be a profile that will show an improvement in leadership effectiveness through training. Given the relatively fixed nature of an individual's personality structure, it would seem that behavioral changes can override personality-related tendencies and so address personality-based barriers that limit leadership effectiveness. It is interesting to note that this group is characterized by high neuroticism and low conscientiousness, and that a behavioral program that addresses self-confidence and self-discipline may be enough to counteract these personality tendencies.

Based on the research that we have looked at, we can start to form a personality profile associated with high leadership effectiveness. If we take the original FFM and include the adjectives normally associated with the scores that effective leaders might be expected to have for each facet, we obtain the following profile:

1. Extraversion (generally high to very high):
   a. warmth – warm
   b. gregariousness – social, outgoing and personable
   c. assertiveness – assertive
   d. activity – energetic
   e. excitement seeking – adventurous
   f. positive emotions – high-spirited.
2. Agreeableness (generally high):
   a. trust – trusting
   b. straightforwardness – honest and forthright
   c. altruism – generous
   d. compliance – cooperative
   e. modesty – humble, modest and unassuming
   f. tender-mindedness – empathic, sympathetic and gentle.
3. Conscientiousness (generally very high):
   a. competence – efficient and resourceful
   b. order – organized and methodical
   c. dutifulness – dependable, reliable and responsible
   d. achievement – purposeful, diligent and ambitious
   e. self-discipline – self-disciplined
   f. deliberation – thoughtful, reflective and circumspect.
4. Neuroticism (emotional instability) (generally low):
   a. anxiousness – relaxed and calm

     b. anger hostility – even-tempered
     c. depressiveness – not easily discouraged
     d. self-consciousness – self-assured and charming
     e. impulsivity – restrained
     f. vulnerability – resilient.
5. Openness (generally average to high):
     a. fantasy – practical and realistic
     b. aesthetics – aesthetic interests
     c. feelings – self-aware
     d. actions – versatile
     e. ideas – creative and curious
     f. values – open and flexible.

With the possible exception of aesthetic interests, all of these facets can be seen to have some impact on an individual's leadership effectiveness, particularly in a command situation. While some of them seem self-evident, there is one that stands out as it has been closely associated with extremely high leadership effectiveness: modesty. In analyzing the profiles of chief executive officers (CEOs) who ran or had run the most successful companies in America's Fortune 500 list, James Collins identified two qualities that exemplified the most successful leaders, those who he categorized as being on level 5 of his hierarchy of leadership[19]; first, they had an extremely strong professional will, and secondly, they had considerable personal humility. While this research focused on CEOs, his identification of modesty as a key determinant in high-performing leaders may seem paradoxical. The exceptional leaders that he identified were not necessarily household names because they channeled their motivation and ambition into developing the status of the company rather than their own status. It is unclear whether this observation translates into leadership in the flight deck; however, high scores in the domain of agreeableness associated with effective leadership reinforce the notion of modesty being a useful attribute.

5.2.2 Non-technical skills for leadership

Now that we have identified some personality variants that are associated with high leadership effectiveness (and with the potential for high leadership effectiveness through training), we can relate what we know back to aviation, specifically non-technical skills. There is one domain, comprising four subdomains, in the NOTECHS non-technical skills assessment system that covers leadership, each subdomain having both positive and negative behavioral markers associated with it. The domain, subdomains and examples of positive and negative behavioral markers for each are shown below:

1. Leadership and managerial skills
     a. Use of authority and assertiveness
          i. Positive behavioral markers:
               – takes initiative to ensure involvement and task completion
               – takes command if situation requires.
          ii. Negative behavioral markers:
               – hinders or withholds crew involvement
               – does not show initiative for decisions.

b. Providing and maintaining standards
  i. Positive behavioral markers:
    – ensures standard operating procedure (SOP) compliance
    – intervenes if task completion deviates from standards.
  ii. Negative behavioral markers:
    – does not comply with SOPs
    – does not intervene in case of deviations.
c. Planning and coordination
  i. Positive behavioral markers:
    – encourages crew cooperation in planning and task completion
    – clearly states goals and intentions.
  ii. Negative behavioral markers:
    – intentions not stated or confirmed
    – changes plan without informing crew.
d. Workload management
  i. Positive behavioral markers:
    – delegates tasks to keep individual workload under control
    – plans actions for high-workload phases of flight.
  ii. Negative behavioral markers:
    – does not delegate workload
    – allows secondary operational tasks to interfere with primary flight duties.

Given that not everyone will have a personality variant that will naturally generate these positive behaviors, it is still possible to exhibit them although this may require slightly more effort and, perhaps, some additional training. However, the question remains: What impact will there be on flight safety when someone does not have what could be characterized as "a good leadership personality" and does not address this by adjusting their behavior? Does it affect flight safety?

5.2.3 Leadership, personality and flight safety

Although having a personality profile consistent with effective leadership will make it easier to exhibit these positive behavioral markers, appropriate training can instill the relevant behaviors in individuals with leadership potential. It is fortunate for all of us that we do not have to be slaves to our personality programming, and training can go a long way in developing and refining the necessary skills required for effective leadership. So what happens when the commander of the aircraft does not fit the profile of an effective leader, when his behavior includes some of the negative behavioral markers we have mentioned? The US National Aeronautics and Space Administration (NASA) conducted a study that divided captains into groups depending on the results of various psychometric tests.[20] Unfortunately, as with so much of the scientific literature on personality, a different set of psychometric tests was used instead of one based on the FFM we have been using so far. However, based on the adjectives used to describe the captains in each group, you will be able to estimate for yourself how these personality variants correlate with what you know about the FFM. The definitions of "instrumentality" and "expressivity" given by the Personal

Attributes Questionnaire, one of the psychometric tests given to the captains, cover many of the facets across all domains of the FFM. High scores for instrumentality implied independence, competitiveness, easy decision making, persistence and self-confidence. High scores for expressivity were associated with emotionality, gentleness, helpfulness, empathy and warmth. Using the original psychometric assessment terms, the groups were as follows:

- Group 1: "Positive Instrumental/Expressive" – Captains in this group had high scores on instrumentality, expressivity, mastery (preference for challenging tasks), work (desire to work hard), achievement striving and competitiveness (specifically, an enjoyment of interpersonal competition).
- Group 2: "Negative Instrumental" – Captains in this group had high scores for negative instrumentality [arrogance, hostility and interpersonal invulnerability (boastful, egotistical, dictatorial)], verbal aggressiveness, impatience/irritability and mastery. They also scored very low for expressivity. This cluster of traits comprises a more authoritarian orientation.
- Group 3: "Negative Expressive" – Captains in this group were negatively expressive and not at all instrumental relative to other pilots. This group also included captains who scored below average on three of the following: instrumentality, achievement striving, mastery, work or impatience/irritability. This cluster of traits is associated with expressing oneself in a negative fashion, for example by complaining and below-average achievement motivation.

All the captains in the study were paired with a randomly selected first officer and undertook a 1.5 day simulated trip consisting of five sectors. Crews led by captains from group 1 (positive instrumentality/expressivity) were consistently effective and made the fewest errors. Crews led by captains from group 2 (negative instrumentality – authoritarian) were less effective on day 1 but improved on day 2, possibly owing to increased familiarity between the crew members. Crews led by captains from group 3 (negative expressive – low achievement motivation and high complaining tendency) were consistently less effective and made more errors.

After expanding on the idea that there is a personality profile that facilitates effective leadership and that behavior stemming from these positive personality traits (or behavior that is engendered through training) is consistent with positive behavioral markers in the NOTECHS system, we now have some evidence that demonstrates how these positive behaviors lead to a lower error rate and a more effective, safer operation. Before we consider another significant variable that affects leadership and teamwork in the flight deck, we should consider another set of personality traits associated with negative performance. Unlike the negative clusters of traits defined in the NASA study, there may be another, more significant cluster of traits that pose a critical risk to safety because they are the polar opposites of those we have identified for effective leadership. Consider a personality profile that is the opposite of the profile that correlates with effective leadership. This individual would have low scores for extraversion, agreeableness, conscientiousness and openness, and a high score for neuroticism. While this personality profile certainly exists in the general population and is not, per se, indicative of any disorder, in a flight deck environment this personality type may pose a significant safety risk, particularly if this person is in command.

It is also worth noting that pilots are not immune to mental illness such as a personality disorder. There are likely to be some pilots flying who could be classified as having a personality disorder given that an estimated 4.4% of the UK population suffer from some form of this mental illness.[21] The prevalence is likely to be significantly lower in the pilot population owing to psychological and psychometric screening as well as a lower incidence of personality disorders in people currently in employment. It is also worth noting that the diagnosis of a personality disorder requires repeated assessments by an expert in psychiatric medicine and cannot be based purely on psychometric assessment.

Even if we disregard the possibility of a captain having a personality disorder, we must still consider a captain who has a personality type (and exhibits behavior based on that personality type) that could be defined as having the opposite traits to those we have identified for effective leadership. This person can, therefore, rely only on his authority to run his team as he lacks the ability to effectively use social influence. This phenomenon has been referred to as the "toxic captain".[22]

5.2.3.1 The toxic captain

Although it is not unusual to have pilots who have personalities that do not perfectly conform to the ideal profile of an effective leader, there do seem to be personality profiles that pose a significant risk to safety. The NASA study identified two groups of captain that showed an increased risk of error: the authoritarian kind and the passive complainers. However, there is a more pathological profile that can pose a much greater risk to flight safety.

In the case reported in Box 5.1, the captain's overbearing style coupled with the first officer's non-assertive style meant that this was essentially a single-pilot operation. Robert Baron, who coined the term "toxic captain", defines it as a pilot-in-command who lacks the necessary human and/or flying skills to effectively and safely work with another crew member in operating an aircraft. He also highlights that such a captain can make the flight deck environment so acrimonious that safety may be put

---

Box 5.1 Accident Case Study – Toxic Captain

Kenya Airways Flight 507 – Boeing 737-800
Douala, Cameroon, 5 May 2007[23]

Just before take-off from Douala International Airport in the Republic of Cameroon, the crew of this Boeing 737-800 aircraft informed the tower that they would like to remain slightly right of the runway extended centerline after departure to avoid some bad weather. The captain was the handling pilot and, despite not having received take-off clearance, he commenced the take-off roll. After take-off, the heading bug was progressively moved to the right to remain right of the runway centerline. With a bank angle of 11 degrees right, the captain called for the autopilot to be engaged but the first officer did not respond. Although the autopilot was not engaged, the captain appeared to believe that it was. The aircraft continued to bank to the right but the heading selector was now to the left of the aircraft's actual heading. With the aircraft in a 34 degree right bank on an actual heading of 215 degrees, the heading selector was set to 165 degrees and so the flight director was commanding full left bank. The autopilot was still not engaged. The ground proximity warning system activated and called "BANK ANGLE". The control wheel was turned 22 degrees right, then 20 degrees left and then 45 degrees right, causing the right bank to increase rapidly. It seems that at this point, someone engaged the autopilot, although without making any calls, and inputs to the flight controls decreased. Five seconds later, intense inputs to the flight controls resumed with several applications of right rudder. At this point the captain announced: "We are crashing". The first officer agreed. The bank angle to the right reached 70 degrees, the nose dropped and the aircraft entered a dive. The first officer called for the captain to correct to the left but it was too late. The aircraft crashed into a mangrove swamp, killing all 114 people on board.

Subsequent investigation determined that the cause of the crash was spatial disorientation due to inadequate scanning of the instruments and a lack of crew coordination. The report also highlighted that the captain took over from the first officer when communicating with the tower before take-off, and this may have resulted in the aircraft taking off without clearance. The report also highlighted that no briefing was carried out, autopilot engagement calls were not made and some checklist calls were not made. Although the first officer continued to make heading changes, the fact that the autopilot was disengaged meant that the aircraft was not following the heading changes and he did not highlight the mismatch between the aircraft attitude and the flight director command bar. The first officer did not make any warning calls even when the aircraft was in a highly unusual attitude. The captain's training record and other sources revealed the following:

- During an assessment before his initial command check, the check pilot was dissatisfied with his performance because of his poor systems knowledge and insufficient monitoring of the flight mode annunciators.
- His initial command check was inconclusive and so he was scheduled for a second check flight, which he passed.
- There were repeated cases where the captain was reported to have difficulties with crew resource management, adherence to standard operating procedures and instrument scanning. Because of this, some proficiency checks were found to be unsatisfactory and retraining was required before another check.
- The captain was said to have a strong character and a heightened ego and was authoritative and domineering towards subordinates. This was associated with excessive confidence bordering on arrogance.

The 23-year-old first officer was relatively inexperienced and was said to be non-assertive. The report stated that the captain exhibited a paternalistic attitude towards the first officer and the first officer appeared to be subdued by the strong personality of the captain.

in jeopardy. In his analysis of the Kenya Airways Flight 507 case, Baron identified some further criteria associated with these toxic personalities:

- Multiple first officers not wanting to fly with a particular captain – This suggests that the problem resides with the captain's personality rather than an interpersonal problem between that captain and one particular first officer.
- Multiple cases where deficiencies in crew resource management (CRM) have been identified during training or proficiency checks.
- The captain being unwilling to acknowledge the problem, preferring instead to believe that other crew members are to blame.
- The observed traits of arrogance and egotism may be the result of a coping mechanism in order to deal with personal insecurities. Their sense of control is enhanced by making other people feel weak.
- Captains from a military background may have difficulty adapting to civilian operations because the hierarchical gradient is not so steep. Optimum leadership styles may also differ between military and civil environments.
- A disregard for CRM or a dismissal of its principles – CRM is either not important or it is meant for other people.

It can be very difficult for organizations to manage these extreme personalities. It may be easier for an organization to explain these cases away as being down to simple personality clashes between crew members rather than recognizing that the captain's personality poses a real threat to flight safety. The reality is that our personality structure affects our information processing system and this system dictates every aspect of our behavior: how we perceive ourselves, how we interact with others, how we interact with the aircraft and how we manage risk. Although it is an unpalatable thought, we have to recognize that it is possible to have a personality structure that is not compatible with being a safe captain. Taking an extreme case, it would be highly undesirable to have someone with an untreated personality disorder in command of an aircraft and, in fact, certain personality disorders will disqualify a pilot from holding a medical certificate. However, "normal but borderline" personality variants (i.e. those that may manifest in undesirable behaviors but not meet the diagnostic criteria of a personality disorder) could turn out to be extremely dangerous if they are incompatible with the occupational environment.

As with the difficulty in treating personality disorders, it can be hard to develop strategies for dealing with personality types that could affect flight safety. As we have said, personality structure is relatively fixed and it is usually only behavioral training that can counteract any shortcomings. Baron proposed two strategies for managing these toxic cases once they have been identified (numbers 2 and 3 in the list) and there may also be a further strategy that airlines could consider before these later steps are required:

1. Psychological screening – The most effective way of dealing with this issue is to prevent it being an issue at all. Although psychological screening is not guaranteed to be effective (e.g. it is possible for candidates to give what they consider to be desirable answers in questionnaires rather than honest ones), this can go some way to ensuring that pilots possess the necessary personality traits to operate flights safely. Beyond psychometric testing, psychological interviews can also give a useful insight into an individual's personality.

Screening becomes even more important when an individual puts themselves forward for command upgrade. Any problems recorded in their training files with regard to CRM should be addressed, particularly traits associated with the toxic captain phenomenon, such as arrogant or overbearing traits. Although some personality traits are incompatible with effective leadership, there is an opportunity to manage these traits through training and so they do not necessarily have to prevent a candidate going on to command training as long as these issues are dealt with and resolved. This becomes more difficult when airlines have to employ direct entry captains as there may be no access to that individual's previous employment record.

2. Remediation – Once a toxic captain has been identified, it may be possible to adjust his behavior through training. There are significant barriers to this owing to the nature of his personality. It may be extremely difficult to get the person to accept that there is a problem with their performance and to get them to engage with remedial training.

3. Termination – Although Baron lists this as the last resort, he recognizes that remedial training may not be effective and that some personality traits will always be incompatible with a safe operation. From a safety management perspective, a hazard is anything with the potential to cause harm. Safety management systems are designed to identify and manage these hazards. Unlike hazards that may be located in aircraft maintenance procedures or instrument approach procedure design, this particular hazard may be rooted in the personality structure of an individual. Despite this, once a hazard has been identified, it needs to be managed or the operation will be exposed to significant risk.

In the event that the required organizational steps are not taken, it may fall to the first officer to be the last line of defense. Dealing with this sort of captain is probably the most challenging thing that a first officer will ever need to do, simply because the authority gradient is not in his favor. Below are some things to consider:

- Do not take it personally – A captain who persistently sets a negative tone with the people he works with is likely to be the source of the problem, not you.
- Avoid escalation – If you have managed the difficult task of not taking it personally, you have a good chance of preventing further deterioration. Do not engage in a disagreement on an emotional level.
- Continue operating as you would with anyone else – Maintaining high standards in your role may prove to be important in preserving flight safety.
- Make the SOPs work for you – No matter what tone is set in the flight deck, make all necessary monitoring and alerting calls. If the captain is not happy with this, politely remind him that you have a duty to follow the SOPs. Most companies have guidance for first officer assertiveness in their training manuals or in the SOPs.
- The last resort for a first officer who deems that the captain's actions are putting the flight in serious and immediate jeopardy is to take control. Robert Bescoe, who developed one of the assertiveness tools described in Chapter 6 (Communication), notes that guidance for a first officer taking control of the aircraft is well defined in cases of incapacitation of the captain but less well defined when the captain is putting the flight in immediate jeopardy.[24] The decision to take control of the aircraft is a truly unenviable one because the likelihood is that the aircraft is in an undesired aircraft state and by taking control, the implication is that the captain is incapable of safely managing the flight. Not only is the aircraft in a precarious situation, it has now become a single-pilot operation. Depending on the situation, there could also be resistance on the part of the captain which would only serve to complicate matters further. However, as with all aspects of flight operations, the primary objective is to preserve flight safety. If taking control of the aircraft is the only way of returning it to a

safe condition when the captain has not taken appropriate corrective actions despite your
warnings, it needs to be done.
• Further guidance regarding first officer assertiveness and other communication strategies is
  given in Chapter 6.

Unfortunately, it does not seem that the toxic captain phenomenon is as rare as
we would like. Most professional pilots reading this section will be able to recall flying with someone who has the same characteristics as the captain of Kenya Airways
Flight 507. There is likely to be a generational component given that many captains
who are currently flying may have completed their initial training before the introduction of CRM training. Perhaps the growing awareness of the role of personality
in the flight deck means that these toxic tendencies are corrected earlier on through
training, and so there is a chance that this problem will gradually decrease as the older
generation of captains retire and are replaced by captains who have been trained in
a different way.

5.2.3.2 The toxic first officer

If we recognize the phenomenon of the toxic captain, we should also consider the possibility of the toxic first officer. Imagine a flight deck where the captain embodies all
the behaviors associated with effective leadership, either by virtue of his personality
or through a combination of personality and training. When his non-technical skills
are assessed, he consistently scores highly in the domains of leadership and cooperation. Could a first officer's personality pose a threat to flight safety in the same way as
a toxic captain's personality could? Based on the characteristics of a toxic captain and
the knowledge that personality structure is relatively fixed, it would be safe to assume
that the underlying personality structure that leads to toxicity would be present before
that person becomes a captain; that is, when they are a first officer. Aggressive, overbearing and egotistical first officers are as much of a possibility as aggressive, overbearing and egotistical captains, particularly when you consider that these very traits
may limit the chances of that first officer being promoted to captain. The frustration
for a first officer of not being able to advance in his career may even heighten these
tendencies. The challenge for a captain in managing such personalities is not an easy
one, especially when trying to ensure that the first officer remains a useful resource
in the flight deck. Below are some potential management strategies based on what we
have covered on personality, behavior and non-technical skills:

• Undesirable personality traits may be accentuated owing to circumstances. A first officer
  who may occasionally be difficult to deal with may suddenly become even more difficult
  to deal with as a result of external circumstances, either personal or professional. As soon
  as you identify this, you can anticipate the potential threat to flight safety. In the first
  instance, you may be able to resolve the situation in the briefing room. Taking the first
  officer aside and inquiring if anything is wrong may be a useful first step in defusing the
  situation.
• Do not take it personally – If the problem presents itself during flight, you may not be able
  to defuse the situation. The first thing to realize is that undesirable personality traits or
  behaviors are the responsibility of the individual and although they may be directed at you,

you are not the cause. Although there may be provocation, avoid taking this personally by refusing to become emotionally engaged in the situation.

- In the event that undesirable behaviors are affecting the atmosphere in the flight deck or leading to unsafe acts, recognize that this is happening and deal with it. You have two potential ways of solving the problem:
  - Use leadership: As we have said, leadership employs social influence to enlist the aid and support of others in the accomplishment of a common task. Using social influence may be a more time-consuming strategy but it gives the best chance of achieving an appropriate resolution while still maintaining the first officer as a useful resource.
  - Use authority: In the event that social influence does not work, you are still able to give whatever commands you deem necessary to ensure the safety of the flight and your first officer is legally bound to obey. Giving commands may be necessary in time-critical situations but should be a last resort in other situations because you will potentially diminish any usefulness that the first officer has as a resource.
- Unfortunately, there has been very little research into the best strategies for dealing with difficult behaviors in safety-critical industries. The best strategy will often be dictated by the people involved and the current situation. However, some articles give some general guidance for dealing with difficult personalities and these form the basis of the additional guidance given below[25]:
  - Maintain self-control: The situation will not be helped if you become emotionally engaged.
  - Consider which behaviors require a response: Consider whether the behavior is just annoying to you or whether it represents a more significant risk.
  - Isolate the cause, particularly if the behavior is unusual: There may be a reason for this undesirable behavior, perhaps a problem outside work. If this is the case, this knowledge can help in managing the situation.
  - Acknowledge and validate: The first officer may believe that his behavior is justified; that is, he is well intentioned. Acknowledging that he thinks his behavior is justified but emphasizing that it is, in fact, making things more difficult may prove useful in eliminating the behavior but keeping the first officer as a useful resource.
  - Provide evidence and discuss consequences: If the behavior is leading to operational difficulties or low morale, the next step is to provide evidence for this and state that if the behavior continues, the situation is likely to deteriorate further.
  - Appeal to self-interest: Explain that the day would probably run easier for everyone involved if the undesirable behavior stopped. If this behavior has persisted over multiple duty days, you could suggest that if it does not stop, your only option will be to raise this matter with the training/safety department, a step that you are reluctant to take because of the possible negative light this may put on the first officer.
  - Reward and reinforce: Make an effort to reinforce positive behaviors by thanking or congratulating the first officer when they make an appropriate contribution to the flight.
  - Set limits: If there is still no improvement, make your dissatisfaction with the current atmosphere in the flight deck clear to the first officer, that you have tried to be understanding about his situation/perspective but that you cannot allow it to affect the safe and efficient operation of the aircraft. This is the last stage before having to rely on authority-based strategies such as giving clear, unambiguous instructions to stop behaving in a particular way.
  - Assess the risk: In the extreme case where you are unable to resolve the deteriorating situation through leadership or authority, the final question is: "Is flight safety threatened by this person's presence on the flight deck?" If the answer is yes, your only option may be to offload them at the first opportunity.

* Follow-up: If a first officer continues to cause problems over several duties, it may be necessary to raise this problem with the training/safety department. It is then possible to see if this is just a matter of interpersonal dynamics or a more significant problem with this particular first officer. Further training may help to resolve the situation.

Further guidance regarding communication strategies is given in Chapter 6.

5.3   Flight deck gradient

Toxic captains and toxic first officers are dangerous because of the effect they have on the flight deck gradient. The flight deck gradient could also be called the authority gradient, and it has a significant impact on crew communication as well as being a key determinant of the effectiveness of Threat and Error Management. In a normal flight deck, it would be normal for the captain to have the most authority but for the first officer to be empowered enough to be a useful resource in monitoring, problem solving, error detection and other aspects of Threat and Error Management. In this case, a flight deck gradient exists but is not too steep. An authoritarian captain paired with an unassertive first officer will lead to a steep flight deck gradient, with the first officer being unlikely to question the captain or highlight any unsafe acts that he commits. A passive captain paired with an overassertive, pushy first officer may lead to a very shallow cockpit gradient, or even one where the normal cockpit gradient is reversed. In this case, despite final authority resting with the captain, it may be the first officer who takes it upon himself to make command decisions, either because of his domineering nature or because the captain is too passive to do so.

Everything we have covered so far in this chapter culminates with one captain and one first officer bringing their personalities and behaviors into the flight deck. Their individual characteristics will determine the flight deck gradient and so each pairing will have a unique gradient. Abnormal and undesirable cockpit gradients occur when the combination of the captain's and the first officer's personalities is not conducive to a safe operation. Several factors have an impact on cockpit gradient:

* The personality of the captain and the first officer.
* Company culture – Some companies may not encourage first officer assertiveness and so the gradient is likely to be steeper.
* National culture – The nationality and cultural background of the captain and the first officer can have a significant bearing on the flight deck gradient, and this is explored in more detail in the next section.
* Situation-induced behavior of the captain and the first officer, for example behavioral changes associated with fatigue or anxiety. In this respect, the gradient may change over the course of a duty owing to external factors but there is likely to be a fixed gradient for most of the time.

5.3.1   Culture and flight deck gradient

Now that we have considered the role of personality and the potential of training in determining effective leadership, we need to look at one more significant variable in the leadership/teamwork equation, namely culture. Whether we like it or not, our cultural background has an impact on our information processing, particularly when selecting responses to particular stimuli. To complete the picture, we have an information processing system in our brain and this has two major modifiers that will give a tendency to generate certain types of behavior in response to certain stimuli: personality and cultural programming. Until relatively recently, the largest airlines were usually national carriers such as American Airlines, Air France and British Airways. National carriers tended to be staffed by pilots of the same nationality as the airline, meaning that there was not a great deal of cultural diversity in the flight deck. With the emergence of large low-cost carriers and rapidly expanding airlines in the Middle East and Far East, pilots from all over the world may be working in one airline. Over the course of two decades, cultural diversity in the flight deck has increased significantly. Two pilots from the same cultural background will generally have an insight into each other's cultural programming and be able to behave in a culturally appropriate way. In cases where crew members are of different nationalities, there may be elements of cultural programming in each person that the other is unaware of. Some cultural programs have an impact on communication style and behavior, and having an awareness of this can allow crew members to best utilize their own skills and those of others.

Unfortunately, the science of human factors has not kept up with this massive rise in cultural diversity in the flight deck. It has been noted that the majority of human factors research carried out in the aviation industry focuses on operations in North America and Western Europe. There is much less research in Middle Eastern and Far Eastern operations, which are the regions showing the fastest growth.[26] Aside from differences in cultural programming, it also seems that there are differences in basic cognitive function between different nationalities. One study highlighted three differences between Eastern and Western information processing. In a study of cross-cultural attention mechanisms, participants of different nationalities were shown a scene where there was a focal object set against a background. It was observed that Japanese participants recalled the details of the scene in a more holistic manner, remembering details of both the focal object and the background, whereas American participants selectively attended more to the focal object and less to the background.[27]

Although this study looked only at cross-cultural differences in the attention mechanism, this difference in one of the basic components of human information processing is very significant because it highlights that the findings of some human factors research can be culture specific. Therefore, CRM techniques that are designed for Western pilots based on data collected from Western operations may need to be adjusted to reflect cultural differences when applied to pilots from other cultures. Many of the psychometric tests that we have mentioned so far, such as the NEO-PI-R, are adjusted so that they are more relevant to the cultural environment in which they are being used and the results are usually compared against average values from people with the same cultural background.

Geert Hofstede was a researcher who closely examined how different nationalities approach work situations. He suggested that the culture of a country could be thought to lead to a "collective programming of the mind", an idea consistent with the impact of culture on information processing and the behavior it generates. In a large study, Hofstede collected data from 160,000 employees of a large multinational corporation with offices in 40 countries.[28] He concluded that national culture had more of an impact on the reported work-related values and attitudes than any other variable such as job role, age or gender. He identified four dimensions that varied significantly according to cultural background:

- Individualism/collectivism – This dimension reflects the cultural tendency towards individualism (more likely to be motivated by individual achievement) and collectivism (more likely to be motivated by group achievement). Countries that tended more towards individualism included the UK, the USA, Australia and the Netherlands. Countries that tended more towards collectivism included Taiwan, Thailand, Pakistan and the Philippines.
- Masculinity/femininity – This dimension reflects a culture's emphasis on materialism or quality of life, those being at the masculine or feminine end of the spectrum respectively. Cultures that tended more towards masculine goals included Japan, Australia, Ireland, the UK and the USA. Cultures that tended more towards feminine goals included Sweden, Denmark, Norway, the Netherlands and Finland.
- Uncertainty avoidance – This dimension reflects how easily people from different cultures cope with uncertainty and how likely they are to avoid uncertain situations. Cultures that showed a tendency for high uncertainty avoidance were more likely to generate firm rules and adhere to them strongly. Cultures that tended to be strongly uncertainty avoidant included Greece, Spain, France, Japan and Portugal. Cultures that showed a weaker tendency towards uncertainty avoidance included Denmark, Sweden, the UK, Ireland, the USA and India.
- Power distance – Some cultures with a strong class system will tend to have quite steep hierarchies in their organizations. In cultures that score highly in the power distance dimension, there is greater respect for authority and a lower likelihood of authority being questioned. In cultures that scored low in the power distance dimension, subordinates and superiors were more likely to treat each other as colleagues, with subordinates being more willing to point out mistakes to their superiors. This dimension is very relevant in the flight deck as it has a significant impact on flight deck gradient. Countries that had a tendency towards higher power distance included Singapore, Thailand, the Philippines, Mexico and India. Countries that tended towards lower power distance included Australia, Israel, Denmark, Ireland, New Zealand, the UK and the USA.

Interestingly, when interpreting the results of the cross-cultural study looking at differences in the attention mechanisms of Eastern and Western cultures, Hofstede's work, specifically individualism/collectivism, may suggest a reason. Eastern cultures tend to have more complex, collective social structures and so assessing context is important in ensuring that one's behavior is appropriate. This may lead to a tendency to distribute attention more diffusely in a particular environment so as not to miss any relevant social cues, as opposed to Western cultures that do not have the same emphasis on collectivism and are more individualistic.[27]

Since Hofstede's original work was carried out in a non-aviation environment, a study was published in 2000 to see whether his conclusions were applicable to the flight deck environment.[29] In this study, 9400 commercial pilots from 19 countries were surveyed using the Flight Management Attitudes Questionnaire, which was designed to measure pilots' attitudes towards command, communication and other work-related issues. The analysis showed that there was significant correlation between the results that Hofstede found when looking at employees working in a multinational corporation and pilots working in different countries around the world. The correlation was particularly high for individualism/collectivism and power distance. The author of this report concluded that this evidence suggests that training needs to be adjusted to reflect cultural differences.

In the context of this chapter, the phenomenon of power distance is directly relevant to the flight deck gradient. Differences in individualism/collectivism can be important in goal setting, and differences in uncertainty avoidance have an impact on SOP compliance and potential stress levels when deviations from procedures are necessary. Below are some strategies that can help you to use crew resources effectively, particularly when the crew come from a variety of different cultural backgrounds. It should be noted that in the same way as our behavior is not completely dictated by personality, neither is our behavior dictated by our culture. Behavior is a result of the workings of our information processing system, a system that has more determinants to it than just personality and cultural backgrounds. Training, environment, knowledge and, most importantly, the situation being engaged with have a significant impact on behavior. In short, we are not slaves to our cultural background and it would be inappropriate to generalize these strategies to everyone you work with who has a particular background. These are just some general considerations that may be useful:

- High power distance can limit cross-checking and error detection by increasing the flight deck gradient – Crew members from cultures that tend to score highly on the power distance dimension may be less likely to question or point out errors made by someone who they perceive as more senior to them. In order not to lose this essential error-checking resource, senior crew members, especially captains, should work extra hard to promote open channels of communication in the flight deck and encourage junior crew members, particularly first officers, to highlight any errors they detect and to ask for clarification if they encounter any circumstances or behavior they do not understand. A simple way of doing this is to state that any uncertainty they have about what is going on means that you are either diverging from the SOPs or you have not briefed them properly. Either way, they should make this known to you. When a crew member does this, this positive behavior should be reinforced by recognizing it and thanking them for it. This will make it more likely that they will repeat this positive behavior in the future.
- Low power distance may be mistaken for being overbearing – In the event that the captain is from a cultural background that has high power distance and the first officer is from a cultural background that has low power distance, the captain may misinterpret the more relaxed attitude of the first officer as disrespect. First officers should be aware of this possibility and consider this when pointing out errors or questioning the captain. Although error detection and cross-checking are essential functions, carrying them out in a culturally appropriate manner may help to avoid misunderstandings. Captains should be equally aware of these cultural differences and ensure that the manner in which they communicate to more junior crew members is culturally appropriate.
- Uncertainty avoidance can work both ways – A tendency to follow SOPs may sound like it is advantageous in every circumstance. The drawback of cultures with high uncertainty avoidance may be a tendency to try and develop rules and procedures to cover every situation, a risk in itself as systems with large numbers of procedures may be unworkable for the human operators. It also means that in abnormal situations, there may be a tendency to spend time trying to find some sort of procedure that could be applicable, rather than dealing with the situational complexities as they arise. Low uncertainty avoidance may be associated with less rigid adherence to SOPs but a greater capacity to see beyond them when situations dictate.

5.4   Cooperation and conflict solving

Aside from the characteristics of the team leader, all team members need to be able to cooperate in order for the team to be effective. The NOTECHS framework includes a domain for this skill set; this is shown below, along with some positive and negative behavioral markers:

1. Cooperation
   a. Team building and maintaining
      i. Positive behavioral markers:
         – establishes atmosphere for open communication and participation
         – encourages input and feedback from others.
      ii. Negative behavioral markers:
         – blocks open communication
         – competes with others.
   b. Considering others
      i. Positive behavioral markers:
         – takes notice of the suggestions of other crew members even in cases where he disagrees with suggestion
         – gives appropriate personal feedback.
      ii. Negative behavioral markers:
         – ignores suggestions from other crew members
         – shows no reaction to other crew members' problems.
   c. Supporting others
      i. Positive behavioral markers:
         – helps other crew members in demanding situations
         – offers assistance.
      ii. Negative behavioral markers:
         – hesitates in helping other crew members
         – does not offer assistance.
   d. Conflict solving
      i. Positive behavioral markers:
         – keeps calm in conflicts
         – concentrates on what is right rather than who is right.
      ii. Negative behavioral markers:
         – overreacts in conflicts
         – sticks to own position without compromise.

Team building and maintaining, supporting others and considering others are covered in Chapters 6 (Communication) and 3 (Error Management and Standard Operating Procedures for Pilots). Chapter 6 describes strategies for establishing a positive team atmosphere from the outset, ensuring open channels of communication and assertiveness strategies for overcoming certain barriers. What remains is conflict solving, and this is covered in this chapter because, although it is an important skill for all team members, it is an essential skill for leaders. Conflict in flight should be avoided at all costs because its emotional nature is highly distracting and this can lead to flight safety issues. If a conflict occurs, while it should be the goal of both parties to resolve it, the responsibility for finding a suitable resolution lies with the more senior crew member. This may involve relinquishing his own position and accepting that he was in error. As we covered in Chapter 3, one of the most effective ways of ensuring open channels of communication in the flight deck is to openly admit your mistakes.

5.4.1 Individual differences and conflict-solving strategies

When groups of people work together, even when there is no personality clash, there may be occasions when a difference of opinion has the possibility of turning into a conflict. Various conflict-solving styles have been identified and these can be categorized according to two properties:[30]

- the conflict solver's assertiveness (may be related to their use of authority)
- the conflict solver's cooperation.

Using these dimensions, conflict resolution styles can be categorized into five main types, as shown in Figure 5.1.

Although most individuals will adapt their conflict resolution strategy to the needs of the situation, personality traits can predispose us to use one strategy more often than others. Certain psychometric tests, such as the Thomas–Kilmann Conflict Mode Instrument, pose questions regarding preferred conflict management styles and, based on the results, can assess whether someone is more predisposed to using one style over the others.[30] Below are summaries for each of the styles:

- Competing: high assertiveness and low cooperation – The conflict solver will try to win at all costs and will make every effort to impose his position on the other person, to the extent that they will back down.
- Accommodating: low assertiveness and high cooperation – The conflict solver will yield to the other person's argument readily.
- Avoiding: low assertiveness and low cooperation – The conflict solver will not engage with the problem and will actually avoid dealing with it.
- Collaborating: high assertiveness and high cooperation – Rather than one person's argument taking precedence over the other's, this style of conflict resolution is characterized by

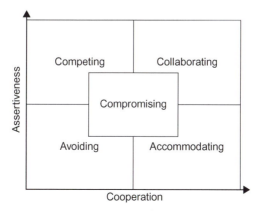

Figure 5.1 Conflict resolution styles.

further discussion of the problem in an attempt to find a common ground that satisfies both parties.
- Compromising: medium assertiveness and medium cooperation – This is a style where elements of all the other styles may be called on to resolve the conflict in a more compartmentalized way. Some elements of the conflict may be avoided and there may be some give-and-take with regard to other elements of the conflict. This is similar to collaboration but rather than exploring the nature of the conflict to try to find common ground, the conflict is split so that some elements are resolved to the satisfaction of the conflict solver and some to the satisfaction of the other person.

5.4.2   Situational variables and conflict-solving strategies

Although our personalities may lead us to have a greater tendency to use a particular conflict-solving strategy over the others, this can cause problems if this style is used in a situation that does not warrant it. For example, collaborative conflict solving would not be appropriate for a disagreement that occurs during the approach phase of a flight. There are too many tasks to perform and engaging in in-depth conflict exploration at that time would be inappropriate. If the nature of the conflict was not directly relevant at the time, it would be preferable to use the avoiding strategy and then deal with the conflict at a more appropriate time. In a similar way, other situations will be associated with an optimum conflict resolution strategy. These are given below[31]:

- Competing:
  - In emergency situations requiring quick action. It helps to have authority in these situations as well.
  - When enforcing unpopular rules or enforcing discipline.
  - When protecting oneself from people who take advantage of accommodating behavior, for example dealing with other people who will only provide the services you need if you are insistent.
- Accommodating:
  - When you are wrong: For successful error management, it is always advisable to admit your errors.
  - If the issue is more important to the other person and you will not lose out in a significant way, being accommodating can help to establish a cooperative relationship.
  - If you realize that you are going to need this person's help in the future, it may be beneficial to be accommodating now in the hope that they will be accommodating in the future.
  - If avoiding disruption is more important than winning the argument, accommodation can bring a swift end to the conflict.
- Avoiding:
  - When an issue is of low importance, there are other tasks to be dealt with and this is not an appropriate time to deal with the conflict.
  - The cost of dealing with the conflict outweighs the benefits of resolving it.
  - To allow people to calm down.
  - To allow for more information to be gathered.
  - When this conflict is symptomatic of a larger issue that needs to be resolved.

- Collaborating:
  - When it is important to find a solution that satisfies both parties, particularly if both parties have genuine concerns.
  - In order to combine different problem-solving approaches.
  - When reaching a consensual decision will make it easier to get the other party's help to action it.
  - When there are more complex reasons underlying the conflict, such as a personality clash, and both parties need to resolve this conflict in order to establish some rules for how the relationship can work in the future.
- Compromising:
  - If a situation does not suggest one particular conflict-solving strategy over the others, it may be necessary to use elements from several of them.
  - If two parties have equal power and are committed to mutually exclusive goals, such as in a labor dispute, both parties will have to give up some demands in order to have demands met in other areas.

Chapter key points

1. Personality is a relatively fixed structure that underlies behavior and gives people a tendency to act in certain ways.
2. The Five-Factor Model assesses personality according to five primary factors: extraversion, agreeableness, conscientiousness, neuroticism and openness.
3. An individual's personality affects behavior by affecting information processing.
4. There appears to be a genetic component to many personality variants.
5. When employers ask potential employees to undertake personality tests, they are looking for people with the right personality type to suit the job.
6. Different personality types may be better suited to different jobs.
7. If certain aspects of our personality are causing us problems, it is possible to override these effects with behavior modification.
8. Personality has a definite role in leadership.
9. Leadership is defined as a process of social influence in which one person is able to enlist the aid and support of others in the accomplishment of a common task.
10. There are certain personality variants that are associated with effective leadership and these are closely correlated with the positive behavioral markers that should be displayed when exercising good non-technical skills.
11. There are some personality variants that have a negative effect on flight safety.
12. Dealing with difficult personalities in the flight deck requires careful management.
13. Flight deck gradient refers to the distribution of power in the flight deck. A flight deck with a high gradient would be characterized by an authoritarian captain and a timid first officer.
14. Cultural background can have an effect on the flight deck gradient as well as many other work-related abilities.
15. Conflict solving is an important part of leadership and teamwork and there are several conflict-solving strategies that can be used.
16. A particular situation may require a particular strategy but problems can arise when people have a tendency to use one strategy more than others, especially in circumstances where it might not be appropriate.

References

1. Cattell H, Mead A. The Sixteen Personality Factor Questionnaire (16PF). In: Boyle G, Matthews G, Saklofske D, editors. *The SAGE handbook of personality theory and assessment. Volume 2: Personality measurement and testing*. London: SAGE; 2008, pp. 135–60. http://dx.doi.org/doi:10.4135/9781849200479.n7

2. Widiger TA, Costa PT. Integrating normal and abnormal personality structure: The Five-Factor Model. *J Pers* 2012;**80**(6):1471–506.

3. Lynam DR, Widiger TA. Using the Five-Factor Model to represent the DSM-IV personality disorders: An expert consensus approach. *J Abnorm Psychol* 2001;**110**(3):401.

4. Raine A, Lee L, Yang Y, Colletti P. Neurodevelopmental marker for limbic maldevelopment in antisocial personality disorder and psychopathy. *Br J Psychiatry* 2010;**197**(3):186–92.

5. American Psychiatric Association. *The diagnostic and statistical manual of mental disorders: DSM-IV*. Washington, DC: American Psychiatric Association; 2000.

6. Glenn AL, Raine A. Antisocial personality disorders. In: Decety J, Cacioppo J, editors. *The Oxford handbook of social neuroscience*. New York, NY: Oxford University Press; 2011, pp. 885–94.

7. Munafò MR, Flint J. Dissecting the genetic architecture of human personality. *Trends Cogn Sci* 2011;**15**(9):395–400.

8. Karimi S, Kangavari MR. A computational model of personality. *Procedia – Social Behav Sci* 2012;**32**:184–96.

9. Leichsenring F, Leibing E, Kruse J, New AS, Leweke F. Borderline personality disorder. *Lancet* 2011;**377**(9759):74–84.

10. Myers IB, Myers PB. *Gifts differing: Understanding personality type*. London: Nicholas Brealey; 2010.

11. Myers IB. *Introduction to type: A description of the theory and applications of the Myers–Briggs Type Indicator*. Palo Alto, CA: Consulting Psychologists Press; 1987.

12. Costa PT. Work and personality: Use of the NEO-PI-R in industrial/organisational psychology. *Appl Psychol* 1996;**45**(3):225–41.

13. Davidson K, Norrie J, Tyrer P, Gumley A, Tata P, Murray H, et al. The effectiveness of cognitive behavior therapy for borderline personality disorder: Results from the Borderline Personality Disorder Study of Cognitive Therapy (BOSCOT) trial. *J Personal Disord* 2006;**20**(5):450.

14. Leichsenring F, Salzer S, Beutel ME, Herpertz S, Hiller W, Hoyer J, et al. Psychodynamic therapy and cognitive–behavioral therapy in social anxiety disorder: A multicenter randomized controlled trial. *Am J Psychiatry* 2013;**170**(7):759–67.

15. Kohn R, White CN. Royal aeronautical society flight operations group specialist document: So you want to be captain? Retrieved from: <http://aerosociety.com/Assets/Docs/Publications/SpecialistPapers/So_You_Want_to_be_a_Captain.pdf/>; 2010.

16. Chemers MM. *An integrative theory of leadership*. Philadelphia, PA: Psychology Press; 1997.

17. Judge TA, Bono JE, Ilies R, Gerhardt MW. Personality and leadership: A qualitative and quantitative review. *J Appl Psychol* 2002;**87**(4):765.

18. O'Neil DP. Predicting leader effectiveness: Personality traits and character strengths. PhD thesis, Duke University. Retrieved from: <http://dukespace.lib.duke.edu/dspace/handle/10161/385/>; 2007.

19. Collins J. Level 5 leadership: The triumph of humility and fierce resolve. *Harv Bus Rev* 2001;**79**:66–76.

20. Chidester TR, Kanki BG, Foushee HC, Dickinson CL, Bowles SV. Personality factors in flight operations. Volume I: Leader characteristics and crew performance in a full-mission air transport simulation. NASA technical memorandum 102259; 1990.

21. Coid J, Yang M, Tyrer P, Roberts A, Ullrich S. Prevalence and correlates of personality disorder in Great Britain. *British J Psychiatry* 2006;**188**(5):423–31.

22. Baron RI. The toxic captain. *Aerosafety World* 2012(March):39–42.

23. Technical Commission of the Republic of Cameroon. (n.d.). Technical investigation into the accident of the B737-800 registration 5Y-KYA operated by Kenya Airways that occurred on the 5th of May 2007 in Douala.

24. Besco RO. PACE: Probe, alert, challenge, and emergency action. *Business and Commercial Aviation* 1999;**84**(6):72–4.

25. Duncum, K. *Confronting the problem personality*. Bloomberg Businessweek. Retrieved from: <http://www.businessweek.com/managing/content/feb2010/ca20100211_356400.htm/>; 2010.

26. Tam L, Duley J, Hamilton BA. Beyond the west: Cultural gaps in aviation human factors research. In: *Proceedings of the Mini-conference on Human Factors in Complex Sociotechnical Systems, Atlantic City, NJ, USA*. California: Human Factors and Ergonomics Society; 2005, pp. 1–5.

27. Nisbett RE, Masuda T. Culture and point of view. *Proc Natl Acad Sci USA* 2003;**100**(19):11163–70.

28. Hofstede G. *Culture's consequences: International differences in work-related values*. Newbury Park, CA: Sage; 1980.

29. Merritt A. Culture in the cockpit: Do Hofstede's dimensions replicate? *J Cross Cult Psychol* 2000;**31**(3):283–301.

30. Thomas KW. *Thomas–Kilmann conflict mode instrument*. Tuxedo, NY: Xicom; 1974.

31. CPP. TKI profile and interpretive report. Retrieved from: <https://www.cpp.com/Pdfs/smp248248.pdf/>; 2012.

Communication

Chapter Contents

Introduction   163
6.1 The Sender–Message–Channel–Receiver model of communication   163
   6.1.1 Barriers to communication   166
   6.1.2 The NITS briefing   167
6.2 Communication between pilots   168
   6.2.1 Introduction to transactional analysis   168
   6.2.2 Ego states   169
   6.2.3 Complementary and crossed transactions   170
   6.2.4 Ulterior transactions   176
6.3 Establishing a positive team atmosphere   177
6.4 Communication strategies for effective briefings   178
6.5 Communication strategies for assertiveness   180
   6.5.1 First officer assertiveness at the threat management stage   182
   6.5.2 First officer assertiveness at the unsafe act management stage   183
   6.5.3 First officer assertiveness at the undesired aircraft state management stage   185
   6.5.4 How captains can encourage assertiveness   185
Chapter key points   186
References   187

Introduction

Pilots, when they are introduced to the NOTECHS system, often ask why there are no specific non-technical skills listed under the heading of Communication. When you consider the four domains in the NOTECHS system (cooperation, leadership and managerial skills, situation awareness and decision making), communication is implicit in all of them. Despite it being a widely used skill, one that the vast majority of humans develop as they grow, there is still considerable variability in the effectiveness of communication skills between individuals and research has provided some useful strategies for improving this. This chapter will consider communications research as applied to three scenarios: normal crew communications, briefings and emergency situations. By the end of the chapter you will be able to apply contemporary research to improve the quality of your communication.

6.1   The Sender–Message–Channel–Receiver model of communication

Before we delve into specific limitations of communication and the strategies that we can use to overcome them, we should first establish what we mean by communication

Practical Human Factors for Pilots. DOI: http://dx.doi.org/10.1016/B978-0-12-420244-3.00006-6

and use this as the basis for the rest of the chapter. Although communication may seem to be a subject that can be intuitively understood, it is always helpful to establish precisely which aspect of human behavior we are talking about so that we can frame the topic more clearly. The best definition of communication for this purpose is as follows:

> *Any act by which one person gives to or receives from another person information about that person's needs, desires, perceptions, knowledge, or affective states. Communication may be intentional or unintentional, may involve conventional or unconventional signals, may take linguistic or non-linguistic forms, and may occur through spoken or other modes.[1]*

Based on this definition, in order for communication to occur there must be a sender, some sort of message and a receiver. It is worth noting that neither the sender nor the receiver needs to be consciously aware that communication is occurring. The principles behind the definition of communication that we have given are encapsulated in Berlo's Source–Message–Channel–Receiver (SMCR) model.[2] This has been slightly adapted to include some other important elements and, for clarity, "Source" is replaced by "Sender". The modified version of the SMCR model is given in Figure 6.1.

You will notice that the same five factors affect the sender and the receiver. There needs to be some degree of alignment between the five factors in order for successful communication to occur. For example, no matter how good the sender's communication skills, if the receiver is completely unengaged, it is unlikely that successful communication will occur. Similarly, if the sender is trying to communicate a new piece of knowledge, the receiver will need to have some contextual framework in order to absorb that piece of knowledge. For example, the sender may be an excellent teacher who can communicate Heisenberg's Uncertainty Principle with considerable

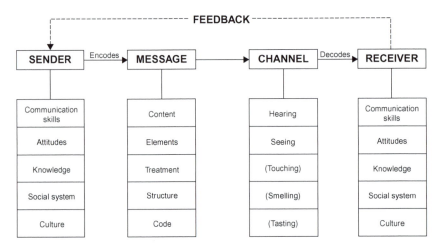

Figure 6.1 Modified SMCR model of communication: Sender–Message–Channel–Receiver.

clarity, but if the receiver has absolutely no knowledge of physics, it is unlikely that the information will make any sort of sense. Attitudes, culture and the social systems in place at the time can also either facilitate or limit communication. When the sender composes the message, the content is the entirety of the message, the elements are the verbal and non-verbal techniques used to convey the content, the treatment is how the sender understands how best to present the message (e.g. the sender's perception of his own message), the structure is the arrangement of the content, and the code refers to the correct combination of content and elements for the message to have the desired effect. For example, if a doctor were to deliver bad news to a patient using an upbeat tone with lots of gesticulation, although the content may be delivered successfully, the choice of elements and the treatment would be inappropriate. That is, the doctor does not understand the appropriate way to deliver such a message and, because of this failure, the code is wrong and the patient may be confused by the style of the message or not fully appreciate the gravity of the situation.

The channel refers to the five senses that can be used to deliver and receive messages. Although the primary means of communication is normally through the hearing channel (given that the spoken word is the most common type of communication), the other channels, particularly the visual channel, have a role. In order to understand this, let us look at types of verbal and non-verbal communication. Verbal communication is determined by the content and structure of the message; in other words, if the message was simply transcribed onto paper, the verbal element would be the words on the page. The non-verbal element comprises many more features and can have a significant effect on how the verbal element is interpreted by the receiver. Non-verbal elements include:

- paralinguistics:
  - speaking rate
  - tone of voice
  - pitch
  - volume
  - prosody: rhythm, stress and intonation of speech
  - paralinguistic respiration: sighs, gasps, groans, etc.
- facial expressions
- eye contact
- body language and posture
- hand gestures.

These non-verbal elements can have a significant impact on the receiver's interpretation of the sender's message, and their importance should not be underestimated, particularly for messages that have an emotional content. A common misconception is that only 7% of information is communicated using verbal elements, with other 93% being transmitted through non-verbal elements. If this were true, you could travel to a country where you did not speak the language and yet be able to understand 93% of what people were "saying" to you by interpreting their non-verbal language. This is a misinterpretation of the original research by Albert Mehrabian, which found that only 7% of the receiver's understanding of the *emotional* content of a message is based on what is said, with 38% being based on the tone of voice used and 55% being based on the body language of the sender.[3] Based on this, if someone says some words that would ordinarily be interpreted as being friendly but their tone and body language suggest unfriendliness, the receiver is more likely to base their emotional reaction to the message (and the sender) on tone and body language. These statistics were based on experiments carried out in a controlled environment but it is likely that the relative magnitudes of the importance of verbal, tone and body language are valid in real-world communication.

One final modification to the SMCR model is the idea of feedback. Ideally, when time allows, to ensure that the message has been successfully communicated, it is useful for the sender to receive feedback from the receiver. This will ensure that the receiver has received the message as intended by the sender.

6.1.1 Barriers to communication

Based on the SMCR model, we can now describe some barriers to communication that can lead to communication difficulties and some of the strategies that can help to overcome them:

- Sender deficiencies – If the sender has impaired communication skills, this will act as a barrier to communication. This impairment can be alleviated by communication skills training and, sometimes, enhanced communication skills on the part of the receiver.
- Mismatches between sender and receiver – If there is a significant disparity in the attitudes, knowledge, social system or culture between the sender and the receiver, these may all limit the effectiveness of communication. For example, a difference in culturally perceived status, particularly if the receiver is of a higher status than the sender, may mean that the sender is less likely to issue a message and the receiver is less likely to take notice of it. In a flight deck environment, this phenomenon would occur when there is a steep flight deck gradient for either cultural or personal reasons. It is up to the higher status individual to ensure that an appropriate tone is set to allow for a more junior crew member to communicate as necessary. Junior crew members should also not be limited by any perceived cultural or social mismatches.
- Mismatches in language – If the sender and the receiver do not share a common language, or the operational language of the sender is more jargon heavy than that of the receiver (as in the case of pilots communicating with cabin crew), the onus is on the sender to ensure that the message is communicated in a language that can be understood by the receiver. Jargon should be avoided when a disparity exists.
- Message deficiencies – A message is deficient if it is not encoded in a manner that reflects the situation. For example, a carefully crafted, respectful message that employs all the required social graces would be a deficient message if the aim was to communicate an imminent threat to safety. The treatment of the message when it is encoded should be a reflection of the sender's perception of the goal of the message. It is, therefore, acceptable to ignore social niceties if the sender perceives the message to be of vital importance to the receiver. A message that is overly long may lead to loss of understanding, particularly if there is a knowledge gradient between the sender and the receiver. Chunking the important parts of the message using a framework such as NITS (Nature, Intentions, Time, Special instructions) can aid the encoding of the message, and details of this framework are given in Section 6.1.2. Feedback is a useful tool for assessing whether the message was appropriate. The sender can ask the receiver to read back the salient points of the message to confirm that these have been successfully transmitted.

- Barriers affecting transmission/reception channels – There are various physical barriers that can limit the auditory and visual channels that are important for sending and receiving messages. Some of these barriers and some strategies for overcoming them are given below:
  - ◦ Noise: Increase volume of transmission, make use of interphones or communicate through an alternative channel.
  - ◦ Oxygen masks: Communications can decrease drastically should the crew have to don their oxygen masks. Given that this only tends to happen during emergencies, this has the added effect of limiting communications at a time when it is most important. Communication should be kept as brief and clear as possible and crew should be prepared for the fact that communication between the flight deck and the cabin may become impossible.
  - ◦ Flight deck door: This limits the visual channel and so any emotionally relevant content of the message may not be transmitted or may be misinterpreted because the receiver is unable to see the facial expressions or body language of the sender. Should time allow, critical communications should be made face to face to allow for feedback to ensure understanding, and for both the sender and the receiver to be able to judge message content and importance as completely as possible.
- Receiver deficiencies – As well as mismatches between the sender and the receiver, it is possible that the sender has encoded an appropriate message through an appropriate channel but that, for reasons that are not apparent to the sender, the receiver is limited in his ability to receive or decode the message. A common reason for this would be that the receiver is selectively attending to a particular stimulus and the message is unable to break through this tightly focused attention. This is one of the reasons that ground proximity warning system (GPWS) messages sometimes do not prompt the reaction that they were designed to elicit from the pilots, even though the message is valid. In fact, pilots often report not having heard the GPWS warnings despite them being clearly heard when cockpit voice recorders are replayed. To avoid this, feedback is crucial as it proves that the message has been received.

6.1.2   The NITS briefing

The NITS format is a standardized way of structuring a message. It can be used in any abnormal situation where it is important to pass on certain information regarding the situation to one or more people. It works best when all crew are trained to use it. The standard procedure is as follows:

1. Confirm that the intended receiver is ready to receive a NITS briefing – If NITS is used in your company then the receiver will know that they are listening out for four key pieces of information and that they will be expected to read these back afterwards. Some companies encourage the receiver to take notes as required. If the NITS briefing is being used to pass information on to air traffic control (ATC), it may be useful to ask them to take careful note of what is to follow as you will want them to read it back at the end.
2. Conduct the NITS briefing – Each point should be the verbal equivalent of a bullet point; it should only include what is relevant. Every additional piece of information decreases the chances that it will be recalled. When covering each point, use the relevant heading; for example, "The Nature of the problem is that we've had an engine failure. Our Intention is to divert to Geneva". Cover the following points:
   a. Nature of the situation – a brief, jargon-free summary of the situation. It should only contain information that will be relevant to the receiver (e.g. it is probably not necessary to tell cabin crew about every technical ramification of the electrical failure that has happened, only the elements that are relevant to them).

    b. Intentions – a brief summary of what is going to happen now (e.g. we are going to divert to Geneva).

    c. Time – how long before landing (for cabin crew) or how long before the crew will be ready to make an approach (for ATC).

    d. Special instructions – anything that you need the receiver to do or to arrange (e.g. check both wings for anything unusual, arrange for the fire service to meet the aircraft after landing); also, anything else that the receiver needs to know that has not been covered already.

3. Ask if they have any questions – The receiver may not have fully understood one or more points or may think that something important has not been covered. Give them an opportunity to ask any questions they have so that they have a complete understanding of the message.

4. Ask them to read it back – This will ensure that the message that they have received is the one that was intended.

5. The receiver may then become the sender for other people – In order to promulgate the message to those people for whom it is important, the original receiver may have to give a NITS briefing to other crew members.

Cabin crew can also give NITS briefings to pilots regarding abnormal situations in the cabin, for example, medical emergencies. Pilots may also use NITS briefings when communicating with ATC, particularly after an initial Mayday or PAN-PAN ("Possible Assistance Needed") call has been made. Some ATC units have adopted the ASSIST framework when it comes to communicating with aircraft that have made a Mayday or a PAN-PAN call[4]:

- Acknowledge the call.
- Separate the aircraft from other traffic. Give it room to maneuver.
- Silence – on the frequency. Provide separate frequency where possible. This prevents unnecessary clutter for the pilots.
- Inform those who need to know and those who can help; inform others as appropriate.
- Support the pilots in any way possible – start to think of alternative routings, etc.
- Time – give the pilots time to collect their thoughts; do not harass them for information.

The last letter of the ASSIST framework is very important. It may take some time for the pilots to go through checklists, obtain weather reports and complete a TDODAR (time available, diagnose, options, decide, assign tasks, review) before being in a position to brief ATC or cabin crew. ATC will generally wait to hear from pilots regarding the problem and, at an appropriate time, the NITS format can be a very useful way of passing this information on.

6.2 Communication between pilots

6.2.1 Introduction to transactional analysis

In developing the approach known as transactional analysis for use in psychotherapy, the psychiatrist Eric Berne attempted to impose a structure on human communication. The foundations of this are given below[5]:

- Humans have a stimulus hunger; that is, they need emotional and sensory stimulation to remain biologically healthy. If they are deprived of this, there is deterioration in the individual's psychological and biological health, enough to cause death.

- As an individual matures, this stimulus hunger becomes more specific insofar as the individual will develop a need for social recognition, for example, as a way of differentiating the individual's own qualities from those of others based on social feedback. This is known as recognition hunger.
- The unit of social action that counts in order to satisfy this recognition hunger is known as a stroke. A simple stroke could be as basic as someone acknowledging your presence. Strokes can be positive or negative.
- An exchange of strokes constitutes a transaction and this is the basic unit of social intercourse.
- The most effective way of satisfying stimulus hunger and recognition hunger is intimacy. Berne's prototypical example is the act of making love with a view to procreation: a highly intimate act between two people who love each other and are demonstrating this both through this act and through their desire to procreate. It is worth noting that this act needs to be differentiated from non-loving sex which, although pleasurable for both people, may not satisfy their deeper stimulus and recognition hungers in the same way.
- Given that intimacy must at times give way to the practicalities of day-to-day living, it seems that there are substitutes, albeit inferior ones, that can partly satisfy our stimulus hunger and recognition hunger. These are known as games.
- It should be noted that the use of the word "games" in this context does not suggest that these games are always fun. It simply implies a set of transactional patterns that usually follow a sequence based on unspoken rules and regulations, often social in origin. There are ways that many of these games seem to play out naturally, but when one player diverges from the rules the social consequences can be surprisingly serious.
- To understand games in the context of communication between pilots, we need to understand the idea of an ego state and, more so than achieving an understanding of games, understanding the different ego states and the types of transaction that occur when different ego states are used offers the greatest chance of improving communication in the flight deck.

6.2.2   Ego states

Consider the different roles you play in your life. To your children, you are a parent; to your colleagues, you are another adult; and to your parents, you may still be a child. However, have you ever reacted in a childish way when your partner has said something to you, or used the tone of a critical parent with a work colleague who is not performing as well as you would like? Berne describes an ego state as a set of feelings accompanied by a related set of behaviors. While we may have many of these ego states, they can be broadly divided into three types:

- Parent (P) – The supervisory, instructional and authoritative ego state as would be needed to bring up a child. It is also activated in other social situations.
- Adult (A) – An ego state that permits objective information processing with relatively little alteration by the effects of personality. This ego state operates predominantly on logical information.
- Child (C) – This ego state is a vestige of the sets of feeling and behaviors we used during childhood. This ego state is driven by impulse and knowing what it wants. However, it is not necessarily an undesirable ego state for adults to enter into because it is also an outlet for creativity, pleasure seeking and playfulness, all of which have a place in adult relationships.

It should be noted that there are further subtypes of these ego states but the basic, three-state structure given here is suitable for the examples that follow.

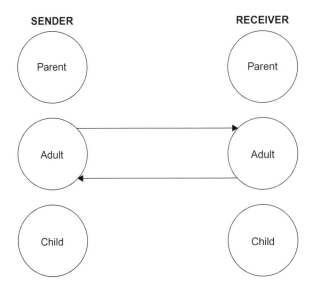

Figure 6.2 The parent, adult and child ego states of a sender and receiver, and an (A–A) (A–A) complementary transaction.

Now that we have established the three ego states that adults can occupy, we can begin to see how complex transactions between adults can play out, given that both parties may change ego states either spontaneously or as a result of a transaction initiated by the other party.

6.2.3 Complementary and crossed transactions

Figure 6.2 represents two people, the sender and the receiver, each having three ego states: parent (P), adult (A) and child (C) – this is a PAC figure. The transaction shown by the arrows is one that predominates in everyday life, whether it be at work or at home. This is known as the adult–adult complementary transaction (A–A)(A–A).

Below is an example of an A–A transaction in the flight deck. The ego states are given in parentheses at the end of each statement.

Captain	First officer
Please could you check my performance calculations? (A)	
	Yes, of course. (A)

Because the transactions complement each other, that is, the adult response is appropriate to the adult request, this is a socially acceptable transaction. Interactions that follow a complementary pattern can be sustained indefinitely.

Another type of complementary transaction is shown in Figure 6.3. In this case, the first person is in a parental ego state and the nature of transaction they initiate makes it likely that the second person will generate a response based on their child ego state. This transaction is (P–C)(C–P).

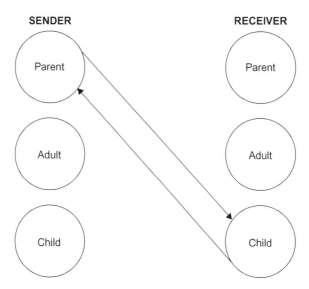

Figure 6.3 The parent, adult and child ego states of a sender and receiver, and a (P–C)(C–P) complementary transaction.

Do (P–C)(C–P) transactions occur in the flight deck? Let's consider a less likely (but still possible) transaction in which the captain starts off in a parental ego state, perhaps because of a personality tendency or as a result of his perceived dim view of the first officer's ability.

Captain	First officer
I checked your performance calculations and you made a serious mistake. What's wrong with you today? Why are you getting so much wrong? (P)	
	Why can't you just let me get on with my job instead of constantly interfering and distracting me? I might have more chance of getting things done if you didn't keep butting in! (C)
Listen, I'm the one in charge here. It's my neck on the line if this stuff is wrong. You clearly can't be trusted to carry out even the simplest task so I have to waste time checking on you. (P)	
	You're the only captain who acts like this with me. I really hate seeing you on my roster. (C)
And I'd be much happier if they rostered me with a first officer who knew what the hell he was doing. (P)	

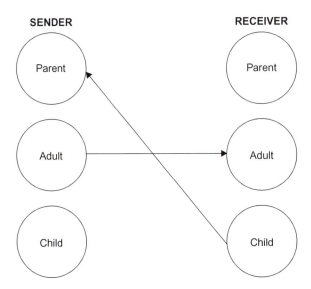

Figure 6.4 The parent, adult and child ego states of a sender and receiver, and a (A–A)(C–P) crossed transaction.

Although it may not seem it, this conversation is a sequence of complementary transactions. As you can see from the PAC figure, the transactions run parallel to each other and so, in the same way as (A–A)(A–A) can be sustained indefinitely, (P–C) (C–P), (P–C)(C–P), (P–C)(C–P) … can also be sustained indefinitely. The result of this in a flight deck setting is a rapid steepening of the flight deck gradient with the captain taking on an increasingly authoritative role. Lots of resources are now lost as the first officer is probably going to be very reluctant to point out threats or unsafe acts for the captain, either because he does not want to be criticized or, more worryingly, because he hopes that the captain will make an error that he can seize upon in order to score points. Although they could continue this argument indefinitely, it may be that it is occasionally broken by a tense, painful silence, only to start up again when someone initiates a transaction again. Berne uses complementary transactions as the basis of his first rule of communication: as long as transactions are complementary, communication can, in principle, proceed indefinitely. However, especially in light of the second example given, although the communication may continue indefinitely, it may turn out to be severely damaging to crew interaction.

The corollary to the first rule of communication is that the communication sequence can be broken in the event of a crossed transaction. Figure 6.4 shows a crossed transaction which starts with the adult ego state of the first person initiating a transaction directed at the adult ego state of the second person. In this case, for some reason, the second person reacts using their child ego state and directs their response to the parent ego state of the first person. This is an (A–A)(C–P) transaction.

The second rule of communication states that in the event of a crossed transaction, one person is going to have to change ego states for the conversation to continue.

In reality, this sort of crossed transaction will temporarily stop communication until the transactional vectors can be realigned. The easiest way, but not necessarily the best way, of turning this into a complementary transaction is for the first person to then respond using their parent ego state and direct it to the second person's child ego state; that is, to re-establish a complementary transaction by entering into the sequence given in the previous example of the argument in the flight deck. The transactional sequence might go as follows:

(A–A)(C–P)–[pause in communication]–(P–C)(C–P), (P–C)(C–P), (P–C)(C–P) …

The pause in communication occurs because the first person had been expecting an adult response from the second person but this has not happened. The first person now needs to make a quick assessment of why this is so and what he is going to do about it. In this example, they get back into a similar (P–C)(C–P) argument as before. Let's see how this could play out in the flight deck. In this scenario, the first officer is the handling pilot:

Captain	**First officer**
Based on our distance from the runway, I think we're a bit high on profile (A)	
	Who's flying this approach? I know what I'm doing and *some* of us are capable of flying fuel-efficient approaches. (C)
Pause in communication due to unexpected crossed transaction	
Who the hell do you think you're talking to? All I said was that you're high and now you're getting snappy. (P)	
	Seriously, just get on with what your doing and let me fly this approach. (C)
Listen to me. You're too high. I'm now *telling* you to slow down, start configuring and increase your rate of descent. (P)	
	You're really not capable of letting first officers do their job. You always have to interfere so that we're just an interface between you and the autopilot, even when we're the handling pilot. (C)
Yeah, that's really interesting! For now, just count the stripes on my shoulders and those on yours and then do exactly what I'm telling you. (P)	
	Whatever you say … captain! (C)

There may be a natural break in the argument at this point as they fly the approach but, once they have taxied on to their parking stand, you can probably imagine this pattern continuing as they try to "discuss" what transpired during the approach. The argument can continue indefinitely and could permanently damage their working relationship, all from one crossed communication early on [the (C–P) response from the first officer after the (A–A) enquiry from the captain]. Before we move on, we should consider what led the first officer to make this crossed transaction. Some reasons are given below:

- The captain's perceived criticism could have been the last straw; perhaps the captain had been constantly pointing out things that were wrong ever since they departed. Although this may have been done with the best intentions, this constant stream of suggestions and corrections may wear down a first officer to the extent that he makes a crossed transaction.
- In a high-workload environment, the first officer may have vocalized a thought that would otherwise be suppressed or at least modified before being expressed. Social discourse relies on us passing our initial thoughts through an internal filter to ensure that we will not be breaking social rules. This is one of the functions of the frontal lobe of the brain. When we are under high workload or when we are tired, this filter mechanism is less effective, meaning that we are more likely to let slip some of these "unfiltered" thoughts.
- It could have been intentionally provocative. Making a crossed transaction is a fairly reliable way of initiating an argument.

There are a few points that we have covered so far that are worth summarizing before we move on:

- Complementary transactions can continue indefinitely.
- (A–A)(A–A) transactions are what we should always aim for in the flight deck. They are complementary and they engage an ego state in both the captain and the first officer that is objective and logic driven.
- The (P–C)(C–P) transaction is complementary and can be sustained indefinitely but it is characterized by an argument, an undesirable change in flight deck gradient and diminishing crew resources as the atmosphere becomes more unpleasant. Sustained (P–C)(C–P) transactions seriously limit the effectiveness of crew interaction and, consequently, pose a significant risk to flight safety. They can also cause long-lasting damage to working relationships.
- A crossed transaction can end a sequence of complementary transactions. In the case of (A–A) transactions, a crossed response (C–P) may lead to an argument.
- People may make crossed transactions for several reasons; for example, they inadvertently express what they are feeling rather than modifying it to fit the social rules in force at the time. This may be more likely under high-workload conditions, conditions where it is even less desirable for the communication between pilots to deteriorate into an argument.

Understanding these concepts brings us to a crucial strategy to try and maintain reciprocal (A–A) transactions in the flight deck: in the same way as a crossed transaction can break an (A–A) sequence and cause an argument, another crossed transaction can recover the situation. The only way of having any hope of restoring an (A–A) sequence is for one person to override their desire to adopt the undesired ego state that the other person's communications are clearly aimed at, and to keep communicating from their adult ego state. In non-transactional terms, this could be called "not rising to the bait". One person's negative tone and manner is trying to force the other person to adopt a complementary ego state that will result in an argument. By resisting this provocation repeatedly by responding with a crossed transaction that is aimed at restoring the (A–A) sequence, it will soon be clear to the provoking party that the only way to restore a sequence of complementary transactions is for them to switch back to their adult ego state. Consider the following dialogue, which uses this strategy:

Captain	First officer
I've flown this approach about a thousand times. I know what I'm doing. Just let me get on with it. (C)	I think we're still a bit fast on this approach. Just for my information, could you tell me when you plan to start configuring the aircraft so we can be stable by 1000 feet? (A)
Pause in communication due to unexpected crossed transaction. The first officer is now in a difficult position because he is being forced into a parent ego state that will be quite counterintuitive because of his rank. It is better for him to maintain the adult ego state.	
Why don't you make yourself useful and monitor your shoelaces while I get on and complete this flight? (P)	I appreciate that you've flown this approach before, but I'd just like to know what your configuration plan is so that I can carry out my role as monitoring pilot. (A)
Another pause in communication. The captain has switched his own ego state from child to parent but this is still a crossed transaction. The captain is trying to force the first officer into a child ego state while the first officer is trying to force the captain into an adult ego state.	
OK, OK. I'd been aiming for 500 feet but 1000 feet might be better. Reducing speed to 190 knots. (A)	Captain, please. We're 8 miles from the runway and you're still at 250 knots. We should already have slowed down and started putting out flaps. You do recall that we need to be stabilized at 1000 feet and not 500 feet. If we're not stable, I'll have to call go-around. (A) We've got about 1500 feet to get stabilized so we'll have to configure quite quickly. Standing by with flaps. (A)

The first officer's tenacity in not giving in to the provocation from the captain and switching to a parent or child ego state meant that he could keep using (A–A) crossed transactions to restore the reciprocal (A–A) sequence. This is the type of transaction that both pilots in the flight deck should be aiming to use all of the time.

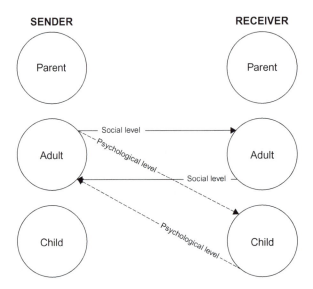

Figure 6.5 The parent, adult and child ego states of a sender and receiver, and an ulterior transaction.

6.2.4 Ulterior transactions

Some transactions engage more than one ego state in an individual. In these cases, there are usually two separate transactions at work. One is carried out on a social level (the words used) and a simultaneous one is conveyed on a psychological level (the implicit, additional meaning of the words). These are known as ulterior transactions. Berne uses the example of a salesman using an ulterior transaction with a customer. For example, the salesman could say, "This model, of course, is much more luxurious but you can't afford it". The salesman is, technically, obeying social convention and has initiated an (A–A) transaction on a social level. However, the psychological transaction is directed towards the child ego state of the customer, as "this is much nicer … but you can't have it". The customer can respond on an (A–A) by saying, "Yes, it is lovely and yes, you're correct in saying that I can't afford it". In so doing, he has resisted the psychological provocation implicit in the salesman's ulterior transaction. However, he may let his child ego state dictate his response and reply, "Actually, I think I will take this model". Although this is an (A–A) response on a social level, there is a (C–A) psychological motivation, which is "Don't tell me what I can and can't have. I'm as good as any of your customers". This sort of transaction is shown in Figure 6.5.

This brings us to the third rule of communication: the outcome of an ulterior transaction is based on the psychological level rather than the social level.

Ulterior transactions may happen in the flight deck, often associated with one-upmanship or grandstanding (trying to elicit admiration or applause from observers, often during the exercise of a skill). Consider the following transaction:

Captain	First officer
So you're not going to fly this approach manually? You're probably right. It is quite a tricky approach and you're probably out of practice. [This is socially (A–A) but psychologically (A–C) by implying that the first officer is not skillful enough to manually fly the approach.]	
	Actually, on second thought, I will hand fly this approach. [This is socially an (A–A) response but psychologically is (C–A), as if to say "of course I can do it!"]

In the example given above, the first officer's response was determined at the psychological level and, in some scenarios, this motivation could draw people into risky behaviors. Maybe it was inappropriate to manually fly this approach but because of the implicit provocation in the captain's statement, the first officer falls foul of the third rule of communication and undertakes a risky maneuver that he would not have carried out otherwise. It is possible to resist the third rule but it requires resolve because the provoking party may take this as a sign of weakness or evidence of a lack of skill. This is why any sort of point-scoring in the flight deck is a bad thing. It is bad for morale and may encourage pilots to take more risks than they would otherwise.

We can now summarize the three rules of communication with regard to some of the strategies that they offer us in managing flight deck communications:

1. As long as transactions are complementary, communication can, in principle, proceed indefinitely – We want to maintain a continuous sequence of (A–A) transactions in the flight deck in order to work together effectively.
2. In the event of a crossed transaction, one person is going to have to change ego states in order for the conversation to continue – A crossed transaction that disrupts the (A–A) sequence should be countered with another crossed transaction to restore it.
3. The outcome of an ulterior transaction is based on the psychological level rather than the social level – Do not conceal intentions or use psychological provocation. Always try to continue with (A–A) transactions at the social and the psychological levels.

6.3 Establishing a positive team atmosphere

In some smaller airlines, when the crew meet, they will probably all know each other. Some may even be good friends. In other larger companies, it may be that every duty is with different crew and it would be rare to work with the same people twice in the course of a year. Both of these scenarios present their own individual problems for team building. In the first case, it may be that one of the crew members who you know is someone that you do not particularly get on with. In the second case, you have to start the team-building process from scratch every time you start a new duty. One consideration that is important to both of these cases is the use of strokes. When we introduced transactional analysis, we defined a stroke as a unit of social action that can partly or temporarily satisfy the recognition hunger of someone else. A stroke could be as simple as saying hello, non-verbally acknowledging someone's presence or inquiring whether that person enjoyed their time off. Different actions will have different stroke values and our relationship with someone is partly predicated on how many mutual strokes we might expect from each other when we interact. It may be sufficient to politely smile at someone who you recognize from your apartment building. From then on, when you meet, it is likely that you will both smile politely. If, one day, that same person ignores you in the hallway, it is likely that you will feel slightly aggrieved. You had been expecting a stroke and you did not receive it. However, if that person stopped you in the hallway and began a spontaneous conversation inquiring about your health and that of your family, you might also think that strange. In general, the increasing intimacy of a relationship over time involves gradually elevating the number of strokes that occur at each encounter. If you had been away on holiday and, when you returned, your neighbor remarked on your tan and said how this explained why he had not seen you for a while, this could be perceived as a friendly, socially appropriate step-up in the relationship. The next time you meet, you may exchange a few pleasantries rather than just smiling politely at each other.

This idea of strokes can be useful when trying to integrate individual crew members into an effective team. The psychiatrist Leonard Zunin hypothesized that the first 4 minutes after two or more people meet will have a significant effect on the interpersonal relationships between the people involved.[6] If the appropriate strokes are not transacted in this time, this can leave one or more crew members feeling distant from the group. If the captain notices a crew member reporting to the briefing room and makes no effort to engage in a social transaction with them, this may lead to the crew member feeling socially short-changed. Given that many of the other strategies given in this book rely on empowering less senior crew members, the responsibility is on the more senior crew member to take the earliest opportunity to meet the rest of the team and establish a positive tone. After 4 minutes, this becomes a lot more difficult and it may be that a negative tone has already been set that will last for the entire duration of the work trip.

6.4 Communication strategies for effective briefings

Several psychological observations are worth considering before we look at effective briefings, especially with regard to threat management during briefings:

- Primacy/recency effect – Humans have better recall for information that is presented at the start and at the end of a briefing.[7]
- Levels of processing – Humans have better recall for information that has required a deeper level of processing. Thus, being told something requires relatively little processing on the part of the receiver whereas having to engage with that information on a semantic level (the meaning) in order to answer a question will require a deeper level of processing and so is more likely to be recalled.[8] For example, rather than telling the monitoring pilot that in the event of a go-around, he needs to set up a VHF omnidirectional radio range (VOR) radial that the handling pilot can track, it would be better to ask him what he thinks would be the best navigational aid to set up in the event of a go-around. In so doing, the monitoring pilot

has to check the instrument approach chart, ascertain the appropriate radial and then talk through how and when he would need to set this up. All of this requires a deeper level of processing and it is more likely that this information will be recalled later.

- Context and recall – We have better recall of information that was encoded in the same environment where it will need to be retrieved.[9] Fortunately, most briefings occur in the flight deck and so we can take advantage of this phenomenon.
- Chronological recall – Except for introducing and re-emphasizing important points at the start and the end of the briefing, the rest of the briefing should follow a structure that matches the chronology of the sequence of actions that the briefing refers to. For example, a departure brief should cover pushback, engine start, taxi routing, take-off (and rejected take-off) and then the standard instrument departure.
- Temporal contiguity and touch drills – Combining visual and verbal information at the same time aids recall, i.e. touching the relevant controls as you talk through a drill or maneuver that you may need to perform will make it more likely that it will be carried out correctly should it need to be.[10]

One particular guidance document concerning briefings stated that a thorough briefing should be performed regardless of how familiar the airports and procedures are or how often the crew members have flown together.[11] Based on what we know about human attention, this statement, although well intentioned, may be flawed. An effective briefing must strike a balance between being comprehensive and holding the listener's attention. Does an approach briefing for an approach into the home base of both pilots on a day when they have flown and briefed the approach already have to be as detailed as an approach into an airport that neither pilot has been to before? Carrying out a highly detailed approach briefing just for the sake of carrying out a highly detailed approach briefing is unnecessary and, ironically, introduces more risk into the flight deck because both crew members have now taken on an additional, unnecessary task. If this briefing is being carried out according to a standard script, if the other pilot has heard it before or determines that he already knows everything that is going to be said, his attention is likely to drift. To address this question of brevity versus comprehensiveness, we must first consider what the purpose of a briefing is. In short, a briefing is meant to synchronize the mental models of both pilots so that they have a shared understanding of how the aircraft is going to be maneuvered on the ground and in the air, what threats exist, how those threats are going to be managed, and what each pilot's actions are going to be under normal and (select) abnormal conditions. If both pilots already have a shared mental model, as demonstrated by an earlier departure or arrival, cannot identify any threats, and have already talked about each pilot's actions under normal and abnormal conditions, a departure/arrival brief could be fairly short. It may just require confirmation that the instrument approach charts are correct, the preprogrammed FMC data are valid and the relevant speed and altitude bugs are set and, for an arrival brief, a review of the go-around procedure. It is also worth noting that a briefing can be beneficial for the person leading it. By talking through the predicted sequence of events in chronological order, this can be an opportunity to mentally rehearse the actions that will be needed. Mental rehearsal has been shown to be highly beneficial in promoting recall of skilled-based actions.[12]

Based on the amended Threat and Error Management model (TEM2) proposed in Chapter 3 (Error Management and Standard Operating Procedures for Pilots), a departure or arrival briefing should focus on threat management and, taking account of the psychological phenomena listed in this section, a briefing should be carried out as follows:

1. Given that briefings often require decisions to be made, ensure that System 2 is able to dictate these decisions by carrying out briefings during low-workload periods.
2. Distractions and interruptions should be avoided wherever possible.
3. Ensure that the other pilot is ready to participate in a briefing.
4. Start by discussing any of the key threats that both of you have identified and back this up with the threat identification framework described in Chapter 3: low visibility, meteorological, NOTAMs, other traffic, performance, QNH, radiotelephony, systems, terrain and unfamiliarity (L-M-N-O-P-Q-R-S-T-U). Remember, this threat identification framework is just an example and can be revised and amended to include threats that are specific to your operation.
5. At this stage, management strategies do not need to be covered as they will be discussed when the impact of each threat is related to particular operational phases. For instance, there is little point talking about management strategies for windshear after take-off before discussing the potentially confusing taxi routing out to the runway.
6. Now that the major threats have been highlighted, the discussion that follows can refer back to them and to the strategies necessary to stop them affecting flight safety.
7. Discuss the forthcoming phases of flight in chronological order. For example, a departure briefing would cover pushback, engine start, taxi, take-off (or rejected take-off), climb and cruise; and an arrival briefing would cover descent, approach, landing (or go-around), taxi, park and shutdown.
8. To take advantage of the levels-of-processing phenomenon, every opportunity should be taken to make the other pilot think about what is being said. This could take the form of questions or asking him to describe his actions if a particular threat is encountered. Any procedures or maneuvers that may be needed should be rehearsed using touch drills; this will take advantage of temporal contiguity and make it more likely that these actions will be accurately carried out if the need arises.
9. When threats relating to a particular phase are relevant, use the avoid/buffer/contingency plan system to ensure that a threat management strategy is agreed upon.
10. By discussing actions that may be required in abnormal situations, the relevant production rules in procedural memory are partly activated (primed), making it more likely that they will be activated quickly should the need arise.
11. At the end of the briefing, take advantage of the primacy/recency effect by recapping the key points of the brief, especially the threats and the strategies that will be used to limit their effect on flight safety.
12. Check for understanding by asking questions and encouraging the other pilot to ask questions as well.
13. If time allows, unusual emergencies can also be discussed using the "What would you do if ...?" format given in Chapter 2 (Information Processing).[13]

6.5   Communication strategies for assertiveness

One of the recurring themes across several of the chapters in this book is that a flight deck crew can only function effectively if there is open and honest communication between both pilots. The unsafe act management section of TEM2 relies on either pilot being able to highlight an unsafe act to the other. These sorts of strategies are more difficult when they run counter to the flight deck gradient; that is, when they have to be directed upwards in the chain of command. The person with the hardest job in an airline is the new first officer who thinks that his overbearing captain has committed an unsafe act. Unfortunately, despite the advances we have made in human factors training in aviation, we still have a significant way to go in empowering first officers to be able to do this confidently, particularly in cultures where there is a high power-distance regard for authority as described in Chapter 5, Section 5.3.1. The list of accidents that could have been prevented by first officer (or other crew) assertiveness is disconcertingly extensive. Some of these cases are given in Box 6.1 along with brief descriptions.

First officer assertiveness is important at all three stages of threat and error management and there are different strategies that can be used depending on the stage.

Box 6.1 Accident Case Studies – Assertiveness

KLM Flight 4805 – Boeing 747-200
Tenerife, Spain, 27 March 1977[14]

During the take-off roll in low visibility at Los Rodeos Airport in Tenerife, the flight engineer expressed concern that there was another aircraft on the runway. The captain continued the take-off and the first officer did not intervene. The aircraft collided with another Boeing 747.

Birgenair Flight 301 – Boeing 727-200
Puerto Plata, Dominican Republic, 6 February 1996[15]

Owing to a blocked pitot tube, the crew received multiple contradictory speed warnings. The captain was the handling pilot and maneuvered the aircraft into a nose high attitude. The relief crew member in the flight deck made some subtle suggestions that the captain should refer to his attitude directional indicator (ADI) when the stick shaker activated to signify that the aircraft was near its stalling angle of attack. The first officer also did not counteract the captain's back pressure on the control column. The aircraft stalled and crashed.

Korean Air Flight 801 – Boeing 747-300
Guam, 6 August 1997[16]

During an approach into Guam, the crew of this aircraft ended up significantly lower on the approach than normal. Multiple ground proximity warning system warnings activated and the first officer and flight engineer both told the captain to go-around. The captain did not react in time and the first officer did not execute the go-around. By the time the captain elected to go-around, it was too late and the aircraft crashed into Nimitz Hill, killing 228 people. The report into the accident cited the first officer's and flight engineer's failure to challenge the captain's performance as causal to the accident.

Korean Air Cargo Flight 8509 – Boeing 747-200
Essex, UK, 22 December 1999[17]

Shortly after take-off, the captain's ADI malfunctioned and this caused a comparator warning as there was now a discrepancy between the two primary ADIs.

The captain was the handling pilot. The first officer canceled the comparator warning but did not comment on its significance. The report noted that there was a considerable difference in experience levels between the captain and the first officer and the captain had criticized the first officer a number of times before take-off. The captain proceeded to overbank the aircraft and the first officer did not comment or intervene. The flight engineer repeatedly expressed his concern over the increasing bank but neither pilot responded. The aircraft crashed into the ground with approximately 90 degree left bank.

Crossair Flight 3597 – RJ-100
Bassersdorf, Switzerland, 24 November 2001[18]
The captain of this aircraft continued to descend below minimum descent altitude despite not having visual contact with the runway. The first officer made no attempt to prevent the continuation of the flight below minimum descent altitude. The report noted the significant age and experience gap between the captain and the first officer. During the flight, the captain had lectured the first officer about the interpretation of a runway report, thereby placing him in the position of a pupil despite the fact that explanation was unnecessary. In the 24 seconds after descending through the minimum descent altitude, no speech or actions on the part of the first officer are documented. The aircraft crashed into the ground, killing 24 people.

Garuda Indonesia Flight 200 – Boeing 737-400
Adisucipto International Airport, Indonesia, 7 March 2007
See the accident case study presented in Box 3.1 (Chapter 3).

Air India Express Flight 812 – Boeing 737-800
Mangalore International Airport, India, 22 May 2010
See the accident case study presented in Box 7.1 (Chapter 7).

6.5.1 First officer assertiveness at the threat management stage

Threat management predominantly occurs during briefings. Ideally, both pilots will collaborate to identify threats that are relevant to the day's operation and suggest strategies for dealing with them. If it is the first officer who is leading the brief, there should be no difficulty for him to identify threats and suggest strategies to overcome them. If the captain is leading the brief or, more worryingly, does not conduct a brief, the first officer needs to find some way to draw attention to the threats that he considers relevant and obtain agreement from the captain about potential management strategies. Todd Bishop suggests using the following strategy to raise concerns in a non-accusatory fashion, and this has been adjusted slightly to fit in with the TEM2 model.[19] Ideally, flight deck communications should not require this sort of managed intervention but occasionally it may be required:

1. Attention – Get the commander's attention by using his name or title.
2. Concern – State that you are concerned.

3. Threat – State the threat that is concerning you.
4. Strategy – State which management strategy you think is best.
5. Agreement – Obtain agreement regarding your strategy.

For example:

1. Attention – David.
2. Concern – I'm concerned.
3. Threat – That thunderstorm is on our departure track.
4. Strategy – I think we should negotiate an early right turn before departure.
5. Agreement – How does that sound to you?

Remember, if a threat is successfully managed at the briefing stage, there is a much smaller chance that it will cause problems later on. In this case, if the first officer had not suggested negotiating an alternative departure routing prior to take-off, the crew could find themselves in a position of taking off into bad weather, having to accelerate the aircraft and retract the flaps, being switched over to departure control and waiting for a gap in the radio communications before they could ask for radar vectors around the weather. By that point, they could be very close to the bad weather or may have to take avoidance headings without approval from ATC and potentially come into conflict with other aircraft.

6.5.2 First officer assertiveness at the unsafe act management stage

If the first officer considers that an unsafe act has occurred, he may now need to escalate the intensity of his inquiry in case the unsafe act leads to an undesired aircraft state. This seems to be discomforting for some first officers, particularly if there is a steeper than normal flight deck gradient for either cultural or personality-based reasons. However, when we consider the situation practically, there are only three reasons that a first officer might think that an unsafe act has been committed:

• The captain has actually unintentionally committed an unsafe act.
• The captain may have intentionally committed an unsafe act but this was a rule-based violation aimed at trying to satisfy competing goals (such as an efficiency–thoroughness trade-off as mentioned in Chapter 3). In this case, the captain is diverging from the standard operating procedures (SOPs) for a potentially legitimate reason but has not briefed the first officer.
• The captain has not made an error and is complying with the SOPs but the first officer is, for some reason, unaware of this particular SOP or has simply made an error himself.

No matter which of these three reasons is behind the first officer's discomfort, raising his concern can only have a positive outcome:

• In the case of an unintentional unsafe act, the captain can correct it immediately.
• In the case of an intentional rule-based violation, the captain will have to justify his decision to diverge from the SOPs. The first officer might then be in a position to suggest an alternative, safer, more procedurally compliant course of action.
• If the captain has not made an error, the first officer will either be made aware of the relevant SOP or can have his own error highlighted to him, thus minimizing the risk of repeating it.

There is no way of knowing in advance what the reason behind the first officer's discomfort is. The only things that would stop him pointing out an error would be:

- Fear of embarrassment at not knowing a procedure
- Being concerned that the captain will not like having his mistakes pointed out to him.

The first reason is a legitimate concern but one that reflects the increasing complexity of the systems within which we work. Every single pilot will, at some point, have missed something and it is highly unlikely that there is a pilot who knows every single procedure, rule, amendment and temporary technical instruction in their company. In the same way as humans will always be prone to making errors, not always being 100% up to date with every single procedure is simply a fact of life. There is no justification for a captain being upset at having his mistakes pointed out. It is the reason we have two pilots in the flight deck and is at the very heart of effective Threat and Error Management. It is up to the captain to set the tone whereby a first officer can raise his concerns without fear of embarrassment or recrimination, and some strategies for achieving this are covered later, in Section 6.5.4. For the time being, we can consider another strategy that a first officer can employ to clarify the nature of the captain's action (or inaction) that has made him uncomfortable. Besco suggests an escalating strategy, based on the acronym PACE, that can be useful at the unsafe act management stage[20]:

1. Probe – Ask questions to gain a better understanding of what is happening. This will help to determine which of the three reasons has caused the first officer's discomfort.
2. Alert – Warn the captain of the anomalies that have occurred.
3. Challenge – Challenge the suitability of the current course of action.
4. Emergency – Give an emergency warning of critical and immediate dangers.

Hopefully, the situation will be resolved at the probing stage. This will clarify whether an error has been made and who has made it. If this is disregarded, there is an escalating sequence of statements that can further draw the captain's attention to the deteriorating situation. Unlike the previous strategy, PACE is an escalating strategy and the first officer would move up a step if he is not satisfied by the response to the previous step. For example:

1. Probe – David, I notice that you're flying this approach 30 knots faster than I was expecting. Is there a reason for this?
   Captain: Trust me, OK?
2. Alert – David, we're well above the standard speeds for this approach.
   Captain: Seriously, it's fine.
3. Challenge – David, we're far too fast. If you don't slow down, I'm going to have to call for a go-around.
   Captain: We're fine. I'm just doing this because we're late.
4. Emergency – We're too fast and now we're unstable. Go-around!

If unsafe act management has reached the stage where the first officer has had to issue instructions to the captain, something has gone badly wrong. However, if the captain decides not to listen to the first officer and does not execute a go-around, the aircraft is now in an undesired aircraft state and the first officer needs to do something about it.

6.5.3   First officer assertiveness at the undesired aircraft state management stage

If the situation has reached the stage where the aircraft is in an undesired aircraft state because the captain has failed to respond to the escalating warnings from the first officer, there is now only limited time to avoid the undesired aircraft state resulting in a negative outcome (an accident, an incident or a near-miss). If the captain has not responded appropriately, the first officer will be forced into a position where he needs to take action. This may be giving a clear command or, in dire situations, taking control of the aircraft. If we consider the last scenario from Section 6.5.2, where the first officer was using PACE to alert the captain about his inappropriately high approach speed, if his emergency warning is not heeded and the aircraft's approach is now unstable, the first officer is now faced with a dilemma. If he backs down and allows the captain to continue an approach which he deems to be unsafe, his function as a safety resource is now compromised. While it is clearly the captain's fault if he ignores the first officer's warnings and allows the aircraft to land too fast and potentially run off the end of the runway, the question is, what more can the first officer do? Below is a brief summary of what we have considered so far plus some additional strategies if the aircraft is in an undesired state:

*   If the first officer has identified a threat that could lead to an unsafe act, he can use attention–concern–threat–strategy–agreement to mitigate this risk.
*   If this does not work, and the captain makes an unsafe act, the first officer can manage this using the probe–alert–challenge–emergency (PACE) framework.
*   If the captain ignores the emergency warning, the first officer can:
    *   Repeat the warning.
    *   Give a command: "go-around", "increase thrust", "pitch down", etc.
    *   Perform an action that, while not jeopardizing flight safety, will allow recovery from the unsafe state; for example, informing ATC that the aircraft is going-around or increasing the selected speed on the mode control panel.
    *   Take control: This is the last resort as it implies that the captain is, in some way, incapacitated and incapable of performing his duty. The job of recovering the aircraft to a safe flying condition is now in the hands of the first officer and he may not get any support from the captain.

These emergency actions are only appropriate if the aircraft is in an undesired aircraft state and the intervention that the first officer makes is the most conservative course of action. They would be entirely inappropriate if they did not represent the safest, most conservative course of action; for example, the captain wanting to fly a go-around and the first officer intervening to continue the approach.

6.5.4   How captains can encourage assertiveness

It is the captain's duty to facilitate the first officer in his role as second-in-command. He can do this at all three stages of the Threat and Error Management process. During the threat management stage, he can accomplish this by leading or facilitating an interactive briefing. At the unsafe act management stage, he can ensure that a tone has been set that allows the first officer to speak up any time he thinks he detects a problem. Based on the two most common reasons why a first officer might feel uncomfortable (the captain unintentionally committing an unsafe act or intentionally committing a rule-based violation without briefing the first officer), the captain can encourage the first officer's input at the briefing stage by explaining that any uncertainty that he may have is down to the captain either making an error or not briefing him sufficiently. This sets a tone whereby the first officer will feel comfortable to highlight any doubts that he has. The captain should always be sensitive to any expression of concern on the part of the first officer. As we covered in Chapter 2 (Information Processing), specifically, the effect of the overconfidence heuristic on decision making, a captain may mistakingly feel entirely justified in his chosen course of action and it may only be an alert from the first officer that triggers a re-evaluation of the situation.

If none of this has worked and the aircraft is now in an undesired state, the captain may find himself in a very difficult position. The first officer may now feel that he has no option but to give commands. He may even feel the need to take control of the aircraft. In most cases, this will be an extremely difficult decision for the first officer and requires a great deal of courage. The captain must now re-evaluate how bad the situation must be for the first officer to take such drastic action. Although there is likely to be some embarrassment on the part of the captain, if the first officer has had to take this course of action, the captain should respect this and try not to make the situation worse. For better or worse, the captain is now in the subordinate position but should be professional enough to assist and recover the aircraft to a safe and stable condition.

Chapter key points

1. The Sender–Message–Channel–Receiver (SMCR) model explains the dynamics of communication.
2. Although the majority of the information of a message is encoded verbally, the non-verbal aspect is primarily used to interpret the emotional meaning behind the message.
3. Feedback is a useful way of confirming that a message has been communicated correctly.
4. The SMCR model can also be used to identify barriers to communication, many of which can be problematic in aviation.
5. The NITS (Nature, Intentions, Time, Special instructions) briefing is a useful way of overcoming several of these barriers because of its simple format.
6. Communication between two people can also be analyzed using transactional analysis.
7. People can display three ego states: parent, adult and child.
8. In aviation, it is preferable for both pilots to remain in their adult ego states.
9. A complementary transaction is one where the message is generated from the sender's adult ego state and directed to the receiver's ego state, and the receiver responds accordingly.
10. A complementary transaction may also occur when the sender generates a message using their parent ego state and directs it to the receiver's child ego state, and then the receiver's child ego state directs their response to the original sender's parent ego state. In the flight deck environment, this would probably lead to an argument.
11. Complementary transactions can continue indefinitely and a crossed transaction is needed to stop them.

12. Ulterior transactions occur when two messages are conveyed at the same time: one on a social level and one on a psychological level. The outcome of such a transaction is usually based on the psychological level of the message rather than the social one.

13. The concept of strokes in transactional analysis is important when establishing and maintaining a positive team atmosphere.

14. When meeting members of a team, the first 4 minutes will have an important role in determining how effective that team will be.

15. There are several psychological phenomena that we can use to our advantage in carrying out an effective briefing, especially with regard to threat and error management.

16. Many crashes have occurred where the first officer was aware of the deteriorating situation but was not assertive enough to intervene.

17. First officer assertiveness is vital in ensuring flight safety and there are specific techniques that can be used during threat management, unsafe act management and undesired aircraft state management.

18. Captains should make an effort to encourage first officers to be assertive.

References

1. National Joint Committee for the Communication Needs of Persons with Severe Disabilities. Guidelines for meeting the communication needs of persons with severe disabilities [Guidelines]. Retrieved from: <http://www.asha.org/policy/GL1992-00201.htm/>; 1992.

2. Berlo DK. *The process of communication: An introduction to theory and practice*. New York: Holt, Rinehart, & Winston; 1960.

3. Mehrabian A. *Silent messages: Implicit communication of emotions and attitudes*. Belmont, CA: Wadsworth; 1981.

4. Eurocontrol. Guidelines for controller training in the handling of unusual/emergency situations. Retrieved from: <http://www.skybrary.aero/bookshelf/books/15.pdf/>; 2003.

5. Berne E. *Games people play: The psychology of human relationships*. New York: Ballantine Books; 1964.

6. Zunin LM, Zunin N. *Contact: The first four minutes*. New York: Ballantine Books; 1989.

7. Ebbinghaus H. *Memory: A contribution to experimental psychology*. New York: Teachers College, Columbia University; 1913.

8. Craik FI, Lockhart RS. Levels of processing: A framework for memory research. *Journal of Verbal Learning and Verbal Behavior* 1972;**11**(6):671–84.

9. Smith SM, Vela E. Environmental context-dependent memory: A review and meta-analysis. *Psychonomic Bulletin & Review* 2001;**8**(2):203–20.

10. Ginns P. Integrating information: A meta-analysis of the spatial contiguity and temporal contiguity effects. *Learning and Instruction* 2006;**16**(6):511–25.

11. Airbus. *Flight operations briefing notes: Standard operating procedures – conducting effective briefings*. Retrieved from: <http://www.airbus.com/fileadmin/media_gallery/files/safety_library_items/AirbusSafetyLib_-FLT_OPS-SOP-SEQ06.pdf/>; 2004.

12. Driskell JE, Copper C, Moran A. Does mental practice enhance performance? *Journal of Applied Psychology* 1994;**79**(4):481.

13. Martin WL, Murray PS, Bates PR. What would you do if … ? Improving pilot performance during unexpected events through in-flight scenario discussions. *Aeronautica* 2011;**1**(1):8–22.

14. Netherlands Aviation Safety Board. (n.d.). *Final report and comments of the Netherlands aviation safety board of the investigation into the accident with the collision of KLM*

flight 4805, Boeing 747-206B, PH-BUF and Pan American Flight 1736, Boeing 747-121, N736PA at Tenerife Airport, Spain on 27 March 1977. (ICAO Circular 153-AN/56).

15. Flight Safety Foundation. Erroneous airspeed indications cited in Boeing 757 control loss. *Accident Prevention* 1999;**56**(10):1–8.

16. National Transportation Safety Board. *Aircraft accident report: Controlled flight into terrain, Korean Air Flight 801, Boeing 747-300, HL7468, Nimitz Hill, Guam, August 6, 1997* (NTSB/AAR/-00/01 PB00-910401); 2000.

17. Air Accidents Investigation Branch. *Report on the accident to Boeing 747-2B5F, HL-7451 near London Stansted Airport on 22 December 1999* (Air accident report 3/2003); 2003.

18. Aircraft Accident Investigation Bureau. (n.d.). *Final report no. 1793 by the Aircraft Accident Investigation Bureau concerning the accident to the aircraft Avro 146-RJ100, HB-IXM operated by Crossair under flight number CRX 3597, on 24 November 2001 near Bassersdorf/ZH.*

19. International Association of Fire Chiefs. Crew resource management: A positive change for the fire service. Retrieved from: <https://www.iaff.org/06news/NearMissKit/6.%20 Crew%20Resource%20Management/CRM.pdf/>; 2003.

20. Besco RO. PACE: Probe, alert, challenge, and emergency action. *Business and Commercial Aviation* 1999;**84**(6):72–4.

Fatigue risk management

7

Chapter Contents

Introduction   189
7.1 Introduction to sleep   190
7.2 Fatigue   192
   7.2.1 Cognitive effects of fatigue   193
7.3 Role of sleep in managing fatigue   195
   7.3.1 Physiology of sleep   196
   7.3.2 Sleep inertia   199
   7.3.3 Homeostatic sleep drive and sleep need   200
   7.3.4 Biological clock, circadian rhythms, chronotypes and sleep urge   201
         7.3.4.1 Biological clock and circadian rhythms   202
         7.3.4.2 Chronotypes   203
         7.3.4.3 Melatonin   203
         7.3.4.4 Sleep urge   204
   7.3.5 Sleep debt   206
7.4 Fatigue risk management strategies   210
   7.4.1 How to achieve sufficient high-quality sleep   211
         7.4.1.1 Planning your sleep   211
         7.4.1.2 Sleep hygiene   212
   7.4.2 How to nap effectively   213
   7.4.3 How to deal with insomnia and sleep disorders   214
         7.4.3.1 Chronic insomnia   215
         7.4.3.2 Obstructive sleep apnea   215
   7.4.4 How to mitigate risk if you find yourself fatigued   216
   7.4.5 How to use sleep medications   218
   7.4.6 How to deal with jet lag   219
   7.4.7 Organizational strategies for fatigue risk management   222
Chapter key points   222
Recommended reading   223
References   223

Introduction

Unless you are under the influence of alcohol or drugs, nothing will affect your performance as much as fatigue. Of all the human factors subjects in this book, fatigue is probably the most misunderstood and yet the one where improving our knowledge of it can have a massive impact on safety. Despite the fact that we spend approximately one-third of our lives asleep, the reasons for this and the strategies that can be used to achieve high-quality sleep are widely misunderstood. For many, sleep seems to be one of those annoying chores that gets in the way of everyday life. Given the nature of the

Practical Human Factors for Pilots. DOI: http://dx.doi.org/10.1016/B978-0-12-420244-3.00007-8

industry within which we work, having a deeper understanding of the science of sleep is a crucial step in creating a safe and efficient operation, and this is the goal of this chapter. As you will notice, this chapter is quite long but the amount of space given over to this subject is a reflection of how important it is in improving performance. More than any other guidance in this book, if every crew member were to adopt the sleep strategies presented here, there would be a huge improvement in both safety and efficiency in the aviation industry. No pills, no simulators, no computer-based training; just sufficient amounts of healthy, natural sleep.

7.1   Introduction to sleep

Although animals and humans have always needed sleep, the reason for this is still subject to debate. In evolutionary terms, sleep puts organisms at tremendous risk because they are more vulnerable to predators. If sleep were something that we could manage with less of, we would have evolved to manage on less sleep millions of years ago. However, sleep is a biological necessity. It is only really in the past 100 years that we have achieved any understanding at all about the purpose of sleep. But before we get into the reason why we sleep, let's talk about what we know about why we *need* sleep.

The answer to this question may seem deceptively simple. We need to sleep because we get tired. At this point we have to define two physical states of an organism: one state is being awake and the other state is being asleep. The greater the period of time awake, the greater the need to sleep. While this does not seem like a particularly profound statement, it is at the heart of our understanding of the importance of sleep and also leads to the most effective strategies for managing sleep. Let's first look at what happens when we are in the "awake state". The range of activities that humans can undertake when they are awake is broad. When we consider the different jobs that people carry out we can begin to see what the effect on the body might be. Take, for instance, someone who works in the construction industry. If we assume that he starts work at 9 a.m. in the morning and finishes at 5 p.m. in the evening and lives 30 minutes from his place of work, we can start to construct a model of this worker's day. He will probably get up at about 7.30 a.m., spend an hour having breakfast and attending to personal hygiene, and leave for work at about 8.30 a.m. We can imagine that his work is physically strenuous. There may be a lot of moving of heavy materials, climbing up and down ladders, maintaining balance using postural muscles to move bricks into place, and other physical tasks of a similar nature. Even accounting for breaks (including lunch), by 5 p.m. this particular person has undertaken a lot of physical activity.

We know from our understanding of muscle physiology that physical activity requires energy. The muscles, like some other parts of the body, store energy in a readily available form. There are other energy stores that can be used as necessary, although they are not as easily accessible as the first kind. As the day goes on, the energy stores that are readily accessible become depleted. As well as this, depending on the extent of the physical activity, the muscles themselves may incur damage. This is not necessarily noticeable, but takes the form of microdamage (as the name suggests, tiny areas of damage, which are eventually repaired). As time goes by, the worker's muscles will become tired. After an extended period, he may notice that his performance and strength start to decrease as energy stores are used up and microdamage accumulates. During his lunch break, depending on what he has for lunch, he may find that he has some additional energy when he begins working in the afternoon. What has happened is that some of the energy that he had used up has been replaced, some of the areas of microdamage will start undergoing repair and he is now able to continue for longer than if he had not taken a lunch break at all. When work finishes at 5 p.m., his muscles have been used for approximately 7 hours. Energy stores are depleted, microdamage has occurred, and he may have the feeling of muscle tiredness. Fortunately, there is now a break of approximately 16 hours before his muscles are going to be required to exert themselves to the same extent again.

The example that I have just given illustrates some of the phenomena that relate to the purpose of sleep. While sleep certainly has a role in helping muscles to recover from microdamage as it is during this period when they are not being used, a similar thing is happening in the brain. With reference to Chapter 2 (Information Processing), let's talk about what happens in the brain during the period that it is awake.

The brain makes up about 1–2% of the body weight of a normal human being. Paradoxically, every time the heart beats to pump blood around the body, the brain uses 20% of the blood that is pumped out. When you consider the size of the brain in relation to how much blood it requires, it becomes clear that it is a particularly demanding organ. The reason for this is very simple: the brain is extremely active and any form of activity requires energy, energy that we generate using the oxygen and glucose in our blood. As a factor of size, the brain is the most active organ in the body. With approximately 86 billion neurons to feed, the brain has four major arteries that supply it with large quantities of blood. Let's now remind ourselves of the role of neurons in brain activity. The majority of the neurons in the brain are interneurons, neurons that connect multiple other neurons together. In order for these neurons to communicate with each other, electrochemical signals are transmitted from one end to the other by a process of pumping electrically charged chemicals in and out of the cell in a progressive manner that conducts the impulse along the neuron. At the junction between two neurons (the synapse) the normal electrochemical transmission cannot jump the gap between the two cells. Instead, the first neuron will release chemicals known as neurotransmitters across the gap and these neurotransmitters will, in turn, stimulate the second neuron. These neurotransmitters are stored in compartments at the terminal end of neurons and once they are released and have stimulated the second neuron they are broken down to limit their effect. The by-products of this process can accumulate in the brain tissue and must eventually be cleared. All of these processes rely on energy. Because this is happening constantly when we are awake, in the same way as with the construction worker's muscles, as the day goes on energy stores become depleted, by-products accumulate and, over an extended period, performance will decrease. Since the processes we have described occur in every one of the 86 billion neurons in the brain, in the same way that muscles become tired as they are used, so the brain becomes tired and requires time to restore itself. This brings us on to the concept of fatigue.

7.2 Fatigue

Muscles will feel fatigued when they have depleted their energy stores and when they have started to accumulate microdamage. The brain gets fatigued in a similar way. Because the brain is such an active organ it will become fatigued in a more consistent, predictable way, unlike muscle fatigue. For example, someone using their muscles all day, perhaps someone who works in construction or another physically strenuous job, will probably experience more muscle fatigue by the end of the day than someone who works in an office. While there will be some differences in the subjective feeling of fatigue between people who work in different professions, the level of activity of the brain is fairly consistent from person to person and so cognitive fatigue will occur in a similar way and at a similar rate for most people. Cognitive fatigue is the term I will use to describe the subjective feeling of mental tiredness.

As we have said, if the brain has been active for a period of time there will be an accumulation of the by-products of neurotransmission as well as other potentially toxic chemicals (neurotoxins). If this process were allowed to continue, there would be a noticeable drop in the transmission speeds between neurons that would lead to an overall decrease in the processing power of the brain. There is some evidence to suggest that the brain also stores energy in the form of glycogen (a chemical that can be easily broken down into glucose) and that these stores are depleted when the brain is awake, in much the same way as the muscles use their energy stores when they are active.[1] The purpose of sleep is to address this decrease in performance and depletion of energy stores before they reach the stage of affecting the safety of the individual. Before moving on to how sleep accomplishes this, let's talk about the state that we are trying to fix – fatigue.

The International Civil Aviation Organization (ICAO) defines fatigue as:

> *a physiological state of reduced mental or physical performance capability resulting from sleep loss or extended wakefulness, circadian phase, or workload (mental and/or physical activity) that can impair the crew member's alertness and ability to safely operate an aircraft or perform safety-related duties.*[2]

Fatigue is a fact of life. At the end of a busy day, it is acceptable to be fatigued as we know that sleep will fix the problem. The other thing to stress here is that fatigue is a physical state and not a psychological state. If people try to deceive themselves (or, more worryingly, try to deceive others) about the impact that fatigue has on them, then they have misunderstood one of the most basic biological requirements that humans have. While fatigue is most certainly a brain issue, it is not "all in your head"! You cannot remedy fatigue by having a positive mental attitude, no matter how good it is. Humans need only a few things in sufficient quantities to survive: oxygen, food, water and sleep. None of these are optional.

The ICAO definition mentions a few other terms that we will need to address, such as wakefulness, circadian phase and workload. Workload has been addressed in Chapter 2 (Information Processing) from the perspective of cognitive workload and that definition works here as well. Wakefulness simply means when the individual is awake, and extended wakefulness suggests a longer period of being awake than would be normal for that individual. Circadian rhythms will be explained later in this chapter.

Based on the ICAO definition, there are two main contributors to fatigue:

- Sleep loss – If we were expecting to be able to sleep for 8 hours and only slept for 4 hours, it is possible to wake up and still be fatigued.
- Extended wakefulness – If we assume that we had a full and restful night's sleep the night before and have awoken fully replenished, if we attempt to stay awake for a long period of time we will eventually deplete our energy stores and accumulate neurotoxins which will lead to cognitive fatigue and a decrease in cognitive performance. If the day has involved a higher than normal level of workload, either mental or physical, or both, then these effects will be accentuated.

The circadian phase can have an impact on both of these factors, and we will see what this means shortly.

7.2.1   Cognitive effects of fatigue

Given that the brain's activity when we are awake leads to a depletion of energy stores and the accumulation of neurotoxins, why is this a problem for us? As stated in the introduction to this chapter, there is nothing except for alcohol and drugs that will affect your performance as much as fatigue. When the brain has been awake for a prolonged period, energy depletion and toxin accumulation means that every one of the 86 billion neurons in your brain will start to perform less well and this leads to a variety of effects. These effects are wide ranging but we can look at some of them in relation to how they affect other human factor concepts covered in this book. In so doing, you will see that fatigue has an impact on just about every aspect of human performance. Probably the most significant impact is in our ability to process information, with fatigue affecting all five stages described in Chapter 2 (Information Processing):

1. Sensation – Prolonged, fatiguing use of sensory organs in a constant way can lead to some negative effects. For example, asthenopia (otherwise known as eyestrain) may occur after a prolonged period spent focusing on an object a fixed distance away from the eyes, usually an object that is quite close. This is particularly common in people who have to spend a long time in front of a computer screen. Symptoms include dry eyes, headache and blurred vision. While it is not common in flight crew, it can occur in other members of staff who may have to spend extended periods in front of a computer. Fatigue may also cause double vision owing to inaccurate coordination of the muscles controlling the movement of each eye, resulting in a misalignment.
2. Attention – Fatigue will have an impact on all three types of attention: selective, sustained and divided. Because our attentional mechanism requires so much of our cognitive capacity, it is very sensitive to fatigue. A fatigued individual will have more difficulty attending selectively to a particular active task such as carrying out a performance calculation, often finding that his attention will stray from the task. It will also become a lot more difficult to sustain attention, for example, during a monitoring task; as time goes by, there will be an increasing risk of missing a signal that would normally be detected.[3] The multitasking abilities that are possible because we are able to divide our attention will also be reduced as we will find it more difficult to switch from one activity to another while maintaining a mental model of the current state of each of the respective tasks.

3. Perception – It appears that fatigue leads to an impairment in the quality of the mental models that we construct. As mentioned in Chapter 2, perceptual illusions are extremely convincing and we have few strategies available to prevent these from causing us problems. When we are fatigued, our ability to predict and plan for situations where these illusions may occur is degraded.[4]

4. Decision making – Because of the neural mechanics discussed in Chapter 2, it is this crucial part of information processing that is most significantly affected by fatigue. To briefly summarize this mechanism, an aspect of the mental model that requires a decision to be made is selected, then the parameters are matched to stored production rules and declarative chunks of information until a match is found and the decision is made. Depending on the workload at the time, the number of attempts to find the best match between a stimulus and a potential response may be limited. System 1 will attempt to come up with a decision as quickly as possible using preprogrammed heuristics that dictate what information is most likely to be retrieved from declarative memory and which production rules are used. If time and workload allow, System 2 can then modify the output of System 1 using all the processing power of the cerebral cortex to refine the decision and avoid many of the errors associated with the rapid, "rough-and-ready" mechanism of System 1. In the fatigued brain, not only does it take longer for information to be retrieved and for matching to occur, it is also more likely that, in the face of the processing limitations caused by the fatigue, the brain will accept the output of System 1 with little or no modification by System 2. The result of this is that when we make decisions in a fatigued state they are slower and more prone to the systematic errors that are associated with System 1. In short, our brain cannot be bothered to waste time and energy coming up with the best answer and, instead, comes up with an answer it thinks is "not great, but good enough". If we reconsider the heuristic of overconfidence, System 1 wants us to think that our knowledge and the appropriateness of our actions are better than they actually are. If we do not modify this instinct using System 2, we are led into riskier behaviors than we would have performed had we been well rested. Think of it like a lazy colleague who will do the bare minimum to get by, and take ill-advised shortcuts with little thought as to the long-term consequences.

5. Response – Given that neuronal transmission and communication between neurons are slower as a result of fatigue, any response is likely to be slower in the fatigued individual than in someone who is well rested. In many industries, split-second reaction times are not always important but in commercial aviation they may be crucial in maintaining a safe operation. Being fatigued and having to perform tasks during times when it would be normal to be asleep also increases the risk of making skill-based errors.[5]

If we look at some of the other topics in this book, we can see that fatigue will have a big impact on most of them. Owing to the impairment of our information processing capability, we are going to be less able to identify threats, it is more likely that we will commit unsafe acts and our subsequent management of these unsafe acts is likely to be worse than had we been well rested. The overinfluence of System 1 and the increase in the likelihood that we will take more risks may lead to us diverging from standard operating procedures (SOPs). In order to use SOPs, they must be recalled and applied to the situation at hand. The fatigued brain may find it less effort to ignore SOPs and simply respond to a situation in as simple a way as possible. The negative effects of automation complacency, together with impaired monitoring of automation, can lead to abnormal flight conditions. Communication between crew members reduces when one or both of the individuals are fatigued, and this may lead to missing standard callouts or not making the appropriate monitoring callouts that the situation requires. Finally, because fatigue is generally perceived as a negative sensation, the impaired mood of the fatigued crew member may have an impact on his team-working abilities and, if he happens to be in command, this can easily result in a negative flight deck environment and inhibition of other crew members' assertiveness.

In summary, fatigue actually makes you more stupid: your brain works more slowly, you communicate less and your behavior will alter so that you are likely to become someone who other people do not want to work with. You can drink all the coffee you want, you can switch on the radio to try to keep yourself awake as you drive home, you can tell yourself that fatigue is only in your mind, but none of that is going to solve your problem. There is only one cure for fatigue: sleep.

7.3   Role of sleep in managing fatigue

While sleep has some very important functions for the body, such as allowing for the repairing of damage, allowing growth and regulating the immune system, it is the effect on the brain that is the most important. First of all, we will talk about the effects of sleep and then we will talk about how sleep accomplishes these. In the same way as sleep replenishes the body, it also replenishes the brain. Used neurotransmitters are broken down and cleared from the synaptic junctions between neurons, and the brain uses a clever mechanism for clearing itself of these byproducts. Recent research has shown that certain cells in the brain shrink when we are asleep and this permits the intercellular fluid to wash through the brain, clearing out neurotoxins and other unwanted material.[6] The channels through which this occurs are known as the glymphatic system and this cleaning process is thought to be one of the mechanisms that explains the restorative nature of sleep. Interestingly, one of the toxic substances that is removed by the glymphatic system is that responsible for the development of Alzheimer's disease, namely β-amyloid. This cleaning process is the "housekeeping" purpose of sleep.

There is another reason why human beings sleep. Given that during the day we encounter huge amounts of sensory information, learn new things, make many decisions that involve matching stimuli to responses and, in short, go to sleep having processed a lot of information, we do not start the following day back at square one. The wonderful thing about humans is their ability to learn from new experiences. For example, as we discussed in Chapter 2 (Information Processing), when we encounter new stimuli, the mechanism whereby our pattern-matching system retrieves information from declarative memory and attempts to use production rules to generate a response may lead to the combination of multiple production rules into a new, more comprehensive one that is better suited to dealing with that particular stimulus. Using this example, we have generated a new chunk in our declarative memory (e.g. "this combination of warning lights means that there is a fault with a radio altimeter") and a new production rule ("if these warning lights illuminate, disconnect the autothrottle"). Although these chunks and rules become part of their respective memory stores, there is no guarantee about how securely they will be embedded in those stores. We do occasionally learn new bits of information which we then forget. One of the theories of sleep is that memories are reinforced and embedded in their stores by reactivating the respective neurons in a synchronous fashion so that they become more strongly associated with each other (remember, "cells that fire together, wire together") meaning that they can be more quickly and reliably activated and used in the future. This process is known as memory consolidation.[7] Although you might assume that this process is related to dreaming, it actually appears to be a completely separate phenomenon as we will describe later.

In summary, there are two purposes of sleep with regards to the brain:

- housekeeping – clearing out neurotoxins and replenishing energy stores
- memory consolidation.

Let's see how these happen by looking at the physiology of sleep.

7.3.1 Physiology of sleep

Sleep is divided into several different stages based on the level of brain activity. It is possible to measure brain activity by looking at the electrical activity of neurons using electrodes attached to the scalp, a procedure known as electroencephalography (EEG). Although the timing and location of neural activity are based on the task or stimulation being dealt with at the time, there is a certain amount of synchronization between groups of neurons within close proximity to each other in terms of their electrical activity. This synchronous activity results in brainwaves or, more precisely, neural oscillations. Like any other wave, brainwaves have amplitude (the height of the wave) and frequency (the number of waves occurring per second). The frequency of the brainwaves can tell us about the activity of the neurons causing them.[8] Different tasks may lead to different frequencies being seen in the neurons in particular brain regions, but if we take a global view of the brain, the general frequency of the brainwaves detected will show whether the person is awake or asleep and, if they are asleep, which stage of sleep they are in. Figure 7.1 shows some examples of different brainwave patterns associated with different levels of alertness and different stages of sleep.

Before we get into the specifics of what the different stages of sleep mean, let us look at a visual representation of how a normal adult transitions from being awake to being asleep, the different stages they pass through when they are asleep and how they then transition from being asleep to being awake. This is known as a hypnogram, an example of which is shown in Figure 7.2. As you can see from the figure, we normally pass through all of the stages several times throughout the night and spend a different amount of time in each stage depending on how long we are asleep. The general structure of our path through the various stages of sleep is known as the "sleep architecture" and maintaining this architecture is important for the quality of our sleep. Broadly speaking, there are two types of sleep:

- rapid eye movement (REM) sleep
- non-rapid eye movement (NREM) sleep – this type of sleep is broken into several stages.

Let's start by looking at NREM sleep and how we transition from being awake to being asleep. We will look at each of these stages in turn and how they correlate with

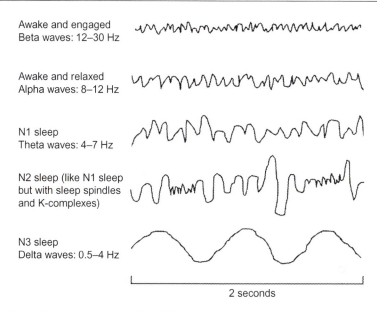

Figure 7.1 Brain waves characterizing different stages of alertness and sleep.

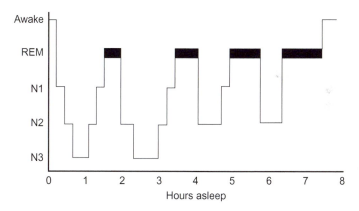

Figure 7.2 A normal sleep hypnogram.
REM: Rapid eye movement.

brain activity. Each of the stages can be given as a number or as a number prefixed with the letter "N" to denote "NREM" sleep. Generally speaking, the brainwaves get slower the deeper asleep you are.

• Awake and engaged in a task – When we are awake, the brain is usually quite busy. The frequency of brainwaves is high (about 12–30 waves per second, i.e. 12–30 Hz) and they are not well synchronized. This is because there is a lot of activity happening in different brain regions with lots of activation of interneurons connecting these different regions. These brainwaves are known as beta waves.

- Awake but not engaged in tasks (relaxed) – If we are awake but are not engaged in a task (or we simply have our eyes closed), brain activity is less and the frequency of our brain-waves is lower, about 8–12 waves per second (8–12 Hz). These brainwaves are known as alpha waves and mark an important, intermediate stage in going to sleep.
- Stage 1 (N1) sleep – This is the transition stage between wakefulness and sleep. Brainwaves slow down even further to the 4–7 Hz range and this marks the onset of sleep. These brainwaves are known as theta waves. It should be noted that in people who are severely fatigued, microsleeps (very short periods of sleep that the person is unable to recall) can occur while the person is attempting to stay awake and these may only be detectable by seeing theta waves in among the alpha waves on the EEG. At this stage of sleep we are still somewhat aware of our environment and it is relatively easy for an external stimulus to wake us up.
- Stage 2 (N2) sleep – During this stage of sleep the body becomes increasingly relaxed, heart rate and breathing rate slow, and we become less aware of our environment, although it is still relatively easy to wake someone from this stage of sleep. Our body temperature also begins to drop and our movements become less. Brainwave activity is similar to that seen in stage 1, but two additional brainwave features are seen in stage 2: sleep spindles (short, high-frequency waves) and k-complexes (short-duration, high-amplitude waves). The dura-tion of this stage of sleep varies but an adult will spend approximately 50% of the night in stage 2 sleep, normally for about 20 minutes at a time.
- Slow-wave sleep (N3; also called stage 3 and 4 sleep or just stage 3 sleep) – As the name suggests, at this point the frequency of the brainwaves has slowed to about 0.5–4 Hz and they are now described as delta waves. Our brainwaves are now about 60 times slower than they were when we were awake and this denotes a significant difference in the amount of brain activity. People in slow-wave sleep are very difficult to wake up. Slow-wave sleep is the most restorative stage of sleep and it is at this stage that much of the housekeeping activity of the brain is carried out and also when memory consolidation occurs. The synchroniza-tion of these slow waves is a way of reinforcing neural connections that have been formed during the day so that they become more firmly established. As we go through the day, our brain has an increasing need to spend time in this slow-wave state and this is thought to be the primary drive that makes us sleep. If someone has been awake for a prolonged period and is very fatigued, when they are allowed to fall asleep they will rapidly enter slow-wave sleep and spend much more time in this stage than someone who is not as severely fatigued.
- Rapid eye movement (REM) sleep – This stage of sleep is sometimes called paradoxical sleep. After the individual has completed the cycle from stages N1 to N3 and then back to N2 or N1, the brain enters a state where the brainwave activity is similar to that which would be seen if the individual was awake (beta waves). An observer would see rapid eye movements (although the eyelids are closed) but with no other muscle movement. It is at this stage of sleep that dreaming occurs. High-frequency brainwaves and increased energy consumption in the brain are paradoxical as they seem to run counter to the idea that the brain slows down when it is sleeping in order to replenish itself. The true purpose of REM sleep remains unclear. The assumption was that dreaming was part of the memory con-solidation process, but we now know that memory consolidation occurs during slow-wave sleep. Strangely enough, even though we spend a significant amount of our time in the REM stage, for people who are unable to enter the REM stage, either because of medication or because of a localized brain injury, there appear to be no negative consequences.[9] For this reason, the true purpose of REM sleep remains a mystery. One theory is that REM sleep is useful for brain development in infants. The dreaming state enables many more opportuni-ties to stimulate the developing brain and refine the neural connections. Paradoxically, the

infant brain has about 10 times as many synaptic connections between neurons as the adult brain (5 quadrillion in infants compared with 500 trillion in adults) and dreaming may allow for "synaptic pruning" to preserve important connections and eliminate non-functional ones to make room for the developing brain.[10] The reason for the paralysis at this stage of sleep is obvious. Given the realistic nature of dreaming, without some form of temporary paralysis, the individual would start acting out their dreams. Experiments in animals have demonstrated that when this mechanism for inducing temporary paralysis is disabled, the animal will begin to behave as if it is interacting with a stimulus that is not really present (e.g. chasing imaginary prey).[11] At all times, though, the animal is asleep.

Ideally, we would be allowed to go to sleep naturally and wake up naturally. Unfortunately, the reality of our daily life is that we are normally forced to be awake at a certain time by our alarm clock. If we look again at the hypnogram of the sleep cycle for a normal person who is not excessively fatigued, we can see that towards the end of the sleep cycle, the person is transitioning through the lighter sleep stages. If the alarm clock wakes them at this point, it will take a relatively short period for their brainwaves to "speed up" back to a frequency that is compatible with being awake and alert, 12–30 Hz. Normally, if you wake having just been in REM sleep, you may feel absolutely fine because your brainwaves in REM sleep are almost identical to those that would be seen when you are awake.

However, let's imagine that this person has gone to sleep in an excessively fatigued state, perhaps due to limited sleep over the previous few nights. When he does go to sleep, the brain is desperate to carry out some repair and replenishment and so will rapidly enter a deep, slow-wave sleep state where all these restorative processes happen. Someone who is excessively fatigued will have to spend more time in slow-wave sleep because their brain has accumulated a lot of neurotoxins and it needs slow-wave sleep to correct this. Rather than just needing a couple of episodes of slow-wave sleep at the start of night, this person may find himself transitioning back into the slow-wave stage throughout the entire night. But what happens when it is time to wake up?

7.3.2 Sleep inertia

As we have seen, slow-wave sleep is characterized by delta waves at a frequency of 0.5–4 Hz. An awake brain has beta waves that have a frequency of about 30 Hz, up to 60 times as fast as delta waves. We know that it is very difficult to wake someone up when they are in slow-wave sleep because the brain that is deeply asleep is in such a radically different mode of operation than an awake brain. This is why we spend less time in slow-wave sleep as the night progresses. Unfortunately, the fatigued sleeper might be spending a lot more time in slow-wave sleep, even towards the end of the sleep period when the alarm clock is set to go off. If the alarm goes off when they are in slow-wave sleep, the brain now needs to reconfigure itself to run 60 times faster. This is not a quick process and all of us will be familiar with that unpleasant feeling of grogginess that sometimes happens when we are awoken from a deep sleep. This can last for up to 30 minutes and is known as sleep inertia. It is the slow process of transitioning from the slow delta-wave activity characteristic of slow-wave

---

Box 7.1 Accident Case Study – Sleep Inertia

Air India Express Flight 812 – Boeing 737-800
Mangalore International Airport, India, 22 May 2010[12]

In the early hours of the morning, this Boeing 737-800 flight from Dubai ran off the end of the runway at Mangalore International Airport in India, killing 158 people. The accident investigation report reveals that the captain had been asleep for approximately 90 minutes during the flight and was only awoken when the first officer gave him the weather at Mangalore. Before this, conversation between the first officer and the cabin crew in the flight deck and several radio calls did not appear to wake him, suggesting that he was in slow-wave sleep. It was just 21 minutes from when he woke until the crash. The captain was the handling pilot for the approach and, although it cannot be conclusively proven, it may have been sleep inertia that led to problems with the descent profile that left them twice as high on the final approach as would be normal. Aside from this, sleep inertia may have also led to the captain disregarding warnings from both the ground proximity warning system and the first officer, which included a call to go-around. The aircraft touched down two-thirds of the way along the length of the runway, then ran off the end and into a ravine beyond. It was the worst commercial aircraft crash in 2010.

---

sleep back to the beta-wave activity that we need in order to be fully awake and alert. (See Box 7.1.)

Sleep inertia has a doubly negative effect for people who are chronically sleep deprived. People who are sleep deprived will spend more time in slow-wave sleep, thus increasing the likelihood that they will be woken when they are in this phase. As well as this, the normal routine of people who are sleep deprived would normally include a shorter than normal time in bed. So as well as spending more time in slow-wave sleep, having to wake up before the end of the natural sleep cycle greatly increases the chance of experiencing sleep inertia on waking. We will return to this subject when we discuss fatigue risk management strategies, as it is particularly relevant to effective napping.

Now that we have discussed the general structure of sleep, including the phenomenon of sleep inertia, we must now ask another question: What is it that makes us sleep? To understand this we must look at two new concepts: the homeostatic sleep drive and the circadian rhythm.

7.3.3 Homeostatic sleep drive and sleep need

Although the term sounds quite daunting, the concept itself is quite simple. All through this chapter we have been talking about the purpose of sleep and now we need to cover the internal mechanism that makes us go to sleep. Given what we know about the restorative nature of sleep, evolution has given us a mechanism that means

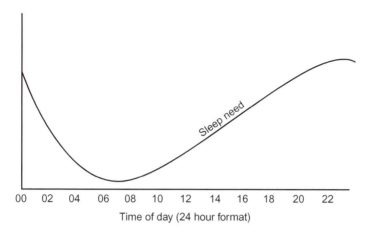

Figure 7.3 The change in sleep need over 24 hours.

that we do not need to wait until the brain is completely exhausted before we are able to go to sleep but, rather, will give us an increasing need to sleep so that we may enter this restorative state before there is any significant decrease in cognitive performance. Nature wants us to spend approximately one-third of our lives asleep and the optimal way to achieve this is to spend approximately one-third of each day asleep. The precise mechanisms through which the brain regulates wakefulness and sleep are quite complex but, in summary, there are two systems operating interactively: one system promotes wakefulness and the other promotes sleep. The chemical adenosine accumulates in the brain while we are awake, and gradually inhibits the system that promotes wakefulness and progressively activates the system that promotes sleep.[13] The reason that we need a system in order to sleep is that, as we have seen, the brain does not simply switch off when it is asleep and, in fact, sleep is a highly organized state that requires coordination. Adenosine accumulates progressively during hours of wakefulness, and this is thought to lead to an increasing need to sleep and is called the homeostatic sleep drive. The homeostatic sleep drive is sometimes called the "sleep need" and Figure 7.3 shows how this need increases during the day and then reduces during the period that we are asleep.

7.3.4 Biological clock, circadian rhythms, chronotypes and sleep urge

We have determined that nature wants us to spend approximately one-third of each day asleep. It has given us a mechanism that will activate the neural machinery that encourages the brain to enter a sleeping state after a period of time has passed (the homeostatic sleep drive, associated with the progressive accumulation of adenosine) and that would seem to be enough. Given that humans have evolved to be diurnal (meaning that we prefer to be active during the day and sleep at night, initially for hunting reasons), we want to make sure that the homeostatic sleep drive will be synchronized to ensure that we are awake during the day and asleep at night. However, the accumulation rate of adenosine may vary from person to person and can be dependent on activity. If our sleep–wake cycle was controlled only by adenosine, we might find ourselves getting sleepy in the early afternoon, sleeping for 8 hours and then waking in the middle of the night feeling fully refreshed. If this were to continue, we could end up completely out of phase with the day–night cycle, and this would have limited our ancestors' ability to hunt and socialize. Although hunting is not really part of modern life, it would certainly inhibit our ability to work if we found that our hours of peak wakefulness were between 6 p.m. and 10 a.m. In order to combat this, we must have another mechanism that keeps the homeostatic sleep drive roughly aligned with the day–night cycle. This is our biological clock.

7.3.4.1 Biological clock and circadian rhythms

As well as managing our sleep–wake cycle, our internal biological clock regulates several other physiological systems such as those responsible for hormone secretion and urine production in a predictable way over a 24 hour period. Any biological process that follows a repeating 24 hour pattern is known as a circadian rhythm. Because these biological processes like to keep to a timetable, it is useful to have an internal mechanism that acts as a timekeeper. Even in the absence of external stimuli, such as daylight, many body systems and cycles (including the sleep–wake cycle) will follow a predictable pattern that repeats approximately every 24 hours. Because the pattern only lasts *approximately* 24 hours, there needs to be a mechanism that realigns our biological clock with the actual cycle of day and night so that we do not go out of phase, much like a slightly inaccurate watch that needs to be reset every day. The body uses several cues (known as zeitgebers – German for "time givers") to achieve this synchronization, the most important being daylight, although meal times, social contact and repetitive schedules also have an impact. To reset the biological clock so that it is synchronized with the real day–night cycle, light passes into the eye and is detected by specialized cells called retinal ganglion cells.[14] These cells have nothing to do with vision and even people who are completely blind can have functional retinal ganglion cells and so will be able to synchronize their body clocks to the day–night cycle.[15] When light is detected by the retinal ganglion cells, this is signaled to the brain's biological clock which is, in fact, a small cluster of neurons (about 20,000) called the suprachiasmatic nucleus (SCN), located near the crossing point of the optic nerves coming from each eye. The neurons in the SCN fire in synchronized pulses; these pulses accelerate to a peak frequency at midday and then decrease in frequency.[16] It is the frequency of these pulses that acts as the biological clock, and the other biological systems that follow a circadian rhythm take their cues from the SCN. Even in the absence of light, the SCN will continue to change the frequency of its neuronal firing in a regular way over approximately 24 hours, even if it is removed from the brain and is being kept alive outside the body. You could, theoretically, wire up the SCN as a watch and it would be fairly accurate over a number of days before it began to drift significantly. Fortunately, the light that is transmitted to the SCN via the retinal ganglion cells keeps it "entrained" to the local day–night cycle.

7.3.4.2   Chronotypes

The fact that the SCN does not always keep to a cycle of exactly 24 hours explains another phenomenon that you may have noticed in your colleagues, your partner or your children: some people can be characterized as "owls" and some people as "larks". The owl/lark distinction is a colloquial way of expressing that there are some subgroups of people who would prefer to wake later in the day and go to bed late (the owls) and there are others who wake early and go to bed early (the larks). This distinction is known as an individual's chronotype and is partly related to the structure of their SCN but is also strongly associated with age. Imagine that you had an SCN that completed its regular cycle of varying the frequency of its neural oscillations in a shorter period than 24 hours. Given that your clock gets reset every day, you would wake up early and from then on, your internal biological clock would run faster than normal. After 14 hours have passed in reality, your body clock may be telling you that 16 hours have passed and it is nearly time to go to bed. If you then go to bed early, you will clear your sleep need in the normal period of time and so then awake early to start the process again. The opposite is true in those people with a biological clock that runs slower than normal (owls). While the majority of people do not necessarily exhibit a strong chronotype and can adjust more easily when they have to rise early or go to bed late, those that do exhibit a strong chronotype will struggle and this may well affect their sleep overall and, consequently, their performance.[17]

As well as individual differences in chronotype, age plays a role in our sleep–wake cycles. Young children and babies tend to awake early (they are experiencing fast SCN cycles – a fast clock) and teenagers are famous for staying in bed late (they are experiencing slow SCN cycles – a slow clock). As these teenagers grow up, their clocks gradually speed up and so adults tend to wake earlier than teenagers. If you have ever shouted at your teenage children to get out of bed and stop being so lazy, it is worth remembering that they are not disobeying you but are rather obeying those 20,000 neurons that form their biological clock. A study exploring the benefits of later school starting times, thus allowing students to stay in bed longer, showed that pupils exhibited significant improvements in mood, attention and overall performance.[18] The growing realization is that chronotypic variation between different people and different age groups is a reality and one that cannot be changed at an individual level. It is the world that needs to adapt to this variation among individuals and age groups, and studies looking at encouraging flexible working times to account for variations in people's chronotypes have given very encouraging results.[19]

7.3.4.3   Melatonin

To complete the picture, there is just one more neurobiological concept to consider, that of the interaction between light and release of a hormone called melatonin. The adjustment of this mechanism is thought to have a therapeutic function in dealing with jet lag and we will be talking more about this later in the chapter. More importantly, the melatonin system is another mechanism to help us keep many of our circadian processes in phase with the local day–night cycle. When the SCN detects a gradual transition from light to dark, it stimulates the production of melatonin from the pineal gland, a cluster of cells about the size of a grain of rice located just behind the thalamus, deep within the brain. The melatonin level starts to rise from about 9 p.m., having been extremely low during the day. It reaches a peak in the early hours of the morning and then decreases until it reaches the low daytime level at about 9 a.m. Along with the SCN, melatonin is another mechanism for keeping certain circadian rhythms, particularly body temperature, in phase with the day–night cycles. As melatonin levels rise, body temperature falls and this fall in body temperature is strongly associated with a desire to sleep.[20] Remember this fact for when we talk about sleeping strategies. One more function of melatonin that is worth considering is its ability to shift our circadian rhythms. This ability evolved in animals to help them to adjust to seasonal differences in light–dark cycles, with long days in the summer and long nights in the winter. For humans, especially those that fly, being able to shift our circadian rhythms is very useful for crossing time zones and melatonin has a role in this process.

7.3.4.4   Sleep urge

We now arrive at a crucial concept. If we already have a mechanism that gives us a biological *need* to sleep based on the amount of time we have been awake, and a biological clock that resets itself to stay in phase with the day–night cycle, what else do we need? It seems that as well as having many biological processes that follow a circadian rhythm, there is a process that is designed to give us a *feeling* of sleepiness that varies over a 24 hour period like any other circadian rhythm. This sense of sleepiness, or sleep urge, is different from the sleep need because it is subjective; thus, you may feel sleepy even though you have not accumulated enough of a sleep debt to actually need to sleep. To illustrate this, Figure 7.4 superimposes the pattern of our circadian sleep urge on top of the varying extent of our sleep need.

From the time we wake up we start to develop a physiological need for sleep as levels of adenosine build up in the brain. But, if you think about your own experience, you do not wake up first thing in the morning and feel absolutely wonderful and then progressively feel more tired as each hour goes by. The circadian rhythm that gives us a sleep urge is designed to peak during late evening when our physiological need for sleep is also peaking. The sleep urge is maintained even while the brain is restoring itself and the physiological sleep need is decreasing. The sleep urge is designed to slow down brain function in order to keep us asleep and so there are periods when our sleep urge would make it difficult for us to carry out certain tasks in the event that we are awake during these periods. The most significant of these periods is the time between 2 a.m. and 6 a.m., and legislation regarding flight time limitations refers to this period as the window of circadian low. At this point in our sleep–wake cycle, everything is telling us that we should be asleep and, consequently, our cognitive performance is lower than normal.[21]

We are probably all familiar with the "post-lunch lull", that feeling of sleepiness we experience in the early afternoon. Although it is commonly thought that this sleepiness is caused by having had a meal at lunchtime, consuming food is entirely unrelated to this small spike in our circadian rhythm of sleep urge. For example,

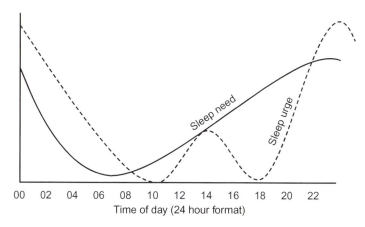

Figure 7.4 The change in sleep need and sleep urge over 24 hours.

you do not necessarily feel tired after breakfast or, indeed, after dinner. Although the reason for this increase in sleep urge during the afternoon is unclear, it may be a way of offsetting some of the sleep need that has been accumulated up to that point rather than carrying it all the way through to the end of the day. Research has shown that there is an increase in performance after an afternoon nap,[22] and some cultures include a siesta in daily life. We will cover effective napping strategies later in this chapter (see Section 7.4.2).

To summarize, we can say that we build up a physiological need for sleep during the day and we address this by obeying our circadian sleep urge, which is there to give us a sense of sleepiness in a circadian rhythm that varies over an approximately 24 hour period. The highest sleep urge occurs during the hours of darkness, with a smaller increased urge to sleep in the afternoon. To keep our sleep urge cycle aligned to the time of day, our biological clock (the SCN) is reset using light and other cues to ensure that we are awake during the day and asleep at night. The biological clock will complete a cycle in approximately 24 hours and this is used to coordinate lots of other body functions that are required to vary over the course of a day (circadian rhythms, including the sleep urge rhythm). For some people, depending on their chronotype and age, this clock runs faster or slower than in others and so gives a tendency to wake earlier or later than normal. In addition, when we detect a transition from light to dark that would normally represent dusk, our pineal glands secrete melatonin, which synchronizes other circadian rhythms, such as body temperature, to the day–night cycle. The reduction in body temperature at night that is caused by melatonin is another major contributor to both our sense of sleepiness and our ability to sleep (think about how difficult it is to get to sleep in a hot room). When we sleep, the brain transitions through several different stages, each characterized by different levels of neural activity. The brain cleans itself of neurotoxins and consolidates memories, and slow-wave sleep is particularly important for these processes. Once these processes have been completed, the sleep cycle comes to an end and the system that arouses the brain becomes more active and wakes us up.

7.3.5 *Sleep debt*

The mechanisms that control our sleep work well provided that we completely satisfy our physiological sleep need every night. The vast majority of people require 8 hours of normal sleep for the brain to restore itself and for the body to do the same, and it is best to consider this 8 hours as a physical limitation. There is no way of shortening this time without some negative consequences. The unfortunate fact is that because of our life-styles, it is becoming increasingly difficult to consistently get 8 hours of sleep per night. The question we must now consider is this: What happens to us if we do not respect this physical limitation? Let us imagine that John starts the week on Monday morning having completely cleared his physiological sleep need. He is now fully refreshed and restored and ready to start work. Because he is so busy, he goes to bed at 10.45 p.m. and gets up at 6 a.m. and so only gets 7 hours sleep (the extra 15 minutes being the time it takes to fall asleep) for the next four nights of the week (Monday to Thursday night inclusive). The nature of his work also means that he cannot take any afternoon naps. Seeing as John normally needs 8 hours of sleep a night, by Friday he has missed 4 hours of sleep; that is, he has accumulated a sleep debt of 4 hours. How can John pay back this debt? If John decides to socialize on Friday night and gets to bed at midnight, he may be able to get his required 8 hours of sleep that night but it is unlikely he will be able to sleep for an additional 4 hours because his circadian sleep urge will decrease during the early hours of the morning and will still lead him to wake up. This is why even if we are really fatigued, we cannot just sleep for as long as we need to pay back all of our sleep debt. Our circadian sleep urge will not let us. If he is lucky, he may be able to pay back 2 hours of his sleep debt but will have to carry the other 2 hours over into Sunday. Again, if he plans it well, he may be able to get his 8 hours of sleep plus an additional 2 and so pay back the remaining 2 hours of sleep debt that he owes in order to start the week on Monday being back to normal with zero sleep debt.

This example, although simplistic, was meant to illustrate the nature of sleep debt. John was lucky that his weekend schedule allowed him to pay back the relatively small sleep debt he had accumulated during the week. What we have not shown in this example is that even though 7 hours of sleep would seem to be acceptable for most people, if we measured John's performance objectively from Monday to Friday there would be a measurable decrease due to his increasing sleep debt. Imagine, though, if John had been even busier and had only been able to get 6 hours of sleep per night or had been unable to repay his sleep debt over the weekend. In these cases, as well as having a measurable decrease in performance during the week, John will then start week two with a sleep debt, perhaps as much as 8 hours. This is the equivalent of miss-ing a night's sleep. There is no way that he can pay back this debt in one go because his circadian sleep urge rhythm will not allow it. At some point in your life, you may have found yourself in this sort of situation. Although you will probably feel tired, it is unlikely that you will feel as exhausted as if you had missed one night of sleep. The reason for this is that it seems that the brain can adapt to increasing sleep debt by downgrading its performance overall and operating at reduced level. This means that even though you may feel okay, perhaps a little tired, your brain is working measurably slower than a brain with no sleep debt. This has been demonstrated experimentally.

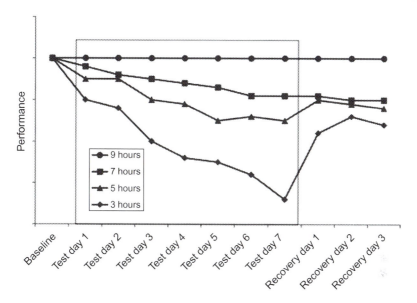

Figure 7.5 Performance changes associated with normal and restricted sleep.

In this experiment,[23] 66 volunteers were divided into four groups and each group was allowed to spend either 3, 5, 7 or 9 hours in bed each night for a period of a week. It should be noted that time in bed does not necessarily mean time asleep. The group that was allowed 9 hours in bed slept for 8 hours, the group that spent 7 hours in bed slept for 6.5 hours and the groups that were restricted to 5 and 3 hours in bed slept for almost the entire time. Each day the reaction times for each group were measured, as well as the number of lapses that the participants made during certain tests. After the 7 day test period, there was a 3 day period where all of the groups were allowed 8 hours in bed and, again, their reaction times and number of lapses they made were measured each day. Figure 7.5 shows a simplified version of the results.

As you can see from the graph, only the participants who were allowed 9 hours in bed maintained their normal level of performance throughout the week. Even though 7 hours in bed might seem quite reasonable for most of us, this experiment demonstrated that this leads to a measurable decrease in performance. Performance clearly suffers in the groups that were restricted to 5 hours and 3 hours in bed. It seems that for the groups that were restricted to 7 hours and 5 hours in bed, their decreasing performance did stabilize eventually, although at a reduced level. This reduced performance continued into the recovery phase and suggests that the brain is adapting to sleep debt by operating at reduced capacity with slowing of neural transmission and alteration of neurotransmitter levels. Other research has supported this idea by showing that there is an alteration in how genes are expressed in the fatigued brain,[24] along with increased destruction of neurons.[25]

Another alarming feature of sleep debt, especially for pilots, is the tendency to take microsleeps. Microsleeps occur in the fatigued brain and, as the name suggests, are brief period of sleep that the sufferer does not recall but can be seen as episodes of theta-wave activity on EEG recordings. They are the brain's way of addressing severe fatigue by taking multiple, brief sleep episodes. They normally last for 5–10 seconds but may last as long as 30 seconds. A NASA study noted microsleeps in flight crew, not only during the relatively quiet cruise phase, but also during descent.[26] During the same study, nearly half of one of the groups of pilots (those who did not sleep during a scheduled in-flight rest period) experienced an involuntary sleep episode of 2–12 minutes, despite the presence of two NASA researchers in the flight deck. While these episodes are not technically microsleeps, they do highlight that sleep will occur whether it is consciously desired or not. The brain's capacity to sleep without it being consciously triggered evolved long before we put our brains and bodies into complex machines traveling at speeds greatly in excess of anything our primitive ancestors could have experienced, machines that require very little input from us to keep moving.

Microsleeps are one of the most alarming phenomena in this book, but understanding microsleeps is not primarily about improving the safety of flight. It is about improving the safety of your drive home from work. Given that these episodes usually occur in the fatigued brain during periods of low stimulation (like driving a car on a motorway), the brain might normally be expected to show alpha-wave activity, normally associated with a relaxed state of wakefulness. During the microsleep, theta brainwaves (normal for stage 1 sleep) replace alpha waves. The brain is technically asleep. At 70 miles per hour (113 km/h) on the motorway, you will travel 157 m during a 5 second microsleep. The problem is, those 157 m may not be traveled in a straight line; they could be traveled across a couple of lanes, through the central reservation or off the side of the road. Unfortunately, you will not realize it is happening until it is too late.

The implications of these experiments into cumulative sleep debt, rest and microsleeping are profound. How many of us are operating at reduced capacity because our brains have adapted to chronic sleep debt? What about those people who consistently get less than 8 hours of sleep and may have accumulated a sleep debt measuring hundreds of hours? How much processing ability have their brains lost to cope with this chronic sleep debt? What will happen if a brain that is running at a reduced capacity because of chronic sleep debt is then subjected to a night of particularly bad or reduced sleep?

The National Transportation Safety Board (NTSB) voted two to one against not including "fatigue" as a contributory factor in the accident described in Box 7.2 because the fatigue level of the crew could not be proven (although the report noted the high likelihood of fatigue on the part of the crew but could not suggest to what degree this affected their performance). Owing to the exclusion of "fatigue" as a contributory cause in the formal report, the Chair of the Safety Board, Deborah Hersman, added a separate statement at the end of the report that expressed her views. This was an extremely insightful précis of the current state of fatigue risk management in the aviation industry, and some of the key points are given here:

1. Fatigue risk management has been in the Top Ten List of Transportation Safety Improvements since the NTSB started the list in 1990.

Box 7.2 Accident Case Study – Fatigue

Colgan Air Flight 3407 – Bombardier DHC-8-400
Buffalo, New York, USA, 12 February 2009[27]

During the hours of darkness, this aircraft was conducting an instrument approach to Buffalo-Niagra International Airport. As the aircraft was being configured for landing, the airspeed was progressively decreasing and neither pilot detected the impending stall until the stick shaker activated. The captain did not initiate the appropriate stall recovery maneuver and instead pulled back on the control column. The stick push system also activated but the captain continued to apply significant back pressure on the control column. The report notes that this could have been due to a startle reaction. Unfortunately, the first officer retracted the flaps and the landing gear, thus increasing the stalling speed of the aircraft. The aircraft entered an aerodynamic stall and went through several roll oscillations. The aircraft reached 105 degrees bank to the right, then rolled left and then to the right again. With the flaps at 0, the aircraft began a rapid descent and crashed into a house, killing all 49 people on board and one person on the ground. Based on the activity of the pilots over the preceding 48 hours, the fact that they both had significant commutes, and both had ended up spending the previous night sleeping in the pilots' lounge at Liberty International Airport, Newark, it is highly likely that they were fatigued.

---

2. The errors made by the crew are consistent with those that would be caused by fatigue (failure to monitor airspeed, inappropriate reaction to stick shaker and other monitoring failures).
3. Over the past 40 years, we have made progress in addressing the problems of alcohol impairment and non-adherence to SOPs but fatigue is still not given the same respect as a real cause of accidents.
4. Impairment due to fatigue is the same as impairment due to alcohol – being awake for 16 hours impairs your performance to the same extent as a 0.05% blood alcohol content [slightly below the drink-drive limit for the USA (0.08%) but well above the 0.01% limit for bus drivers].
5. Although we cannot retrospectively measure the fatigue levels of the pilots in the same way as we can assess alcohol levels, we can easily calculate the likelihood and the degree of fatigue based on their activities before the crash and on the scientific data regarding fatigue.
6. The failure of the NTSB to include fatigue as a contributing factor is symptomatic of the Board's inconsistent approach to addressing fatigue in transportation accidents.
7. While the NTSB agreed that fatigue was likely, examining the evidence *establishes* the presence of fatigue.
8. The captain spent the night before his duty sleeping in the company crew lounge (a place open to other crew members). It is known that he did not sleep for 8 hours uninterrupted because he logged on to a company website three times. He spent two out of the last three nights in the crew lounge and would have accumulated between 6 and 12 hours of sleep debt.
9. At the time of the accident, he had been awake for 15 hours.

10. The accident happened at a time in the evening when the captain would normally go to sleep.
11. The first officer had had a maximum of 8.5 hours sleep in the preceding 34 hours, 3.5 of which were when she was commuting in the jumpseat from her home in Seattle to her base in Newark. This sleep was broken because she had to change planes in Memphis. The other 5 hours sleep were in the crew lounge, which was busy and well lit.
12. Fatigue was identified as a major contributory factor in the investigation report into the Exxon Valdez oil tanker grounding in 1989.
13. Commuting is a significant problem, although it may be explained by the financial challenges of being a pilot in a regional airline and being unable to live near one's base owing to it being in a part of the country where property prices and rental costs are expensive.
14. We are never going to change the debate on fatigue unless we face it head on.

From what you have seen already in this chapter, you could probably make an estimate of what percentage of normal brain processing capacity this crew was operating at, given their fatigued state. Deborah Hersman is correct that it would be possible to calculate their level of impairment due to fatigue based on the duration and quality of their sleep in the days preceding the crash. This impairment, on top of a generalized reduction in performance caused by chronic sleep debt, would easily account for the specific unsafe acts made by the crew. While some fatiguing aspects of daily life are beyond our control, there are scientifically proven strategies that will allow individuals to manage their fatigue levels and give them the best chance of avoiding impairment of brain function that could reduce safety during flight operations.

Before we get on to fatigue risk management strategies, let's look at one more piece of research that Hersman referred to in her statement. Fatigue and alcohol have both been found to decrease performance in a dose-dependent way; that is, the greater the amount of fatigue or alcohol, the greater the decrease in performance. A study was conducted to compare the performance impairment due to alcohol with the performance impairment due to fatigue[28] and the simplified results are shown in Figure 7.6.

As you can see, after 17 hours of sustained wakefulness, performance has decreased to a level equivalent to having a blood alcohol concentration of 0.02%, the legal flying limit. After about 20 hours, performance has decreased to the equivalent of having a blood alcohol concentration of 0.05% (the drink-drive limit in most Western countries). After 25 hours awake, the decrease in performance is equivalent to approximately 0.10% blood alcohol concentration. This decrease in performance will be even more pronounced if extended wakefulness occurs in someone with a pre-existing sleep debt.

7.4 Fatigue risk management strategies

This section is divided into seven topics:

• how to achieve sufficient high-quality sleep
• how to nap effectively
• how to deal with insomnia and sleep disorders

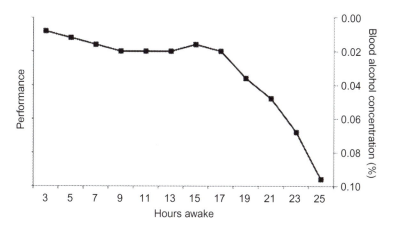

Figure 7.6 Performance changes associated with time awake matched to the equivalent amount of alcohol in the body required to cause such a performance change in someone not suffering from fatigue.

- how to mitigate risk if you do find yourself fatigued
- how to use sleep medications
- how to deal with jet lag
- organizational strategies for fatigue risk management.

7.4.1   How to achieve sufficient high-quality sleep

Guidance in this section is divided into two parts:

- planning your sleep
- sleep hygiene.

7.4.1.1   Planning your sleep

With sensible planning, for most people working in aviation, there is no reason to accumulate a sleep debt. Remember, the vast majority of people need 8 hours sleep although there are some individual variations. This may require 8.5 hours in bed, and the time you normally take to fall asleep should be factored into your sleep planning. Here are some strategies that will help you to achieve your 8 hours sleep per night:

- Sleep is best obtained in one block – If you have the opportunity to sleep for at least 8 hours, take it. It will do you the most good.
- If there is no reason to get up at a specific time the next day, let yourself wake up naturally as this will help to clear at least part of any residual sleep debt. Always remember that planning a wake-up time includes considering the wake-up time for the day after. If you have an early start on the day after, it might not be a good idea to stay in bed until midday as you will find it difficult to go to sleep in the evening.
- Bearing in mind the time that you are planning to go to bed that evening, consider supplementing your normal block of sleep with a nap. See Section 7.4.2 on how to nap effectively.

If you know you are going to have a restricted night's sleep, why not take a nap during the afternoon prior to that? This will clear some of the sleep need you have accumulated up to that point in the day.

- Any time asleep is quality time – If your sleep is disrupted, take lots of small naps or even some big naps. Do whatever you need to do to get your 8 hours of sleep. Given that sleep is best taken as one block, you may actually need more time asleep if you are taking multiple naps as the quality of that sleep is not as good.

7.4.1.2 Sleep hygiene

Now that we have planned when we are going to sleep, we need to look at how we are going to get to sleep. Below is a list of strategies to ensure that you optimize your time in bed and that will allow you to have high-quality, restorative sleep[29,30]:

Sleep hygiene before sleeping

- No caffeine in the 4 hours before going to bed – Caffeine is found in high levels in coffee, tea, cola and many energy drinks. While caffeine does have a role in fatigue risk management, it is a significant disrupter of good-quality sleep.
- No smoking in the 1 hour before going to bed – Although nicotine is not as strong a stimulant as caffeine, it will still make it more difficult to fall asleep.
- Avoid alcohol before bed – We all know that a couple of drinks will make us fall asleep (and even more will make us pass out, which is very different from falling asleep). However, although alcohol is a good sleep initiator, its subsequent breakdown in the body leads to stimulant effects that disrupt sleep by affecting the sleep architecture (the structured transitions between the different stages of sleep). Alcohol has a dose-dependent effect on the quality of sleep. This means that there will always be a decrease in sleep quality with alcohol in your system, and the more you have, the worse your sleep will be. If you are going to drink before bed, one drink will do a lot less damage to your sleep than three, but if you can manage without it you will feel a lot better in the morning.
- Do not engage in aerobic exercise within 2 hours of going to bed - Exercising within 2 hours of going to bed will delay your sleep onset although exercising before that may help to fatigue the muscles and aid the onset of sleep. The slight exception to this rule is sex. Sex releases a chemical in the brain called oxytocin that has a calming effect and encourages sleepiness.
- Wherever possible, keep your schedule consistent – To stabilize your circadian rhythms, try to go to bed and wake up at a similar time each day, ensuring that this gives you at least 8 hours of sleep (at least 8.5 hours in bed).
- Establish a regular bedtime routine – By having a repetitive routine you can train your body to know that carrying out this routine means that it will shortly be time for bed. Perhaps plan that in the hour before you go to bed you will make a list of five things you need to do tomorrow (meaning you will not lie in bed thinking about what you need to do the next day), have a cup of hot chocolate, read something non-work related and relatively easy for 15 minutes and then take a hot bath or shower. A routine like this gives a finality to the day's mental exertion by establishing goals for tomorrow and includes a basic motor task in the preparation of a pleasant, soothing drink. Reading something non-taxing is a means of separating the activities of the day from the anticipated relaxation of bedtime. Finally, the bath or shower has a relaxing effect but, more importantly, once finished, leads to a decrease in body temperature once out of the hot water that acts as a powerful inducer of sleepiness (remember, an increase in melatonin during the hours of darkness causes a drop in body temperature).

Sleep hygiene in your bedroom

- Your bedroom is your sanctuary – Use your bedroom only for sleep and sex. Ban all computers, laptops, iPads and televisions from the bedroom. They will only serve to keep your brain active and prevent the onset of the vital, sleep-inducing alpha brainwaves. Also, the powerful, shortwave light these devices emit will disrupt the function of your sleep mechanisms by making it think that it is still daylight and so inhibiting the release of melatonin, which would normally help to reduce your body temperature and send you to sleep.[43] Limit your use of these devices in the evening and, if you do use them, adjust the display settings to minimize blue light, i.e. set the screen to have a "warmer" appearance. If you must have your smart phone in the bedroom, use it only as an alarm clock.
- Ensure that your bedroom environment is conducive to high-quality sleep – Your bedroom should be completely dark and as quiet as possible, and your bed should have a good-quality, comfortable mattress. The use of an eye mask and earplugs may be required when a dark, quiet environment is not possible, for example in some hotel rooms. Earplugs do take some getting used to, but once you have acclimatized they can radically improve the quality of your sleep by minimizing disturbances caused by noise.
- Keep your bedroom cool – To facilitate the drop in body temperature that is associated with sleep, keep your bedroom cool. The optimum temperature will depend largely on what type of bed linen you use and what you wear in bed. Most researchers agree that the bedroom should generally be on the cool side and a starting point for those trying to work out the best temperature is about 16–20 °C (60–68 °F).
- Do not watch the clock – No clocks should be visible from your sleeping position and you should also remove your watch. If you are unable to sleep, you may become fixated on the time and this concern will inhibit your ability to sleep. If you should wake in the night, you do not want to see anything that reminds you of the fact that you will need to be getting up soon as it increases the psychological pressure to try and get back to sleep.
- Beware of the light – If you do need to get out of bed during the night to use the toilet, try to minimize the amount of light you are exposed to. Some people suggest using a torch but you may find that other strategies work better, such as having a dim night light in the bathroom.
- Sleep hygiene if you cannot sleep – It is perfectly normal to experience insomnia from time to time and even longer bouts of it can be brought on by stress and will pass once the stress resolves. If you find that you have not fallen asleep after about 30 minutes and are not feeling any urge to do so, do not continue to lie in bed. Go to another room, sit quietly in a chair in a dimly lit room and either listen to some music or engage in a low-workload task. Remember, you do not want to overstimulate your brain or you will not get the sleep-inducing alpha waves.

7.4.2 How to nap effectively

If you use them correctly, naps can be a useful supplement to night-time sleeping, but always remember that sleep is best taken in a single block and so naps should not become a replacement for normal, night-time sleep. Let's look at some guidance for taking naps:

- Plan your naps carefully – Naps can be taken prophylactically (e.g. before reporting for a late duty that would ordinarily only start after being awake for an extended period). They may also be taken after a duty to offset some of the sleep need that has been accumulated.

If a duty starts early in the morning and finishes at midday, a nap in the afternoon will improve mood and performance during the evening and shouldn't affect that night's sleep. Naps should be timed to commence when the circadian sleep urge is high. This is normally highest during the night but there is also a smaller peak from 2 to 4 p.m. This is an ideal time to have a nap as it will coincide with your natural urge to sleep.

- Time the length of your naps carefully – There are a lot of myths about the optimum length for a nap, many of them based on the avoidance of sleep inertia, that unpleasant feeling of grogginess that occurs after being woken from the slow-wave stage of sleep. In reality, the optimum length of nap you should take is based on how much sleep debt you have accumulated. If you are very sleep deprived, you will probably enter slow-wave sleep very shortly after the onset of sleep and so there is a high chance that you will have to wake up when you are in this stage. Given that we cannot guarantee that we will not have sleep inertia upon waking, the best way to mitigate this risk is to factor in at least 30 minutes after your nap to fully recover from the sleep inertia; therefore, do not plan a nap immediately before you are planning to drive to work. Caffeine has been shown to shorten the period of sleep inertia and so a cup of coffee will help with the grogginess.[31] In planning the length of a nap, you should consider how many hours of sleep debt you have accumulated. It may be that a 3–4 hour nap will go a long way to clearing some of the sleep debt and still allow a restful night's sleep afterwards. The evidence shows that naps will improve performance and that the longer the nap, the longer the improvement in performance. An analysis of several studies into napping suggested that a 15 minute nap could give 2 hours improvement in performance while a 4 hour nap could improve performance for 10 hours.[32] However, the performance benefit does decrease over time after waking from a nap.
- Treat a nap like normal sleep – Sleep hygiene is as important for napping as it is for your main sleep during the night. By following the guidance given for sleep hygiene, you can make your naps as effective and restorative as possible.
- If you are not particularly tired and you do not think you are carrying a sleep debt, feel free to take an afternoon nap but keep it short. To avoid sleep inertia, limit naps to about 20 minutes as this will give an improvement in performance without having to endure 30 minutes of sleep inertia unnecessarily.
- If you want to limit the length of your nap, try drinking a cup of coffee before the nap. Caffeine reaches its maximum concentration in the bloodstream within about 30–60 minutes and so will lighten your sleep during the time the concentration is rising and facilitate an easier transition to wakefulness at the end of the nap. We will cover more on caffeine later in this chapter (see Section 7.4.4).

7.4.3 How to deal with insomnia and sleep disorders

You may find that even if you are following all of this guidance, you are still not getting enough sleep or that you seem to be getting enough sleep but you still wake up feeling exhausted. Being unable to sleep or being unable to sleep effectively robs you of one of the essential things that you need in order to be a fully functional human being, and you should address the problem as soon as you notice it. The longer it goes on, the more your brain will adapt to operating at reduced capacity. A large range of sleeping disorders have been classified and these vary from a persistent inability to sleep to more complex disorders that involve the sufferer exhibiting waking behaviors but being in an asleep state (these "parasomnias" include sleep walking, eating, drinking and many others). Discussing the entire range of sleep disorders is beyond the scope of this book but we will cover two of the more common ones, ones that can and will continue to cause problems for the pilot population:

- chronic insomnia
- obstructive sleep apnea.

7.4.3.1   Chronic insomnia

It is not unusual for adults to have an occasional bout of insomnia that normally resolves by itself. These episodes are often associated with stress and we will look at stress management strategies in Chapter 8. After a few nights of disturbed sleep, the sufferer will notice a distinct change in performance level and mood. If the insomnia continues, the difficulty now is that the sufferer has accumulated such a significant sleep debt and has such impaired brain performance that another biological mechanism needs to be used to keep the person even partly functional throughout the day. Having to be awake and having to carry out activities when carrying an extreme sleep debt leads to activation of the body's stress-response system.[33] The brain signals the release of cortisol and other stress hormones to keep the body and brain in a state of arousal so that it can keep functioning. It is the equivalent of your body opening up its own little coffee shop on top of your kidneys (the location of the adrenal gland, where many of these stress hormones are secreted) and spending the day drinking double espressos to keep going. The problem is that by evening there are such high levels of stress hormones in the body that sleep is almost impossible. You are now trapped in a vicious cycle of needing stress hormones to keep you going during the day, only to have them block the thing you really need at night – sleep.

Treatments for chronic insomnia depend on the cause of the condition and require expert advice. In cases where it is a transient stressor causing the problem, a short course of sleeping pills may help. In cases where the stressor is more long term, cognitive–behavioral therapy (CBT) may help.[34] CBT is useful for those people where a short period of insomnia has led to chronic insomnia because of an accumulation of bad sleep habits and when an association has formed between being in bed and being frustrated by the inability to sleep. In all cases, it is important to get expert advice early on, before the habits become too ingrained, and revisit Section 7.4.1.2 on sleep hygiene techniques to make sure that the sleep environment is as conducive to healthy, restful sleep as possible.

7.4.3.2   Obstructive sleep apnea

Obstructive sleep apnea is another fairly common medical condition that can result in sleep loss. The Federal Aviation Administration (FAA) estimates that 4–7% of adults, and 30% of obese people, suffer from it.[35] It is accompanied by snoring and occurs when the soft tissues at the top of the throat fall backwards during sleep and obstruct the airway. This happens because muscle tone is reduced during sleep. Another thing that can cause this decrease in muscle tone is alcohol, which is why snoring tends to be worse after an evening of drinking. Once the soft tissues have fallen back and blocked the airway, the sufferer is unable to breathe but remains asleep. Oxygen levels

---

Box 7.3 Incident Case Study – Sleep Apnea

Go! Flight 1002 – Bombardier CL-600
Hilo, Hawaii, USA, 13 February 2008[36]

The crew of this Bombardier CL-600 flight from Honolulu International Airport to Hilo, Hawaii, were out of contact with air traffic control for 18 minutes and ended up flying past their destination airport for 26 miles. On waking, they turned back to the airport and landed safely. It is estimated that they had been asleep for 18–25 minutes and, fortunately, they were carrying fuel for the return trip and their period of sleep started roughly halfway through the 51 minute flight. Had they not been carrying the extra fuel and had they slept for longer or begun sleeping further into the flight, then there would have been a significant risk of fuel exhaustion before being able to return to land. It was subsequently found that the captain had undiagnosed severe obstructive sleep apnea.

---

in the blood fall, blood pressure and heart rate increase until, eventually, a neurological reflex is activated which partially arouses the brain and may either wake the person up or shift them to a much lighter stage of sleep so that they can unblock their airway. This may sound like a very loud snore followed by a gasp and can occur hundreds of times in the course of a night. The result is greatly disturbed sleep, together with all the feelings of tiredness and reduced performance that this will cause (see Box 7.3).

Chronic obstructive sleep apnea can cause further cardiovascular problems, and sufferers have a significantly increased risk of heart attacks and strokes.[37] As well as this, the chronic sleep disturbance will lead to chronic impairment of brain function and mood alteration, including depression. Treatment requires expert medical intervention and may involve medication, adjustment of sleeping position, surgery, or a mask that is worn while sleeping that keeps the airway open by providing a continuous stream of pressurized air to prevent obstruction.

7.4.4   How to mitigate risk if you find yourself fatigued

By now, you can probably predict the most important recommendation for what to do if you are fatigued – sleep. It is the only effective cure and it is what your body is demanding that you do. Wherever possible, you should respect this urge. If you are driving, you must take notice of the warning signs of fatigue because you have no way of knowing when your are going to start microsleeping or even that you are doing it at all. There are just too many accidents where the driver has fallen asleep at the wheel and, unlike flying, drivers always have the opportunity to pull over and rest. We have talked about the neuroscience of sleep and the incredible lengths that your brain will go to in order to ensure it gets sleep. We have talked about microsleeps and how they are impossible to notice once they start happening, by which point you may be traveling at 70 miles per hour (113 km/h) while being essentially unconscious. You can switch on the radio, you can drink coffee, you can open a window but, unfortunately, there is still a reasonable chance that you are going to have an accident shortly after you start noticing symptoms of fatigue. None of those strategies is going to keep you safe. The only thing to do is to actively search for a place to stop and rest. This act of having to pull off the motorway and search for an appropriate site may stimulate the brain sufficiently until you find a place, but then you must have a nap before continuing your journey. Long journeys may need several naps and, in some cases, a nap may not be enough. You may need to book into a roadside hotel for the night and get some proper sleep. The importance of this cannot be overstated because fatigue is far more likely to kill you and others on the road than it is in the air.

Controlled rest in flight follows a similar rationale to normal napping and so I am not going to go over the subject except to say that it must be adapted to the procedures in your own airline. The key risk with controlled rest in flight is the sleep inertia that can occur just after waking. Duty should not be planned to resume until 15–30 minutes after waking to allow the pilot to recover from any sleep inertia.

If we are not driving and we do not have the opportunity to nap, for example, we are just beginning the descent and approach to an airport, what else can we do to mitigate the risks associated with fatigue?

Caffeine is one of the most commonly used stimulants, especially in aviation, and its effects on cognition were reviewed in a scientific paper in 2011.[38] It plays a role in managing fatigue and there are strategies that will allow the user to have this control without having a detrimental effect on a subsequent period of sleep. When we covered the physiological nature of sleep, the chemical adenosine was identified as a molecule that accumulates in increasing quantities in the brain during periods of wakefulness. It is the accumulation of adenosine that leads to the progressive activation of the neural systems that promote sleep. Caffeine acts by temporarily blocking the adenosine receptors in this sleep-promoting mechanism. This means that although adenosine continues to accumulate, some of the receptors are temporarily insensitive to it and so there is a slight decrease in the homeostatic sleep drive. The extent of this effect is largely dependent on the degree of sleep need (and any pre-existing sleep debt) and the dose of caffeine taken. Caffeine taken in liquid form is absorbed in the gut and reaches a maximum concentration in the bloodstream within 30–60 minutes.

Studies have shown that caffeine can improve human information processing and error detection. In one study, habitual caffeine users were given doses of caffeine and it was noted that there was an increase in their ability to detect errors in their own actions, a phenomenon known as error-related negativity.[39] Although two different doses of caffeine were used in this study, there did not appear to be a difference in outcomes between the two groups. Continual and habitual use of caffeine leads to an increase in tolerance to its effects as the brain adapts to the continuing high concentrations of caffeine by altering the number of adenosine receptors. The practical outcome of these studies is that caffeine can be useful in improving performance but that the benefit decreases if caffeine is continually consumed. A dose of 3 mg/kg body mass (about 210 mg for a 70 kg adult) will give an improvement in performance. This is the equivalent of a double espresso or two mugs of instant coffee. The improvement seen with lower doses is more difficult to categorize as the improvement may be counteracted by the tolerance that has been formed due to continued caffeine exposure. In essence, a hit of caffeine larger than that which the person is used to can lead to an improvement in performance, although high doses are also associated with palpitations, increased production of urine and laxative effects. The half-life of caffeine is 3.5–6 hours (the amount of time taken for the concentration of caffeine in the blood to reduce by half) and the effects will last during this period but will decrease steadily after the peak concentration is reached 30–60 minutes after consumption. The relatively slow decrease in the concentration of caffeine in the body needs to be taken into account so that it will not affect sleep later on. Avoiding caffeine 4 hours prior to sleeping is a good idea, but there are individual differences in sensitivity and elimination times for caffeine and so you may have to change this figure to suit you. Also, having four double espressos just over 4 hours before going to bed will mean that you will still have a sufficiently high dose of caffeine in your system to delay sleep onset by the time you are looking to go to bed, and so you need to factor in the amount of caffeine you are consuming as well.

Caffeine has one other useful role in fatigue risk management that is worth covering. A study in 2001 concluded that caffeine prior to napping decreases the chance of waking with sleep inertia and that caffeine shortly after waking improves sleep inertia.[31] Although this study was carried out with sleep-deprived participants, it is likely that this effect will be the same in people who are not sleep deprived.

To summarize, if we assume that peak caffeine concentration occurs 45 minutes after consumption, one or two cups of coffee towards the end of the cruise phase should give an increase in alertness and performance for the approach phase in a brain that has been awake for an extended period. To minimize the effects of tolerance and so maintain the usefulness of this handy strategy, consider switching to decaffeinated coffee on days when you do not need to use it as a fatigue countermeasure. Caffeine affects different people in different ways, so you may need to adjust your dose and time of consumption to maximize its effectiveness. Caffeine taken before a brief nap can aid with the subsequent transition from sleep to wakefulness at the end of the nap and decreases the chance of experiencing sleep inertia upon waking.

7.4.5 How to use sleep medications

The appropriate use of sleep medications is a complicated subject and these medications should only be used after receiving expert medical advice. You should also refer to the regulations in force in your own country regarding the use of such medications and your subsequent fitness to fly. Some medications may not be taken before flying duties, and pilots should seek specific guidance from their medical examiner. What follows is a simple explanation of the uses of some of the more common sleeping medications, but it is for information only and does not take precedence over any laws, rules, procedures or any medical advice you receive.

Sleep medications can be broadly divided into two categories[29]:

- Sleep initiators – These medications will aid the transition from wakefulness to sleep. They have a limited ability to keep the individual asleep and work on the principle that the internal sleep mechanism will take over. They may be used in cases of insomnia where the difficulty is thought to lie in sleep initiation, or to promote short periods of daytime sleep (4–7 hours) when, for operational reasons, these periods need to be scheduled at times when sleep urge is not high. Like any other sleep medication, they should not be used long term but they do not appear to cause any significant disruption to sleep architecture (the normal pattern of transitioning through the various stages of sleep). Drugs of this class are known as non-benzodiazepines or, sometimes, "Z-drugs", and include zaleplon, zolpidem and zopiclone.

• Sleep maintainers – These have a longer sedative effect and can be used to maintain sleep in conditions that would normally not be conducive to sleep or when the sleep period will be out of phase with the body's circadian sleep urge rhythm. In the latter circumstance, it may be easy to initiate sleep but it will be difficult to maintain sleep for the required 8 hours owing to a circadian decrease in sleep urge. These drugs can cause postdose drowsiness because of their long half-life. As with sleep initiators, sleep maintainers should not be used long term as they can lead to tolerance and dependence. Sleep architecture is disturbed when using these drugs and so the quality of sleep is not as good as naturally maintained sleep. Drugs of this class tend to be benzodiazepines, the most common of which is temazepam.

It is essential that you check with your medical examiner which medications you may legally use in your state, as some of these cannot be taken when the user wants to exercise the privileges of their pilot's license. These medications should only be used rarely, as prolonged use may cause further problems and the need for long-term use suggests that there is a more serious underlying sleep disorder rather than transient insomnia.

7.4.6   How to deal with jet lag

If there is one thing that evolution did not prepare our bodies for, it is for transiting several time zones in one go, possibly several times in the space of a few days. Our biological clock can adapt to seasonal changes in the light–dark cycle but is not designed to adapt to moving into a different time zone. It seems that there is a genetic block that limits the extent to which light exposure at different times can phase-shift our biological clock. Given that we did not evolve to cross time zones in short periods, this limitation may be there to allow us to maintain our normal 24 hour circadian patterns in conditions where we may not be exposed to the same levels of light. Studies have found that when mice have this genetic block deactivated, they can immediately adjust to their local time zone.[40] Although we are still somewhat limited in our ability to adapt to new times zones, because the SCN can reset itself based on light levels the body can eventually align itself to the new time zone, but this takes time. For normal travelers, long-haul travel usually suggests that a long period of time will be spent in the new time zone, enough to give the body a chance to align itself to the local day–night cycle. A commonly used rule of thumb is that the number of hours difference in the time zones from the point of departure to the point of arrival gives the number of days required to fully adapt to the new time zone. For example, a flight from London to New York results in a 5 hour difference in the time zones from departure to arrival and so 5 days will be required to fully adapt to the day–night cycle in New York. For long-haul pilots, it is a more complicated story as they may not have the luxury of spending a long time at their destination and may have to return to their point of departure or, more disruptively, make another long-haul flight to a time zone that is even further removed from that of their home base. Successful management of jet lag involves all of the techniques and strategies we have discussed so far. Above all, it requires careful planning, and below are some things you may want to consider when planning your sleep:

- Is it worth adapting to the new time zone? – If you are only going to spend 1–3 days in the new time zone, it may be best to maintain your normal sleep–wake cycle in the knowledge that you will be able to fit back into your normal pattern when you get home; therefore, keep your watch set to home time and maintain your schedule accordingly.
- Take into account your next flight – If you do decide to maintain your normal sleep–wake cycle (i.e. you fit your activities, meals and sleep times to the time at your home base, not to local time), will your return flight be during a period where you would normally be awake or asleep? If your normal sleep urge would be telling you to be asleep during the flight (or during the periods where you are at the controls in flight, and not in the crew rest area) it may be better to attempt to change your cycle to ensure that you are awake and alert when you have to operate the flight.
- Be aware of daylight – If you are planning to keep your body clock in phase with the day–night cycle at your home base, possibly because you will be returning shortly and your flight is scheduled for a period when you would normally be awake, your goal is to *avoid* adjusting to your new time zone. As we have said, light is the most powerful zeitgeber (cue that the body uses to detect the local day–night cycle). If you plan to be asleep during the day, make every effort to avoid strong light, especially sunlight, in the period before you go to bed. Wearing dark wrap-around sunglasses if you have to be outside, keeping the curtains drawn and keeping the room dimly lit can trick the brain into thinking it is night-time. Conversely, when waking attempt to get as much light exposure as possible from natural and artificial sources. This will suggest to the body that it is still daylight and will give you the best chance of maintaining your circadian rhythm that was set at your home base, even though you are now in a different time zone. Remember, iPads and laptops emit a strong, shortwave light that can mimic sunlight. Use this to your advantage.
- Take advantage of your circadian sleep urge rhythm to plan naps – If you are planning naps in your new time zone, take advantage of your knowledge about your circadian sleep urge rhythm to time the nap for a period where you would ordinarily have a high sleep urge. For example, if your normal sleep urge starts to rise at 8 p.m. and peaks at midnight before steadily decreasing (and your afternoon sleep urge is between 2 p.m. and 4 p.m.), if your new time zone is 5 hours behind, on the first day you will have an increasing sleep urge at 3 p.m. in the afternoon and at 9 a.m. in the morning. The 9 a.m. sleep urge might be counteracted by the strong sunlight associated with that time of the morning but a decent nap starting at 3 p.m. could be a useful supplement to your sleep. Remember, the longer you spend in the new time zone, the more your circadian rhythms will shift to try and get into phase with the local day–night cycle.
- Remember the light and the other zeitgebers – If you are planning to fully or partially adjust your new time zone, try to match your mealtimes and activities to the local time as well as getting as much exposure to light during the day as possible. This will increase the rate at which your circadian rhythms will phase-shift to match the new time zone. A review of jet lag in the medical journal the *Lancet* suggested the times where light should be sought or avoided based on the nature of the time-zone change.[41] This is shown in Table 7.1.
- Use medications if required and permitted – In consultation with your medical examiner, you may wish to use certain sleep medications to help with jet lag. If you are attempting to sleep during the day, a period where sleep urge is not normally high, it may be useful to

Table 7.1 **Times to seek and avoid light exposure to promote circadian phase shift**

Time zones to west (hours)	Good local times for light exposure	Bad local times for light exposure
3	18:00–00:00	02:00–08:00
4	17:00–23:00	01:00–07:00
5	16:00–22:00	00:00–06:00
6	15:00–21:00	23:00–05:00
7	14:00–20:00	22:00–04:00
8	13:00–19:00	21:00–03:00
9	12:00–18:00	20:00–02:00
10	11:00–17:00	19:00–01:00
11	10:00–16:00	18:00–00:00
12	09:00–15:00	17:00–23:00
13	08:00–14:00	16:00–22:00
14	07:00–13:00	15:00–21:00

Time zones to east (hours)	Good local times for light exposure	Bad local times for light exposure
3	08:00–14:00	00:00–06:00
4	09:00–15:00	01:00–07:00
5	10:00–16:00	02:00–08:00
6	11:00–17:00	03:00–09:00
7	12:00–18:00	04:00–10:00
8	13:00–19:00	05:00–11:00
9	14:00–20:00	06:00–12:00
10	Can be treated as 14 hours to the west	
11	Can be treated as 13 hours to the west	
12	Can be treated as 12 hours to the west	

take a sleep-maintaining agent such as temazepam that could permit 8 hours of sleep under circumstances where it would not normally be possible. Short-acting sleep initiators may be useful in order to take a prophylactic nap before commencing a flight duty.

- Consider melatonin – The evidence surrounding the use of melatonin in treating jet lag is complicated and sometimes confusing. As stated earlier, melatonin is a hormone secreted by the pineal gland during the hours of darkness and is associated with a decrease in body temperature and a feeling of sleepiness. It has a role in altering our circadian rhythms due to seasonal changes in the light–dark cycle. There is some evidence to suggest that melatonin can be useful when trying to phase-shift your circadian rhythms to match the local day–night cycle. A comprehensive review in 2002 of several major studies found that eight out of 10 trials demonstrated that 5 mg of melatonin taken at the target bedtime at the destination (10 p.m. to midnight) decreased jet lag in flights crossing five or more time zones.[42] It had to be taken just before going to bed and only when the bedtime was "normal" for the local time zone (i.e. 10 p.m. to midnight). This dose also led to people falling asleep faster and sleeping better than at a lower dose. Short-lived, high-peak doses of melatonin were more effective than slow-release doses. The benefit seems to be greater when more time zones are crossed and when the flights are eastward. Because of individual differences between people, discuss your own experience of jet lag with your medical examiner to see whether melatonin could be of use.

- Do not forget sleep hygiene – As with normal sleep and napping, your sleep environment and presleep routine in whichever time zone you are in should be optimized according to the strategies laid out earlier in the chapter.
- Make use of any sleep opportunity if you are extremely fatigued – If you are suffering from serious jet lag and have accumulated a significant sleep debt, be aware of the stress response that may occur when you are in this state. In order to function, you may start secreting stress hormones and this will disrupt your sleep even further. Before this happens, start taking any opportunity to sleep in order to pay back your sleep debt, and then try to adapt your sleeping pattern to the local day–night cycle.

7.4.7 Organizational strategies for fatigue risk management

While the individual has to take some responsibility for managing his own fatigue risk, the organization should be equally proactive in managing the fatigue risk on behalf of its employees. Many companies are now implementing a formal fatigue risk management system (FRMS) as part of their safety management system. The principles of an FRMS in aviation are well established and the ICAO has published a lot of guidance on the subject. The implementation details of an FRMS in an airline are beyond the scope of this book but, briefly, it moves beyond looking at flight time limitations and uses biomathematical models based on circadian rhythms to suggest when duties may be susceptible to fatigue-induced risk. These models take account of both cumulative fatigue and the scientific data pertaining to performance levels during the periods of reduced arousal that occur early in the morning in the window of circadian low. With enough feedback from crews, other airline-specific details can be factored into these models, including the difference in workload between sectors, commuting times and fatigue reports received from individual crew members. A fully operational FRMS can have a significant impact on safety, as well as being able to increase employee satisfaction and productiveness.

Chapter key points

1. Unless under the influence of alcohol or drugs, there is nothing that will affect your performance as much as fatigue.
2. Sleep is required because neurotoxins accumulate during the day and these can only be cleared from the brain when we sleep. Energy stores are also replenished.
3. Most people need 8 hours of sleep per night although there is some variation between individuals.
4. Fatigue is defined as "a physiological state of reduced mental or physical performance capability resulting from sleep loss or extended wakefulness, circadian phase, or workload (mental and/or physical activity) that can impair the crew member's alertness and ability to safely operate an aircraft or perform safety-related duties".
5. Fatigue is a physical, not a psychological, state.
6. The two main contributors to fatigue are sleep loss and extended wakefulness.
7. Fatigue has a negative effect on every aspect of our cognition and behavior.

8. The only remedy for fatigue is sleep.
9. Sleep allows the brain to clean itself and replenish energy stores, and also allows memory consolidation, a vital aspect of learning.
10. There are different stages of sleep, which are related to the level of brain activity that is occurring.
11. Slow-wave sleep is the deepest level of sleep and is the most important.
12. Sleep inertia occurs when we are woken from slow-wave sleep, and can have a negative effect on information processing abilities for approximately 30 minutes.
13. There are several factors that contribute to how and when we sleep, including the homeostatic sleep drive (sleep need), our biological clock and its effect on circadian rhythms, and our sleep urge, a circadian rhythm in itself.
14. If we do not get our required number of hours of sleep every night, we accumulate a sleep debt.
15. The greater the sleep debt, the worse our performance.
16. Chronic sleep debt results in the brain downgrading its activity so that it will persistently perform at a reduced level.
17. Sleep strategies and sensible napping can mitigate fatigue risk.
18. There are some strategies for managing jet lag that may be useful to long-haul pilots.

Recommended reading

Martin P. *Counting sheep: The science and pleasures of sleep and dreams*. HarperCollins; 2002. A fascinating introduction to the history and research efforts of those looking at the importance and purpose of sleep and dreams [Difficulty: Easy].

Caldwell JA, Caldwell JL. *Fatigue in aviation: A guide to staying awake at the stick*. Ashgate; 2003. A comprehensive review of fatigue and fatigue risk management in aviation, written by two highly regarded experts in this field [Difficulty: Easy/Intermediate].

References

1. Kong J, Shepel PN, Holden CP, Mackiewicz M, Pack AI, Geiger JD. Brain glycogen decreases with increased periods of wakefulness: Implications for homeostatic drive to sleep. *J Neurosci* 2002;**22**(13):5581–7.
2. International Civil Aviation Organization. *FRMS: Fatigue risk management systems, manual for regulators*. Montreal: ICAO; 2012. Doc 9966.
3. Boksem MA, Meijman TF, Lorist MM. Effects of mental fatigue on attention: An ERP study. *Cognitive Brain Res* 2005;**25**(1):107–16.
4. van der Linden D, Frese M, Meijman TF. Mental fatigue and the control of cognitive processes: Effects on perseveration and planning. *Acta Psychol (Amst)* 2003;**113**(1):45–65.
5. Hobbs A, Williamson A, Van Dongen HP. A circadian rhythm in skill-based errors in aviation maintenance. *Chronobiol Int* 2010;**27**(6):1304–16.
6. Xie L, Kang H, Xu Q, Chen MJ, Liao Y, Thiyagarajan M, et al. Sleep drives metabolite clearance from the adult brain. *Science* 2013;**342**(6156):373–7.
7. Stickgold R. Sleep-dependent memory consolidation. *Nature* 2005;**437**(7063):1272–8.
8. Ward LM. Synchronous neural oscillations and cognitive processes. *Trends Cogn Sci* 2003;**7**(12):553–9.

9. Siegel JM. REM sleep: A biological and psychological paradox. *Sleep Med Rev* 2011;**15**(3):139–42.

10. Mirmiran M. The function of fetal/neonatal rapid eye movement sleep. *Behav Brain Res* 1995;**69**(1):13–22.

11. Sastre JP, Jouvet M. The oneiric behavior of the cat. *Physiol Behav* 1979;**22**(5):979–89.

12. Government of India Ministry of Civil Aviation. *Report on accident to Air India Express Boeing 737-800 aircraft VT-AXV on 22nd May 2010 at Mangalore*; 2010.

13. Porkka-Heiskanen T, Strecker RE, Thakkar M, Bjørkum AA, Greene RW, McCarley RW. Adenosine: A mediator of the sleep-inducing effects of prolonged wakefulness. *Science* 1997;**276**(5316):1265–8.

14. Berson DM, Dunn FA, Takao M. Phototransduction by retinal ganglion cells that set the circadian clock. *Science* 2002;**295**(5557):1070–3.

15. Klerman EB, Shanahan TL, Brotman DJ, Rimmer DW, Emens JS, Rizzo III J, et al. Photic resetting of the human circadian pacemaker in the absence of conscious vision. *J Biol Rhythms* 2002;**17**(6):548–55.

16. Yamaguchi S, Isejima H, Matsuo T, Okura R, Yagita K, Kobayashi M, et al. Synchronization of cellular clocks in the suprachiasmatic nucleus. *Science* 2003;**302**(5649):1408–12.

17. Roenneberg T, Wirz-Justice A, Merrow M. Life between clocks: Daily temporal patterns of human chronotypes. *J Biol Rhythms* 2003;**18**(1):80–90.

18. Wahlstrom K, Dretzke B, Gordon M, Peterson K, Edwards K, Gdula J. *Final report: Examining the impact of later high school start times on the health and academic performance of high school students: A multi-site study.* Center for Applied Research and Educational Improvement, University of Minnesota; 2014.

19. Costa G, Sartori S, Akerstedt T. Influence of flexibility and variability of working hours on health and well-being. *Chronobiol Int* 2006;**23**(6):1125–37.

20. Brzezinski A. Melatonin in humans. *N Eng J Med* 1997;**336**(3):186–95.

21. Blatter K, Cajochen C. Circadian rhythms in cognitive performance: Methodological constraints, protocols, theoretical underpinnings. *Physiol Behav* 2007;**90**(2):196–208.

22. Brooks A, Lack L. A brief afternoon nap following nocturnal sleep restriction: Which nap duration is most recuperative? *Sleep* 2006;**29**(6):831–40.

23. Belenky G, Wesensten NJ, Thorne DR, Thomas ML, Sing HC, Redmond DP, et al. Patterns of performance degradation and restoration during sleep restriction and subsequent recovery: A sleep dose–response study. *J Sleep Res* 2003;**12**(1):1–12.

24. Möller-Levet CS, Archer SN, Bucca G, Laing EE, Slak A, Kabiljo R, et al. Effects of insufficient sleep on circadian rhythmicity and expression amplitude of the human blood transcriptome. *Proc Natl Acad Sci USA* 2013;**110**(12):E1132–41.

25. Zhang J, Zhu Y, Zhan G, Fenik P, Panossian L, Wang MM, et al. Extended wakefulness: Compromised metabolics in and degeneration of locus ceruleus neurons. *J Neurosci Nurs* 2014;**34**(12):4418–31.

26. Rosekind M, Graeber R, Dinges D, Connell L, Rountree M, Spinweber C, et al. *Crew factors in flight operations 9: Effects of planned cockpit rest on crew performance and alertness in long-haul operations.* California: NASA; 1994. NASA Technical Memorandum 108839.

27. National Transportation Safety Board. *Loss of control on approach, Colgan Air, Inc. Operating as Continental Connection Flight 3407, Bombardier DHC-8-400, N200WQ, Clarence Center, New York, February 12, 2009; 2010.* (NTSB/AAR-10/01 PB2010-910401).

28. Lamond N, Dawson D. Quantifying the performance impairment associated with fatigue. *J Sleep Res* 1999;**8**(4):255–62.

29. Caldwell JA, Mallis MM, Caldwell JL, Paul MA, Miller JC, Neri DF. Fatigue countermeasures in aviation. *Aviat Space Environ Med* 2009;**80**(1):29–59.

30. Caldwell JA. *Fatigue in aviation: A guide to staying awake at the stick.* Farnham, UK: Ashgate; 2003.

31. Van Dongen HPA, Price NJ, Mullington JM, Szuba MP, Kapoor SC, Dinges DF. Caffeine eliminates psychomotor vigilance deficits from sleep inertia. *Sleep* 2001;**24**(7):813–9.

32. Driskell JE, Mullen B. The efficacy of naps as a fatigue countermeasure: A meta-analytic integration. *Human Factors: J Hum Factors Ergon Soc* 2005;**47**(2):360–77.

33. Rodenbeck A, Hajak G. Neuroendocrine dysregulation in primary insomnia. *Rev Neurol (Paris)* 2001;**157**(11 Pt 2):S57–61.

34. Morin CM, Bootzin RR, Buysse DJ, Edinger JD, Espie CA, Lichstein KL. Psychological and behavioral treatment of insomnia: Update of the recent evidence (1998–2004). *Sleep* 2006;**29**(11):1398–414.

35. Federal Aviation Administration. (n.d.). *Obstructive sleep apnea.* Retrieved on February 27, 2014 from <http://www.faa.gov/pilots/safety/pilotsafetybrochures/media/Sleep_Apnea.pdf/>.

36. National Transportation Safety Board. *Scheduled 14 CFR Part 121: Air Carrier operation of Mesa Airlines (D.B.A. GO), Incident occurred Wednesday, February 13, 2008 in Hilo, Hawaii, Bombardier, Inc. CL-600, registration: N651BR.* Retrieved from: <http://www.ntsb.gov/aviationquery/brief.aspx?ev_id=20080222X00229/>; 2009.

37. Yaggi HK, Concato J, Kernan WN, Lichtman JH, Brass LM, Mohsenin V. Obstructive sleep apnea as a risk factor for stroke and death. *N Eng J Med* 2005;**353**(19):2034–41.

38. Van Dongen HPA, Kerkhof GA. Effects of caffeine on sleep and cognition. *Human Sleep and Cognition, Part II: Clinical and Applied Research* 2011;**2**:105.

39. Tieges Z, Richard Ridderinkhof K, Snel J, Kok A. Caffeine strengthens action monitoring: Evidence from the error-related negativity. *Cognitive Brain Res* 2004;**21**(1):87–93.

40. Jagannath A, Butler R, Godinho SI, Couch Y, Brown LA, Vasudevan SR, et al. The CRTC1-SIK1 pathway regulates entrainment of the circadian clock. *Cell* 2013;**154**(5):1100–11.

41. Waterhouse J, Reilly T, Atkinson G, Edwards B. Jet lag: Trends and coping strategies. *Lancet* 2007;**369**(9567):1117–29.

42. Herxheimer A, Petrie KJ. Melatonin for the prevention and treatment of jet lag. *Cochrane Database Syst Rev* 2002;(2):CD001520.

43. Cajochen C, Frey S, Anders D, Späti J, Bues M, Pross A, et al. Evening exposure to a light-emitting diodes (LED)-backlit computer screen affects circadian physiology and cognitive performance. *J Appl Physiol* 2011;**110**(5):1432–38.

Stress management and alcohol

Chapter Contents

Introduction   227
8.1 Chronic stress   227
   8.1.1 Prevention of allostatic load due to chronic stress   230
   8.1.2 Management of allostatic load due to chronic stress   230
      8.1.2.1 Management strategies aimed at the stressor   231
      8.1.2.2 Management strategies aimed at the individual   232
   8.1.3 Critical incident stress management   233
8.2 Alcohol   234
   8.2.1 Alcoholism in aviation   238
Chapter key points   240
Recommended reading   241
References   241

Introduction

As well as having an impact on pilot fatigue, chronic stress and alcohol have a significant impact on performance. Both of these topics will be covered in this chapter, with particular reference to their effect on flight safety and how they can be managed.

8.1   Chronic stress

Chronic stress is associated with insomnia and a wide range of other illnesses, from depression to heart attacks. Like fatigue, there is a myth that says that stress is "all in the mind", and while it certainly does have a mental component, the biological effects of stress are quite real. We should start by defining exactly what we mean by chronic stress. At this point, it is important to differentiate chronic stress from acute, situational stress that is related to "workload" and is discussed in Chapter 2 (Information Processing).

Stress is defined by McEwen and Seeman as "a real or interpreted threat to an individual's physiological and psychological integrity that results in adaptive biological and behavioral responses".[1] This definition is consistent with another observation made by Lazarus and Folkman in 1984, that "a stressor is only stressful to the individual if it is appraised as likely to tax or exceed the person's coping skills".[2] While acute, situational stressors fall under these definitions, this section will look at persistent, long-term stressors and their effect on the individual. The recognition that chronic stress causes biological and behavioral changes is nothing new. In 1967, the psychologists Thomas Holmes and Richard Rahe noted a correlation between

Practical Human Factors for Pilots. DOI: http://dx.doi.org/10.1016/B978-0-12-420244-3.00008-X

stressful life events and illness, and went on to develop a list of life events together with a figure reflecting the significance of each event ("life change units").[3] To assess stress levels, an individual works through this list and marks how many times he or she has experienced any of these life events in the past year before tallying up the final score. Scores above 300 suggest a very high risk of developing a stress-related disease. The scale is widely available and some of the life events are shown below, together with their life change units:

- death of a spouse – 100
- divorce – 73
- death of a close family member – 63
- marriage – 50
- retirement – 45
- pregnancy – 40
- death of a close friend – 37
- major mortgage – 32
- trouble with boss – 23
- vacation – 13

This Social Readjustment Rating Scale (SRRS) was one of the early tools that tried to relate stressful life events that people might experience with their chance of developing a stress-related illness. The SRRS has been well validated and consistently demonstrates that high-scoring individuals have a greater risk of developing an illness. High scores suggest that stress management techniques need to be implemented to prevent subsequent illness.

A more contemporary model of an individual's response to chronic stress is based on the concept of allostasis. Allostasis is the ability to maintain stability through change. Allostatic models of illness consider the inputs (stressors), processing (subjective psychological assessment of stressors) and outputs (stress response processes). Allostasis can be differentiated from homeostasis by the timescales involved. Humans have homeostatic mechanisms that keep blood glucose levels, blood pH and a myriad of other things controlled within a narrow range by making short-term alterations to the body that are easily and completely reversible. An allostatic mechanism changes the biology of the body to achieve stability (or near-stability) when a stressor is present and these changes occur over a longer period than homeostatic changes would. These allostatic mechanisms may involve the production of stress hormones, alteration of brain function or gene expression or other physical changes as a response to the stressor. Under normal circumstances, once the stressor is no longer present or the individual has adapted sufficiently to be able to cope with it over the long term, these allostatic responses decrease. We have already seen an example of allostasis at work in the example of someone with chronic insomnia given in Chapter 7. To achieve some sort of stability in the face of chronic, performance-degrading sleep debt, the allostatic response is to activate the hypothalamopituitary axis (the stress response system) to start producing stress hormones in order to give the individual enough chemical stimulation to get through the day. The problem is that, unlike homeostasis, which is short term and reversible, allostasis comes at a price, especially if allostatic mechanisms are repeatedly called upon.

Allostatic load is the "wear and tear" that the body experiences as a result of chronic stress. Allostatic mechanisms are designed to allow the person to maintain stability through change as a response to a stressor. A significant life event may cause an increase in stress hormones and an alteration of brain function, possibly even short-term depression, but over time, these stress responses will reduce until the individual is back to normal. However, there are four allostatic states that can lead to excessive allostatic load and a high risk of illness[4]:

- Repeated hits – An individual has repeated experiences of stressors, often different ones that do not allow for easy adaptation; for example, experiencing several different stressful life events one after the other.
- Lack of adaptation – This is similar to "repeated hits" except that it is more likely that the stressor is the same. Unlike the example given above, if an individual experiences the same stressful life event repeatedly, for example, losing one job after another, there is an opportunity to develop a coping strategy based on this repeated experience. In essence, there should be some sort of adaptation to chronic, repeated stressors, and if coping mechanisms are not used, there will be the same allostatic response to a stressor rather than a slowly decreasing response that would be associated with learning how to manage this stress in a less damaging way.
- Prolonged response – The response to the stress persists long after the stress has passed, for example, post-traumatic stress disorder (PTSD).
- Inadequate response – The response to the perceived stressor is inadequate, meaning that allostasis is not actually happening and so there is no adaptation. The effects of the stressor persist and are met with a low-level response that persists.

All of these states can lead to excessive allostatic load because the stress response persists. Allostatic load can actually be measured as a function of various biomarkers, including stress hormones, metabolic markers (e.g. cholesterol and lipoproteins), immune system markers and cardiovascular measures (e.g. blood pressure). Once these markers are combined, abnormal levels are closely correlated with impaired cognitive and physical performance, decreased immunity and a significant risk of cardiovascular disease.[5] Specific stress-related illness can occur as a result of these biological changes and some of these are given below:

- Stress-induced immune disease
  - worsening of asthma
  - psoriasis
  - increased frequency of cold sores or herpes
  - yeast infections such as thrush
  - impaired immune response leading to respiratory infection
  - higher susceptibility to infection
- Stress-induced gastrointestinal disease
  - diarrhea
  - nausea
  - irritable bowel syndrome
  - abdominal pain
  - gastric ulcers

- Stress-induced cardiovascular disease
  - increased risk of heart attacks and stroke
  - high blood pressure
- Stress-induced musculoskeletal disease
  - tension-induced headaches
  - tension-induced muscle pain
  - bone demineralization
  - increase in abdominal obesity due to high cortisol levels
- Stress-induced neurological disease
  - hippocampal atrophy increasing the risk of neurodegenerative disease
  - memory loss
  - impaired learning
- Stress-induced mental illness
  - acute anxiety disorder
  - depression
  - chronic insomnia

In short, chronic stress exists and is very strongly associated with impaired function, sickness and death. Stressful life events are the main contributor to individuals' stress and sometimes, for reasons beyond their control, multiple events can cause high stress levels that will lead to illness. Having a child die followed by a spouse dying followed by a parent dying is going to be hugely stressful, perhaps to the extent that the person will become immediately unwell from either mental or physical illness. Bear this in mind when we talk about stress management strategies (see Box 8.1).

8.1.1   Prevention of allostatic load due to chronic stress

Research has shown that there are some individual factors that can decrease the risk that a stressor will cause excessive allostatic load. These characteristics are as follows[7]:

- knowing that a stressor will potentially occur and preparing for it
- having a strong social/familial network
- having a regular diversionary activity such as a hobby

8.1.2   Management of allostatic load due to chronic stress

Hopefully, we have some foreknowledge that a stressor is going to occur and have a strong social network that could help to reduce the allostatic load should the stressor affect us. However, in addition to these factors that can reduce the impact of the stressor, we need to have some specific strategies for dealing with chronic stress so that it does not affect safety in the workplace. Before we go on to specific management strategies, we should return to the example of the person who has lost his or her child, spouse and parent in quick succession. There are no stress management techniques that will enable that person to come back to work the day after these events happen and be able to function normally. In some cases, ensuring safety may require keeping the stressed individual away from the workplace until he or she has recovered sufficiently. Chronic stress may go unnoticed in the workplace and may only come to

---

Box 8.1 Incident Case Study – Chronic Stress

Ryanair Flight 9672 – Boeing 737-800
Rome, Italy, 7 September 2005[6]

The crew of this 737-800 flight from Dusseldorf to Rome Ciampino were cleared direct to the outer marker for runway 15 and cleared to descend to 6000 feet (1800 m). The weather at Ciampino was bad, and shortly before the aircraft began the approach to runway 15 the wind changed direction, meaning that runway 33 would now be in use. The crew advised that they could perform a visual approach to runway 33 but to increase spacing with other aircraft, air traffic control (ATC) requested a left-hand, 360 degree orbit. The crew were unable to carry out this maneuver because of a nearby thunderstorm and requested a weather avoidance heading. The crew requested descent to 3000 feet (900 m) but ATC reported that this was below the minimum safe altitude. Eight miles from the runway, the crew were not visual with the airport, entered a zone of turbulence and decided to divert to Rome Fiumicino. The weather was also bad at Fiumicino and the crew began to miss frequency changes, nearly came into conflict with an aircraft on a parallel approach and ended up significantly low on the approach. The captain had disengaged the autopilot owing to turbulence and subsequently reported that they had encountered a microburst that led them to discontinue the approach. The aircraft landed in Pescara 25 minutes later. During the subsequent investigation into this case, the captain reported that he was not in a normal state of mind when he commenced this duty as his 3-month-old son had died a few days previously and this was his first duty since the funeral. The report also notes that he had not reported this fact to the company, despite the operations manual requiring it, because of the "possible temporary removal from operational activity", which he "perceived as a potential condition leading to termination of his employment".

---

light when acute situational stress occurs and the response to this stressful situation is impaired as in the case study given in Box 8.1. For management of other, less devastating stressors, we can implement strategies based on the three parts of the allostatic model of stress:

1. Input – the stressor itself.
2. Processing – the individual's perception of the stressor.
3. Output – the individual's response to the stressor.

Based on this model, our management strategies can focus either on the stressor itself or on the individual experiencing the stress.

8.1.2.1 Management strategies aimed at the stressor

We can manage stress by modifying the stressor itself. Using the same techniques as introduced in the section on threat management (see Section 3.5.1.5 in Chapter 3), there are three management strategies that we can adopt once we have identified a potential stressor:

- Avoid – Some stressors will affect us whether we want them to or not. If you look at the SRRS, you can see that we do not have much influence over the health of our family members or the death of a close friend. We can choose, however, to avoid potential stressors such as taking out a large mortgage or changing to a different line of work. So the first question we need to ask is, "What stressors can I avoid?"
- Buffer – While some stressors cannot be avoided, their effect on us can be buffered. Readjustments going on in the organization that you are working in are bound to cause some sort of stress. We can buffer the effects of this by not becoming too closely concerned with a process that is beyond our sphere of influence. So, although the stress still exists, we limit our engagement with it.
- Contingency plan – If we have identified a potential stressor that we cannot avoid but that we may be able to buffer, we now need to consider what to do if our buffering strategies are insufficient to stop the stress or are affecting us. In the case of a business readjustment, we can make contingency plans in the event that the readjustment leads to us losing our job. While this may not happen, by having a contingency plan in place we have more chance of managing this stress if it does occur.

8.1.2.2 Management strategies aimed at the individual

One of the early definitions of stress given by Lazarus and Folkman recognized that for a stressor to be viewed as negative, it had to be perceived to exceed the individual's capacity to deal with it.[2] We are probably all familiar with that friend or family member who seems to be able to take everything in their stride, and we may ask ourselves why we are not able to do the same thing. The stressors they experience are the same as those that affect us, but their perception of the stressor and their perception of their own ability to cope with it are different. So, if a stressor does affect us, how can we manage this at the perceptual level and in how we respond? Below are some strategies that will allow you to do this[7]:

- Exercise – There is a significant amount of data that demonstrates how effective exercise can be at reducing allostatic load.[8] Unfortunately, a regular exercise routine is often the first thing that we sacrifice when we are under pressure. Thirty minutes of aerobic exercise a few times a week can make a big difference. If you can do 30 minutes every day, even better.
- Do not sacrifice your sleep – Stress can lead to sleeplessness, and sleeplessness worsens stress. To make sure you do not fall into this vicious cycle follow the healthy sleep guidance given in Chapter 7 (Fatigue Risk Management).
- Avoid unhealthy "coping" strategies – When we are under pressure, we may rely more on alcohol, smoking or junk food to get through the day. While these may feel like appropriate "treats" to ourselves, the brief psychological pleasure is outweighed by the more persistent health risks. Alcohol and smoking will also affect the quality of your sleep and decrease your ability to respond appropriately to stressors.
- Social support – If you have a social support network, use it. Talking through your stresses with friends and family members can be a useful first step in the management process. Positive, supportive relationships seem to be associated with a reduction in allostatic load.[9]

- Deconstruct your problems – One of the principles behind cognitive–behavioral therapy (CBT) is breaking large, seemingly insurmountable problems into more manageable chunks and dealing with them one by one.
- Rely on your faith – If you are a person of faith, this can make things easier for you with regard to stress management. The meditative and diversionary aspects of practicing a faith can help in the management of stress and are associated with better health overall.[10]
- Think about your locus of control – The extent to which we believe that we can control what happens to us is related to having an internal or external locus of control. People who think that they can control what happens to them are said to have an internal locus of control and those who feel the opposite are said to have an external locus of control. Being able to switch from internal to external can be useful in managing stress. Successful strategies are reinforced by an internal locus of control: you made this happen and have earned the benefits. When a strategy does not have the desired result, being able to externalize the locus of control will prevent the failure from damaging self-esteem because it can be attributed to external events beyond your control.
- Make sure you get your "me time" – Under stressful conditions, our normal routines are often disrupted. Exercise regimes are sacrificed, social functions are missed and less time is spent with family. We are also very willing to sacrifice quality time with ourselves. When under stress, force yourself to schedule some time for yourself. It could be reading, going to a movie or any other relaxing activity that will divert your mind from stressors.
- Is it time to get some help? – In some cases, self-management strategies may not be enough to deal with the stress in your life. Professional help in the form of CBT or some other form of therapy may be useful. In one study, CBT was given to a group of patients who were recovering from a heart attack. There was a 41% reduction in recurrent heart attacks and a 28% reduction in mortality compared with patients who did not receive this therapy.[11] Stress is a real, biological phenomenon and that stress can sometimes be excessive for reasons beyond your control. There is excellent evidence to suggest that asking for professional help can improve health outcomes significantly.
- Mindfulness-based stress reduction – This form of therapy can be self-directed or can form part of CBT. Mindfulness is training yourself to be able to focus on the present time and your immediate environment by avoiding unrelated thoughts. Although it has its roots in meditation, there is evidence to support its usefulness in stress management.[12]

8.1.3   Critical incident stress management

Critical incident stress management (CISM) is a psychological intervention used with people who have been involved in some sort of critical incident. A critical incident is a situation that causes a strong stress reaction, and in aviation this could be an accident, an incident, a near-miss or any other event with the potential to traumatize the people involved. After a critical incident there can be acute stress reactions resulting in chronic stress disorder that can persist for a prolonged time. About 20% of people who experience a critical incident will go on to develop post-traumatic stress disorder (PTSD). CISM is an intervention designed to limit the stress reactions and avoid chronic stress disorder and PTSD. CISM should be delivered by trained experts and normally follows a set program[13]:

1. Individual intervention – One-to-one discussion of the event takes place with a qualified CISM expert shortly after the event.

2. Critical incident stress defusing – Group discussions regarding the event take place up to 24 hours after it has happened.
3. Critical incident stress debriefing – This individual or group debriefing occurs within 72 hours of the event and focuses on sharing coping strategies.
4. Follow-up – Treatment is given for any residual stress disorders as required.

There is still some debate in the scientific community about the efficacy of CISM, but recognition of the negative effects of critical incident stress and the structure of CISM interventions may provide organizations with a framework for managing post-event stress in their employees.

8.2 Alcohol

The relationship between alcohol and aviation is a complex one but one that deserves discussion. Before discussing this association, let's consider the biological effects of alcohol and how it affects performance. Alcohol is a relatively simple chemical, also known as ethanol. Ethanol binds to multiple neurotransmitter receptors and acts as a depressant of the central nervous system. The results of this depressant action are slowed cognition, impairment of memory formation, impairment of motor and sensory function followed by unconsciousness and, in high enough concentrations, death. As ethanol is broken down in the liver it is converted into acetaldehyde and this causes damage to multiple body systems, for example, cirrhosis of the liver. Other short-term effects of alcohol are as follows:

- It inhibits production of antidiuretic hormone in the brain, causing an increase in urine production and subsequent dehydration.
- The generalized slowing of neural transmissions results in impaired balance and this is worsened owing to thickening of the fluid within the semi-circular canals of the ear that comprise part of our balance system.
- Alcohol stimulates acid production in the stomach and can worsen gastritis and peptic ulcer disease.

The long-term effects of alcohol include:

- heart disease
- cancer – 3.6% of all cancers are attributable to alcohol[14]; the toxic by-product of alcohol metabolism (acetaldehyde) causes the cellular damage that results in cancer
- brain damage
- long-term memory impairment
- depression
- chronic insomnia
- liver disease
- inflammation of the pancreas
- weight gain
- diabetes
- testicular atrophy and impotence

With reference to drinking, alcohol is normally measured in units. One unit is the equivalent of 10 ml of pure alcohol and different drinks have different numbers of

Table **8.1** **Number of alcohol units in common beverages**[15]

Beverage	Strength (ABV)	Volume (ml)	Units
Small glass of wine	12.0%	125	1.5
Medium glass of wine	12.0%	175	2.1
Large glass of wine	12.0%	250	3
Bottle of wine	13.5%	750	10
Pint of lager/beer/cider	3.6%	568	2
Pint of strong lager/beer/cider	5.2%	568	3
Bottle of lager/beer/cider	5.0%	330	1.7
Can of lager/beer/cider	4.5%	440	2
Single, small shot of spirits	40.0%	25	1

ABV: alcohol by volume.

Table **8.2** **Physical effects associated with different blood alcohol concentrations (BACs)**[16]

BAC	Effects
0.01–0.05%	Average individual appears normal
0.03–0.12%	Mild euphoria, talkativeness, decreased inhibitions, decreased attention, impaired judgment, increased reaction time
0.09–0.25%	Emotional instability, loss of critical judgment, impairment of memory and comprehension, decreased sensory response, mild muscular incoordination
0.18–0.30%	Confusion, dizziness, exaggerated emotions (anger, fear, grief) impaired visual perception, decreased pain sensation, impaired balance, staggering gait, slurred speech, moderate muscular incoordination
0.27–0.40%	Apathy, impaired consciousness, stupor, significantly decreased response to stimulation, severe muscular incoordination, inability to stand or walk, vomiting, incontinence of urine and feces
0.35–0.50%	Unconsciousness, depressed or abolished reflexes, abnormal body temperature, coma; possible death from respiratory paralysis

units in them. In the UK, the National Health Service recommends that men should not drink more than 3 units of alcohol a day; for women this limit is 2 units a day. Table 8.1 shows the number of units in some common alcoholic beverages.

Once absorbed into the bloodstream, alcohol can be measured by looking at blood alcohol concentration (BAC), the percentage by volume of alcohol in the blood. High BAC is associated with more profound effects, as shown in Table 8.2.

The effects of alcohol that are most relevant to pilots are obvious. There is impairment of executive function (planning and risk assessment), impairment of balance and a decrease in hand–eye coordination. For these reasons and because of the complex nature of aviation, legal limits for pilots, particularly commercial pilots, are significantly lower than for drivers. For example, in the UK and most of the USA, the legal driving limit is 0.08% BAC. For commercial pilots, the limit is 0.02% in the UK and

0.04% in the USA. It is worth noting that no matter what the state of registration of the aircraft you are flying, you are subject to local laws when abroad. For example, an American pilot reporting for duty in the UK would need to have a BAC of less than 0.02% to be legal to fly.

If we consider the lower limit of 0.02%, how does that relate to drinking activities? Each unit consumed increases your BAC by approximately 0.02%. Alcohol is eliminated from the body at a rate of approximately 1 unit per hour; thus, your BAC will drop by approximately 0.02% per hour.[17] There are several variables that affect how quickly alcohol is eliminated from the bloodstream:

- Gender – Females tend to have quicker elimination than males.[18]
- General health – Any liver function impairment or the use of certain medications will decrease alcohol elimination rates.
- Alcoholic tolerance.
- Duration of the drinking period.
- Factors that slow alcohol absorption (e.g. food) – Drinking on an empty stomach means that the alcohol will be quickly absorbed into the bloodstream and the liver will then begin breaking down the alcohol immediately. When drinking after eating, the alcohol will be absorbed more slowly because of the presence of food in the stomach. This can mean that alcohol will persist in the bloodstream for a substantially longer period.

A commonly used rule is that alcohol must not be consumed within 8 hours of flying. Unfortunately, given the variables that affect alcohol elimination, such as how much alcohol was consumed and whether food was eaten at the same time, there is no way to guarantee that after 8 hours your BAC will be below the legal limit. For example, if in the period between 11 hours and 8 hours before reporting for duty (a 3 hour period) you consume 12 units (4 units per hour), your BAC will be approximately 0.20% at the end of the drinking period. After 8 hours, this will have dropped to about 0.04%, still above the legal limit. However, if food was consumed, alcohol absorption will be slower, meaning that there is a much higher chance of being above the legal limit after 8 hours.

It should be noted that there is considerable variability between people with regard to alcohol elimination,[19,20] but the purpose of that illustration was to show that the 8 hour rule is no guarantee when it comes to ensuring that you are below the legal limit to fly. While some may argue that the legal flying limit is unreasonably low compared to the driving limit, the legal reality is that the limit is fixed and it is enforced. Being found to be above that limit at any time during a duty is a criminal offense and one that carries severe penalties such as loss of license, loss of job and imprisonment. Box 8.2 lists some cases where pilots were found to be above the legal

---

Box 8.2 Incident Case Studies – Alcohol

Japan Air Lines (Cargo) Flight 8054 – McDonnell-Douglas DC-8-62F
Anchorage, Alaska, USA, 13 January 1977[21]
Although staff considered the captain to be under the influence of alcohol, the flight proceeded. The aircraft stalled shortly after take-off and crashed, killing all

five crew members. An autopsy showed that the captain had a blood alcohol concentration (BAC) of approximately 0.3% (the legal limit being 0.04% in the USA).

Northwest Airlines Flight 650 – Boeing 727
Minneapolis, Minnesota, USA, 8 March 1990[22]
All three pilots were found to be above the legal limit after completing a flight from Fargo to Minneapolis. The Federal Aviation Administration had received a tip-off by telephone saying that all the pilots had been seen drinking heavily in a bar the previous night. All three lost their jobs and licenses and spent between 12 and 16 months in jail.

America West Airlines Flight 556 – Airbus A320
Miami, Florida, USA, 1 July 2002[23]
The captain and the first officer left a Miami sports bar at 04:45, having run up a $120 bar tab. They reported for duty at 09:30. Security staff at Miami International Airport suspected the pilots of being under the influence of alcohol and the plane was ordered back to stand after having been pushed back. The captain and the co-pilot were found to have BACs of 0.091% and 0.084%, respectively. They were dismissed from their jobs; the captain was sentenced to 5 years in prison and the first officer was sentenced to 30 months in prison.

United Express Flight 7687 – Embraer 170
Austin, Texas, to Denver, Colorado, USA, 8 December 2009[24]
During the flight, the captain noticed a smell of alcohol from the first officer. The first officer was subsequently found to have a BAC of 0.094% and was sentenced to 6 months in prison.

Delta Airlines – Boeing 767
London, UK, 1 November 2010[25]
The first officer of this flight was stopped at Heathrow Airport when security staff noticed a smell of alcohol from him. He was found to have a BAC of 0.089% and was sentenced to 6 months in prison.

American Eagle Flight 4590
Minneapolis, Minnesota, USA, 4 January 2013[26]
Staff at Minneapolis St. Paul International Airport smelled alcohol on the pilot's breath. He was removed from the flight and subsequently found to have a BAC of 0.107%. He was sentenced to one year in prison.

Pakistan International Airways – Airbus 310
Leeds, UK, 18 September 2013[27]
Staff at Leeds Bradford International Airport noted that the captain was unsteady on his feet and he was removed from the aircraft. He was shown to have a breath alcohol content of 41 micrograms (μg) in 100 ml of breath (the UK limit for commercial flying is 9 μg in 100 ml of breath). He reported that he had stopped drinking 19 hours before the flight was due to leave but because of the amount of alcohol consumed, this was insufficient to allow his alcohol level to reduce to below the legal limit. He was sentenced to 9 months in prison.

limit. These are just a few of the many cases where pilots have been found to be over the legal limit and where they not only have lost their jobs but may never be able to fly commercially again. The prison sentences that are handed down tend to be severe and so conviction brings with it a lot of negative outcomes. However, the reason for including these brief summaries in not to further criticize these pilots. The difficulty we face in our profession is that we often have to spend time away from home and it is a natural tendency to want to socialize with your colleagues. While I cannot comment on the drinking habits of all the people involved, it is possible that some of these convictions resulted from an evening of socializing that got out of hand. As we have seen from the previous paragraphs on alcohol absorption and elimination, it does not take much to end up over the very low alcohol limit that most countries impose on their pilots. The reality is that the legislation surrounding drinking and flying is complex, as are the rules laid out by individual companies. It can be difficult for pilots to know which rules they must work to. For example, the captain of the Pakistan International Airlines flight reported that the rule in his home country was that there should be no alcohol consumed within the 12 hours prior to the flight. He had abided by this rule in the UK but had drunk such a significant quantity before this time that he remained above the legal limit when he reported for duty. Given that some countries, such as the UK, can base a criminal conviction on breath alcohol concentration, it is very difficult to offer practical guidance for remaining below this limit, as there is considerable variability in alcohol metabolism rates between different people. However, based on what we have covered so far, here is some guidance that may help:

- Forget the 8 hour rule – 8 hours is too short. You could consume a moderate amount of alcohol and still be above the legal flying limit 8 hours later. Imposing a 12 or 24 hour rule on yourself is significantly safer.
- Think about your own drinking practices – Think about the number of units you have had. You will need at least that number of hours to eliminate the alcohol, even longer if you have eaten.
- Do not let anyone talk you into anything – For cultural or lifestyle reasons, there can be a social pressure to drink. The psychology and sociology behind this pressure are beyond the scope of this book but it is important to recognize that there can be a powerful, social pressure to consume alcohol. Consider the guidance we have covered so far and do not let anyone talk you into breaking your own rules. If you have decided to adopt a 12 hour rule, make that a limit that is set in stone. Based on this limit, you can politely decline another drink by saying that you do not drink alcohol within 12 hours of reporting for duty.

8.2.1 Alcoholism in aviation

As in any profession, there are pilots out there who are alcoholics. Alcoholism, like other forms of addiction, is recognized as a medical disease and is associated with neurological changes in how the brain processes chemical stimulation. Alcohol and drug addiction are not simply a matter of poor willpower or poor self-control. There are structural and functional changes in the midbrain (an area located close to the thalamus) that lead directly to addictive behaviors and the individual may be completely unable to control these behaviors. A diagnosis of alcoholism is a permanent one. It is incurable, but it is manageable. In many cases, the diagnosis of alcoholism is a terminal one and there is a significant risk of death if it is not properly managed. Unfortunately, professional pilots are used to maintaining a high level of control in their working environments and so the lack of control associated with addictive diseases is often very difficult to deal with. Denial is a very common strategy that is used to deal with the effects of alcoholism and this can be the major barrier to getting professional help to manage this disease.

The fourth version of the *Diagnostic and Statistical Manual of Mental Disorders* (DSM-IV) of the American Psychiatric Association gives diagnostic criteria for alcohol abuse and alcohol dependence (alcoholism).[28] A later version of this manual (DSM-5) slightly adapted the criteria, but the criteria from DSM-IV are more appropriate in the context of this chapter:

- **Alcohol abuse** – A maladaptive pattern of drinking, leading to clinically significant impairment or distress, as manifested by at least one of the following occurring within a 12 month period:
    - Recurrent use of alcohol resulting in a failure to fulfill major role obligations at work, school, or home (e.g. repeated absences or poor work performance related to alcohol use; alcohol-related absences, suspensions, or expulsions from school; neglect of children or household)
    - Recurrent alcohol use in situations in which it is physically hazardous (e.g. driving an automobile or operating a machine when impaired by alcohol use)
    - Recurrent alcohol-related legal problems (e.g. arrests for alcohol-related disorderly conduct)
    - Continued alcohol use despite having persistent or recurrent social or interpersonal problems caused or exacerbated by the effects of alcohol (e.g. arguments with spouse about consequences of intoxication).
- **Alcohol dependence (alcoholism)** – A maladaptive pattern of drinking, leading to clinically significant impairment or distress, as manifested by three or more of the following occurring at any time in the same 12 month period:
    - Need for markedly increased amounts of alcohol to achieve intoxication or desired effect; or markedly diminished effect with continued use of the same amount of alcohol
    - The characteristic withdrawal syndrome for alcohol; or drinking (or using a closely related substance) to relieve or avoid withdrawal symptoms
    - Drinking in larger amounts or over a longer period than intended
    - Persistent desire or one or more unsuccessful efforts to cut down or control drinking
    - Important social, occupational, or recreational activities given up or reduced because of drinking
    - A great deal of time spent in activities necessary to obtain, to use, or to recover from the effects of drinking
    - Continued drinking despite knowledge of having a persistent or recurrent physical or psychological problem that is likely to be caused or exacerbated by drinking.

While these are the clinical criteria for diagnosis, there are many simpler questionnaires that will tell you if you might need to seek help with regard to your alcohol consumption. A commonly used one asks the following four questions[29]:

1. Do you ever feel you should cut down on your drinking?
2. Do you get annoyed when other people criticize your drinking?
3. Do you feel guilty about your drinking?
4. Do you ever need a drink in the morning to steady your nerves or get over a hangover?

Answering "yes" to two or more suggests a problem with alcohol, but question 4 is particularly important as drinking in the morning suggests withdrawal effects and is strongly associated with alcoholism. If you have answered "yes" to two or more questions or if you have answered "yes" to question 4, it is probably time to go and talk things over with your doctor. Do not fall into the denial trap that so many others have fallen into. It is often a fatal one.

A diagnosis of alcoholism does not mean an end to your career as a pilot. Most major aviation authorities recognize that alcoholism is a medical disease beyond the control of the individual. The American Federal Aviation Authority has the Human Intervention Motivation Study, a specialized program that identifies and helps alcoholic pilots with the aim of getting them back into the flight deck.[30] It has helped thousands of pilots so far and has saved not only the careers of the people involved, but also their lives. In the UK, although there is no specialized program for pilots, the Civil Aviation Authority treats alcoholism like any other disease and, provided the disease is managed appropriately and there is monitoring, flying duties may be continued.

It would be inappropriate to go into the management strategies that should be used when alcohol abuse or alcohol dependence is diagnosed. These are difficult diseases to treat and require a lot of professional guidance and personal determination. The goal of this section is to say that admitting to alcoholism does not mean the end of your career. Although it may sound risky to step forward and seek help for this serious illness, the risks of not doing so could be that you end up as a newspaper story or that your health and home life continue to suffer because of alcohol. If in doubt, talk to your medical examiner. Fortunately for all of us, the industry has now recognized the true nature of alcoholism, and help is available so that you can keep flying safely and regain your life outside work.

Chapter key points

1. Stress is defined as "a real or interpreted threat to an individual's physiological and psychological integrity that results in adaptive biological and behavioral responses".
2. Significant life events can cause stress, and the Social Readjustment Rating Scale (SRRS) can quantify these effects.
3. Allostasis is the ability to maintain stability through change.
4. Allostatic load is the wear and tear that the body experiences as a result of chronic stress.
5. Allostatic load can cause a wide variety of illnesses, especially immune-related ones.
6. Managing allostatic load can focus on the stressor itself, the individual's interpretation of the stressor or the individual's response to the stressor.
7. Critical incident stress management (CISM) is a sequence of timed interventions that can be used to limit the chance of developing post-traumatic stress disorder in people who have experienced a significantly stressful event.
8. Alcohol has many negative short-term and long-term effects.
9. The rate at which alcohol is eliminated from the body is very variable and the 8 hour rule is no guarantee that the pilot will be below the legal flying limit when he reports for work.
10. Many pilots have been arrested and imprisoned for being above the legal flying limit.

11. Individuals need to set themselves personal limits regarding their alcohol use.
12. As in any other profession, alcoholism occurs in commercial aviation.
13. There are several signs that show that a person may have problems with alcohol abuse or alcohol dependence.
14. Alcohol dependence is a biological disease affecting the brain, one that is frequently fatal.
15. The nature of commercial aviation means that people who are concerned that they have problems with alcohol abuse or alcohol dependence may not seek help.
16. Most national regulatory authorities have systems in place to help pilots who are suffering from this serious illness and allow them to continue to fly.

Recommended reading

Sapolsky RM. *Why zebras don't get ulcers*. WH Freeman; 2004. A fantastic review of much of the scientific research into stress, allostasis and potential therapies [Difficulty: Easy/Intermediate].

Prouse L. *Final approach: Northwest Airlines Flight 650*. CreateSpace; 2011. Although this is not an academic book, it tells the story of one pilot's conviction and imprisonment for flying while under the influence of alcohol, his journey as a recovering alcoholic and how he eventually made it back into the captain's seat [Difficulty: Easy].

References

1. McEwen BS, Seeman T. Protective and damaging effects of mediators of stress: Elaborating and testing the concepts of allostasis and allostatic load. *Ann NY Acad Sci* 1999;**896**(1):30–47.
2. Lazarus RS, Folkman S. *Stress, appraisal, and coping*. New York, NY: Springer; 1984.
3. Holmes TH, Rahe RH. The Social Readjustment Rating Scale. *J Psychosom Res* 1967;**11**(2):213–8.
4. Juster RP, Bizik G, Picard M, Arsenault-Lapierre G, Sindi S, Trepanier L, et al. A trans-disciplinary perspective of chronic stress in relation to psychopathology throughout life span development. *Develop Psychopathol* 2011;**23**(3):725–6.
5. Seeman TE, Singer BH, Rowe JW, Horwitz RI, McEwen BS. Price of adaptation – Allostatic load and its health consequences: MacArthur studies of successful aging. *Arch Intern Med* 1997;**157**(19):2259–68.
6. Agenzia Nazionale per la Sicurezza del Volo Final report: Serious incident to B737-800, Registration EI-DAV, on approach to Rome Ciampino and Fiumicino Airports September 7th 2005. Rome: ANSV; 2008.
7. Sapolsky RM. *Why zebras don't get ulcers*. New York: WH Freeman; 2004.
8. Tsatsoulis A, Fountoulakis S. The protective role of exercise on stress system dysregulation and comorbidities. *Ann N Y Acad Sci* 2006;**1083**(1):196–213.
9. Brooks KP, Gruenewald T, Karlamangla A, Hu P, Koretz B, Seeman TE. Social relationships and allostatic load in the MIDUS study. *Health Psychol* 2014 Advance online publication, January 20 <http://dx.doi.org/doi:10.1037/a0034528>.
10. Thoresen CE. Spirituality and health: Is there a relationship? *J Health Psychol* 1999;**4**(3):291–300.

11. Gulliksson M, Burell G, Vessby B, Lundin L, Toss H, Svärdsudd K. Randomized con-
    trolled trial of cognitive behavioral therapy vs standard treatment to prevent recurrent
    cardiovascular events in patients with coronary heart disease: Secondary Prevention in
    Uppsala Primary Health Care project (SUPRIM). *Arch Intern Med* 2011;**171**(2):134–40.
12. Grossman P, Niemann L, Schmidt S, Walach H. Mindfulness-based stress reduction and
    health benefits: A meta-analysis. *J Psychosom Res* 2004;**57**(1):35–43.
13. European Organisation for the Safety of Air Navigation *Critical incident stress manage-
    ment: User implementation guidelines*. Brussels: Eurocontrol; 2008. (08/11/03–27).
14. Boffetta P, Hashibe M, La Vecchia C, Zatonski W, Rehm J. The burden of cancer attribut-
    able to alcohol drinking. *Int J Cancer* 2006;**119**(4):884–7.
15. National Health Service Choices. *Alcohol units*. Retrieved from: <http://www.nhs.uk/
    Livewell/alcohol/Pages/alcohol-units.aspx#table/>; 2013.
16. Federal Aviation Administration. (n.d.). Alcohol and flying: A deadly combination.
    Retrieved from 27 March 2014 <http://www.faa.gov/pilots/safety/pilotsafetybrochures/
    media/alcohol.pdf.
17. Wilkinson P. Pharmacokinetics of ethanol: A review. *Alcoholism: Clinical Experimental
    Res* 1980;**4**(1):6–21.
18. Taylor JL, Dolhert N, Friedman L, Mumenthaler M, Yesavage JA. Alcohol elimination
    and simulator performance of male and female aviators: A preliminary report. *Aviat Space
    Environ Med* 1996;**67**(5):407–13.
19. Ramchandani VA, Bosron WF, Li TK. Research advances in ethanol metabolism.
    Pathologie Biologie 2001;**49**(9):676–82.
20. National Institute on Alcohol Abuse and Alcoholism. Alcohol metabolism. *Alcohol Alert*
    1997(No. 35) <http://pubs.niaaa.nih.gov/publications/aa35.htm>.
21. National Transportation Safety Board. (1979). *Japan Air Lines Co., Ltd. McDonnell–
    Douglas DC-8-62F, JA 8054, Anchorage, Alaska, January 13, 1977*. (NTSB–AAR–78–7).
22. Prouse L. *Final approach: Northwest Airlines Flight 650*. CreateSpace Independent
    Publishing Platform; 2011.
23. CBS News. *Drunk airline pilots face prison*. Retrieved from: <http://www.cbsnews.com/
    news/drunk-airline-pilots-face-prison/>; 2005.
24. United States Attorney's Office, District of Colorado. United Express pilot sentenced for
    operating an aircraft under the influence of alcohol. Retrieved from: <http://www.justice
    .gov/usao/co/news/2011/November2011/11_4_11.html/>; 2011.
25. AOL Travel. Drunk Delta pilot jailed in UK. Retrieved from: <http://news.travel.aol
    .com/2011/01/24/drunk-delta-pilot-jailed-in-uk/>; 2011.
26. Associated Press. *Ex-pilot pleads guilty in Minnesota alcohol case*. Retrieved from:
    <http://bigstory.ap.org/article/ex-pilot-pleads-guilty-minnesota-alcohol-case/>; 2014.
27. Telegraph & Argus. *Airline pilot jailed for being three times over limit at Leeds Bradford
    Airport*. Retrieved from: <http://www.thetelegraphandargus.co.uk/news/10828572.Jailed_
    Pakistan_International_Airlines_pilot_was_three_times_over_limit__court_hears/>;
    2013.
28. American Psychiatric Association *Diagnostic and statistical manual of mental disorders:
    DSM-IV*. Washington, DC: APA; 2000.
29. Ewing JA. Detecting alcoholism: The CAGE questionnaire. *J Am Med Assoc*
    1984;**252**(14):1905–7.
30. Werfelman L. When bottle meets throttle. *Aerosafety World* 2006(September):32–6.

Automation management

Chapter Contents

Introduction   243
9.1 Systems of aircraft automation   245
9.2 Flight control laws   246
9.3 Levels of automation and their uses   249
9.4 Flight mode annunciators   252
9.5 Automation, perception and Newton's laws of motion   253
9.6 The ironies of automation   254
9.7 Skill fade and automation dependency   258
9.8 Automation complacency   259
9.9 Automation bias   260
9.10 Automation surprises   262
Chapter key points   265
References   266

Introduction

At a conference on error management held in the Netherlands in 2013, I listened as an aeronautical engineer with no experience of flying commercial airliners stood in front of the assembled group of human factors and safety professionals and bemoaned the era of the "Nintendo pilots". As a product of this era, it was an argument I had heard before: how awful it is that pilots no longer have the basic stick-and-rudder skills; how terrible that pilots have become so reliant on automation that they cannot handle the plane when things go wrong; how remiss it is of modern pilots to let their skills deteriorate like that. This is not just a position held by non-pilots. Senior pilots, often retired, will talk nostalgically about the days when everyone could fly a visual approach without a second thought. They will say how terrible it is that these skills have gone and that a visual approach can now be a source of considerable stress to a pilot. Commentators were quick to report the fact that the captain of the Asiana Airlines Boeing 777 that crashed in San Francisco in 2013 subsequently told the National Transportation Safety Board (NTSB) that he found the idea of a visual approach very stressful but because everyone else had been carrying them out, he was reluctant to say that he wasn't comfortable with conducting a visual approach.[1] The relationship between pilots and automation is a complex one, and one that is difficult both to assess and to manage. Speaking for myself, these criticisms do me no good. They explain nothing, suggest nothing and seem to be based on a nostalgic, rose-tinted view of the past when skies were empty and an approach was stable as long as the aircraft didn't run off the end of the runway. Skies are now busier and more tightly packed than they have ever been,

Practical Human Factors for Pilots. DOI: http://dx.doi.org/10.1016/B978-0-12-420244-3.00009-1

departures and approaches need to be flown with an unprecedented level of precision, we fly instrument landing system (ILS) approaches 99 times out of 100, and every minute action, control input and automation selection made by the pilot is recorded, analyzed and reported by flight data monitoring (FDM) systems. Depending on the safety culture of the company, any deviations picked up by FDM can have significant consequences for the pilot. This is a brave new world of commercial aviation and a nostalgic desire to return to the ways of the past will not address the problems that we face today. Instead, we must tackle the relationship between the pilot and the automation head on, understand it, balance it against what we need to optimize safety and efficiency, and then continually monitor how our strategies work. I can think of no other concept in human factors that is as challenging but I do know this: on a complex, modern airliner, one that pushes the limits of engineering to satisfy the requirements of the commercial air transport world, if systems start failing and computers start flashing up warnings, I want a "Nintendo pilot" who can manage the systems, understand what they are trying to say and can act accordingly. Pilots cannot change the pace of aircraft technology, in the same way as we cannot change the nature of the commercial environment we have to operate in. We just have to make sure that our training is optimized and keeps pace with the operational requirements of this modern era of commercial air transport.

Automation and its increasing role in commercial aviation have created a safety paradox. Although automation seems to improve safety during normal operations, over-reliance on automation means that it can make situations worse in abnormal operations. Dealing with this paradox takes up a large part of this chapter. Many articles written on this subject say that crews must be trained to remain engaged with the aircraft, that crews must be trained not to become overreliant on automation and that crews must be trained to tell when automation is malfunctioning. Sadly, saying that crews *must* be trained to manage automation is far easier than saying *how* crews can be trained to manage automation. Years of research have shown that are no easy answers, only possible strategies for mitigating risks. No other technological advance has changed the world of aviation or the role of the pilot more than automation, to the extent that the question is being seriously asked now whether pilots are needed at all. As air traffic management systems become more automated and the public puts more faith in the reliability of automated systems, could this become a reality? Already, automation has reduced the number of crew required to operate complex commercial airliners from three to two. Will that number reduce even further? Automated aircraft were going to make the skies safer because they could maintain assigned altitudes far more accurately than pilots could. These systems delivered on their promise and the industry reaction was to use this as justification for decreasing the vertical separation of aircraft to fit more into a fixed volume of airspace. It seems that when automation offers the opportunity for a greater level of safety, the industry response is to use this as justification to push the limits even more in order to increase the efficiency of the air transport network by increasing the number of planes in the sky. Aside from these broad, industry-spanning questions, the modern pilot also has to deal with automation on a day-to-day basis. Being able to do this effectively relies on understanding some of the principles behind automation design and considering how it can be best used to permit safe and efficient operation of flights. It is also worth considering how automation has been implicated in accidents and incidents and what we can learn from these cases to keep our operation as safe as possible.

9.1 Systems of aircraft automation

There are seven systems on a modern commercial aircraft that contribute to automation. The combinations used determine the level of automation:

- Flight control laws – In some fly-by-wire systems, during manual flight, the movement of the flight controls does not directly affect the control surfaces. Therefore, there may not be a direct relationship between the control input and the control surface positions. Computers interpret the control input and then command the control surfaces to move according to certain programs and protections. These programs can change according to the phase of flight. Although this may not technically be automation in the traditional way we understand it (i.e. using the autopilot and autothrottle), it does represent a fundamental change in the control principles of an aircraft and needs to be clearly understood, particularly when transitioning from an aircraft without this system where a control input is always matched by a corresponding movement of the control surfaces. At this stage, it is important to differentiate fly-by-wire systems that operate on the basis of flight control laws (such as is common in the Airbus design philosophy) and fly-by-wire systems that do not employ flight control laws but where fly-by-wire systems may be necessary because the strength of the control inputs needed to move large control surfaces is beyond the capabilities of the pilot. A system can still be fly-by-wire and maintain a corresponding relationship between the controls in the flight deck and the movement of the control surfaces. This chapter considers the former type of system that employs flight control laws.
- The autopilot/autothrottle (AP/AT) – This system moves the control surfaces and adjusts the thrust to match the requirements dictated by the flight director and the desired speed or vertical mode.
- The flight director (FD) – This dictates what attitude the aircraft needs to be in to satisfy the requirements coming from the mode control panel.
- The mode control panel (MCP) – This is used to select vertical and lateral flight modes and select speeds. FD position will be affected by MCP selections.
- Flight guidance system (FGS) modes – These are modes that can be selected from the MCP that, for the purposes of this chapter, follow simple control laws that will persist continuously or until a fix end-point is reached, for example leveling off at a preselected altitude. Lateral FGS modes include HEADING/TRACK (HDG/TRK – aircraft will continue on a heading or track) and VOR/LOCALIZER (VOR/LOC – aircraft will follow a VHF omnidirectional radio range radial or localizer). Vertical modes include VERTICAL SPEED (V/S – aircraft will climb or descend at a selected vertical speed), FLIGHT LEVEL CHANGE (FLCH – aircraft will climb or descend as a function of throttle changes while speed is maintained), APPROACH (APP – aircraft will descend on a glideslope) and ALTITUDE HOLD (ALT – aircraft will maintain an altitude).
- Flight management system (FMS) modes – These can send both vertical and lateral guidance to the FD and, by virtue of FD and AP/AT coupling, to the aircraft, provided the relevant MCP mode is selected. Unlike the standard FGS modes, when the FMS is in control of the flight path, there is considerable variability in how it can change the attitude of the aircraft. The programming required for this is far more complex than for FGS modes and resides in the flight management computer (FMC). For example, if the simple FGS mode of HDG is selected on the MCP, the aircraft will continue on the preselected heading indefinitely. The FMS controls the AP/AT in a more complex way and can allow the aircraft to follow complicated routes requiring turns and altitude changes with a high degree of accuracy and virtually no pilot input. The FMS can guide an aircraft during a complex departure,

climb and then step-climb according to weight/efficiency calculations, optimize a descent and fly a complex approach.
- Flight mode annunciators (FMAs) – These display to the pilot which vertical and lateral modes are in use, as well as the status of the AP/AT.

9.2   Flight control laws

In the next section (Section 9.3) we will cover different levels of automation possible using different combinations of some of the systems we have just listed. There is one system that will not be covered in the next section but it is so important that we are going to look at it now. All pilots are familiar with the four basic pilot inputs that determine the flight path of an aircraft:

- pitch
- roll
- yaw
- thrust

If the pilot has control over these variables, he can control the aircraft. There is, however, a fifth input and it is just as important as any of the others insofar as it is one of the basic inputs that determines the flight path of the aircraft:

- flight control laws

Before the invention and integration of these control laws into certain types of fly-by-wire aircraft, a pilot knew that a certain combination of the four original inputs would have a predictable response when flying manually. This changed when Airbus installed programming into their fly-by-wire systems that could modify pilots' manual inputs to ensure such things as stall protection and to limit pitch and bank angles. There are several "modes" associated with flight control laws that change during the flight, and a control input made during the cruise (flight mode) may have a different effect than if it was made during the last stages of the landing (flare mode).[2] While type-rating courses for such aircraft explain these differences and subsequent training ensures that the pilots understand how the control laws will affect the flight path of the aircraft, they nonetheless represent a significant change in the human's relationship with the aircraft, a change that seems to have plenty of advantages but may have some disadvantages as well.

Accidents such as those outlined in Boxes 9.1 and 9.2, though rare, highlight a fundamental irony that is entirely in keeping with those predicted long before the Airbus A320 ever flew, ironies that we will consider later on in this chapter. When certain functions are automated using several sources of data and mathematical formulae (such as those written into the stall protection software), the pilot will not be able to derive the same solution in his head and so will not be able to accurately predict the response of the system. Training a pilot as thoroughly as possible regarding how flight control laws will modify or even ignore his inputs in all manner of different situations is as important as teaching a pilot the basics of pitch, roll, yaw and thrust control.

Box 9.1 Accident Case Study – Flight Control Laws

Iberia Flight 1456 – Airbus A320
Bilbao, Spain, 7 February 2001[3]

During the final stages of the approach, the crew of this aircraft encountered some vertical and horizontal gusts. With a descent rate of 1200 feet (365 m) per minute during the final seconds of the approach, the ground proximity warning system (GPWS) system alerted the crew to the excessively high sink rate. Both pilots attempted to pitch the aircraft up but the control laws in force at the time meant that the angle-of-attack protection system became active and stopped the aircraft from pitching up. The captain applied full thrust but the aircraft landed extremely hard, the nose gear collapsed and one passenger was seriously injured. The investigation found that the combination of vertical gusts and dual inputs from both pilots resulted in the control logic predicting an angle of attack greater than the maximum allowed and so prevented the aircraft from flaring. The report stated that this combination of events was not anticipated in the design and so the next version of the software was modified to prevent this from happening again.

Box 9.2 Accident Case Study – Flight Control Laws

Qantas Flight 72 – Airbus A330
Western Australia, 7 October 2008[4]

While in cruise at 37,000 feet (11,300 m), one of the aircraft's three aid data inertial reference units (ADIRUs) starting sending erroneous data spikes to several aircraft systems. One of the data spikes concerned the angle of attack of the aircraft, one of the parameters that cannot be exceeded due to the control laws. The autopilot disconnected but, even with manual control, the aircraft's control laws caused a 10 degree pitch down movement of the elevators. This resulted in 12 serious injuries and 107 other injuries caused by negative g-forces leading to passengers impacting the aircraft fuselage. Both pilots tried to counter the pitch-down movement but their control inputs had no effect. Fortunately the event was transient and the aircraft was able to land safely. As a result of this accident, Airbus issued specific procedures to deal with this particular problem and issued updated software.

The advantages of flight control laws may be worth the risk. These can be seen when you consider how the control laws work and what levels of protection they provide. Here are some of the protections available in flight mode in the Airbus system[2]:

- Load factor protection – The pilot cannot overstress the aircraft.
- Pitch attitude protection – The pilot can only pitch 30 degrees nose up and 15 degrees nose down.
- Bank angle protection – Maximum bank angle with full control deflection is 67 degrees. Once the control is released, it reduces to 33 degrees.

- High angle-of-attack protection – The pilot cannot stall the aircraft.
- High speed protection – When activated due to a high speed upset, it can roll the wings level and pitch the nose up to recover to normal flight.
- Low energy protection – When low to the ground, low energy protection will give an aural warning to the pilot that aircraft energy is too low and he will need to increase the thrust to recover a positive flight path.

The role of these control laws in one of the most famous aviation success stories of recent times was generally overlooked by the press, but it is worth considering it in the context of looking at the advantages of flight control laws (Box 9.3).

Box 9.3 Accident Case Study – Flight Control Laws

US Airways Flight 1549 – Airbus A320
New Jersey, USA, 15 January 2009[5]

Shortly after take-off from La Guardia Airport, New York City, US Airways Flight 1549 encountered multiple bird strikes that led to almost total loss of thrust from both engines at 3200 feet (975 m). Judging that it was impossible to glide to any nearby airport, the crew of this Airbus A320 elected to ditch in the Hudson River. The Airbus A320 airspeed display shows what is known as "green dot speed", the speed that provides the best lift over drag ratio. V_{LS} is the lowest selectable airspeed that provides an appropriate margin to the stall speed. Slightly below V_{LS} is the α-PROT speed, namely the speed at which one of the flight control laws that prevents the aircraft from stalling becomes active and inhibits or attenuates any pilot action that could lead to a stall. The following text is taken from Section 2.3.3 of the National Transportation Safety Board (NTSB) report into this accident:

… the captain stated during postaccident interviews that he thought that he had obtained green dot speed immediately after the bird strike, maintained that speed until the airplane was configured for landing, and, after deploying the flaps, maintained a speed "safely above V_{LS}", which is the lowest selectable airspeed providing an appropriate margin to the stall speed. However, FDR data indicated that the airplane was below green dot speed and at V_{LS} or slightly less for most of the descent, and about 15 to 19 knots below V_{LS} during the last 200 feet [60 m]. The NTSB concludes that the captain's difficulty maintaining his intended airspeed during the final approach resulted in high AOAs [angles of attack], which contributed to the difficulties in flaring the airplane, the high descent rate at touchdown, and the fuselage damage.

For the last 150 feet (45 m) of the descent, the aircraft was in α-protection mode to prevent it from stalling and so the system disregarded the pilot's rearward control input to the side-stick when he attempted to flare just prior to landing on the Hudson River. While none of this is meant to detract from the quick thinking and excellent teamwork displayed by the crew, the aeronautical engineers and software designers at Airbus played a bigger part than most people realize in the successful resolution of this potentially deadly accident.

The ability to be able to install these protections into commercial airliners marked a turning point in our relationship with the machines we fly and represents an astounding feat of engineering and design. After some initial problems, it seems that this amazing engineering is being paralleled by equally good training to instill the necessary skills in the pilots who will be using this technology. Despite the accident case studies given here, it is worth remembering that many other aircraft that do not have these flight control laws to provide envelope protection have stalled and crashed (e.g. Turkish Airlines Flight 1951, a Boeing 737-800, and Colgan Air Flight 3407, a Bombardier Dash-8 Q400). Other areas of this debate continue, particularly with regard to the limited feedback pilots get through independent side-stick controllers compared to linked control yokes, and whether autothrottle movement is important in keeping the pilots in the loop with regard to what the aircraft is doing. In the final analysis, we do not yet know whether one design philosophy is going to prove itself superior to the other.

9.3   Levels of automation and their uses

Several possible levels of automation are possible using the AP/AT, FD, FGS and FMS modes. Note that AP and AT are assumed to be used together and AP/AT is always assumed to be used with FD:

- level 1 – manual flight with no FD
- level 2 – manual flight with FD and FGS/FMS
- level 3 – AP/AT in FGS modes
- level 4 – AP/AT in FMS modes

Level 1 is clearly the lowest automation level and level 4 is the highest. Although we are accustomed to spending most of our time operating at level 4, this is no more or less legitimate a level of automation than any of the others, provided it is used in the appropriate circumstances. The guidance given by a check airman, Captain Warren Vanderburgh from American Airlines, as seen in a widely circulated video from a training seminar in 1997, was some of the first that recognized that operating at the highest level of automation is not always appropriate.[6] In fact, certain abnormal flight conditions strongly suggest that pilots transition directly from level 4 down to level 1, for example, a ground proximity warning system (GPWS) terrain warning or a traffic collision avoidance system (TCAS) resolution advisory. Table 9.1 shows the different levels of automation with some examples of flight conditions that they would be appropriate or inappropriate for.

Choosing the appropriate level of automation and balancing the workload that it will require of the pilots may be difficult. When we are accustomed to flying using the AP/AT, we may be reluctant to disengage it in light of our knowledge that our manual skills may have deteriorated. Below are some considerations that may help you to decide on the best level of automation for the task at hand:

- Disconnecting AP/AT and FD to try and catch the glideslope from above or to "try and save" an approach that is too high, too fast or at risk of becoming unstable by converting it into a

Table 9.1 **Levels of automation and their uses**

	Good for …	**Bad for …**
Level 1: **Manual, no FD**	• Emergency maneuvers • Well-briefed visual approaches	• Unprepared visual approaches • Spontaneous desire to reduce workload (will increase it)
Level 2: **Manual with FD**	• Practicing manual handling	• Manual handling in busy airspace (traffic or weather)
Level 3: **AP/AT in FGS**	• Temporary ATC instructions	• Emergency maneuvers (AP/AT too slow to react)
Level 4: **AP/AT in FMS**	• Complex procedures • En route navigation	• Late, low clearance changes requiring reprogramming. Level 3 may be better

FD: flight director; AP/AT: autopilot/autothrottle; FGS: flight guidance system; FMS: flight management system; ATC: air traffic control.

visual approach is probably a mistake. If you have not briefed that you will be disconnecting the AP/AT and FD as early in the approach as you are just about to do, it probably means that something unexpected has happened. If you disconnect everything, your workload is going to go up significantly as you now have to control the aircraft manually, as well as having to solve the problem that prompted you to disconnect in the first place.

• Disconnecting the AP/AT and FD should either be thoroughly well planned for (e.g. a visual approach, a relatively unusual type of approach for most pilots that requires more, not less, briefing and preparation than an ILS approach) or be done in reaction to a situation that requires an instantaneous change in flight path, such as a GPWS escape maneuver.

• Manual flying with the FD engaged is probably the most appropriate way of maintaining your manual flying skills in a commercial operations setting. Your company is likely to have guidance regarding this, and none of the recommendations given here are to be followed in contravention of standard operating procedures (SOPs). However, should you have the opportunity to practice your hand-flying skills, the following may help when deciding when to manually fly using the FD:

   ◦ Make sure your colleague is happy: he has to monitor you, so it is best to make sure that this is not going to overburden him during a busy departure.

   ◦ At what point should your colleague alert you? Agree on alerting limits and make sure that your colleague is happy that he now needs to actively monitor your flying and alert you to any flight path or speed deviations beyond the limits you have set. He should also alert you to other unexpected threats such as traffic.

   ◦ Are the weather conditions appropriate? Being able to maintain visual meteorological conditions adds a level of protection, but also consider whether weather avoidance will be required, in which case it may be more useful to use the AP/AT so that sufficient attention can be given to the weather radar without increasing the workload too much.

   ◦ Are the traffic conditions appropriate? Manual flying in an area with lots of traffic, particularly visual flight rules traffic, may be inadvisable as it may be better to maintain a good lookout.

   ◦ Have you briefed your contingencies? If something unexpected happens, have you briefed how the automation is going to be engaged and who is going to do it? Any unexpected occurrence will increase workload. To offset this and reduce workload, consider engaging the automation and confirming the appropriate modes as soon as anything unexpected happens.

⬦ There is no shame in engaging the automation. You may find that during a departure or an approach, the situation changes such that, had you known of it in advance, you probably would not have elected to fly manually. In this case, manage your workload by engaging the automatics and confirming that the correct flight modes are engaged.

⬦ Think about the throttle. The main deficit found when manual flying skills degrade is the ability to control speed manually, and this is also the main cause of accidents during manual flight.[7] Maintaining your manual handling skills is as much, maybe more so, about being able to manage the thrust as it is about controlling the flight path. When flying without the AP, consider flying without the AT as well, as this will keep this highly critical skill of being able to control the speed of the aircraft as well honed as possible.

⬦ Do you want to go back to raw data? This is a controversial topic. Without the FD, you are forced to rely on raw data. Consider how often you are called upon to carry out raw data maneuvers and, if you think it is often enough, you will have to carry out some raw data flying to keep these skills well developed. You will have to think even more carefully about what opportunities you use to practice these skills as there is one less layer of protection and you will have to take one extra step if you decide to engage the automation: switching on the FD and confirming that the appropriate modes are engaged before coupling it to the AP and AT.

• Most modern aircraft will spend their time in level 4 automation. You may temporarily need to drop down to level 3 if given a heading or a late change of clearance. Make sure this change is clearly announced to all crew and, in the case that a heading needs to be followed for a long period, set yourself a reminder that you are operating at the FGS level rather than the FMS level to avoid taking a subsequent "Direct To" and forgetting to re-engage lateral navigation.

• If you are low and you have to change arrivals or approach runway, consider how much workload it will be to change the FMS. FMS changes need to be confirmed by both pilots and if the workload is high, it may not be an appropriate time to divert attention away from controlling the flight path of the aircraft, possibly around active weather, to focus on reprogramming the FMC. If you really do need to change the FMC, consider how you are going to get the workload down so that both pilots can participate in this process: slow down, enter a hold or do an orbit. It may be easier just to drop back to level 3, ask for radar headings and fly the approach using FGS modes. You may be able to get air traffic control to help you out more than they would ordinarily by pointing out that late changes like this make things more difficult for you, and that some headings while under radar control may help you to get the automation reprogrammed for the arrival and approach.

• If you are heads-down, you are out of the loop – Although we have a certain capacity for divided attention, it is best to assume that the pilot reprogramming the FMC is out of the loop entirely. Your self-monitoring responsibilities go up accordingly.

• Decide some personal strategies for cross-checking automation performance – Automation can and does malfunction and can lead to flight path deviations. In Section 9.8 on automation complacency, you will see how easy it is to become overly trustful of automation. The research shows that this is an entirely normal response to a highly reliable system, but we need to consider how we can detect errors that occur in this normally reliable system. These strategies can be based on gross error checks to ensure that automation-related flight path changes are approximately what you would expect.

9.4   Flight mode annunciators

Most modern airliners will provide some feedback to the pilot regarding its current status and the selected automation modes. These are usually displayed in the form of FMAs on the pilot's primary flight displays. These FMA indications are the most reliable way for the pilot to be able to tell what the automation is doing or whether it is even engaged. Most SOPs will dictate that when the pilot is making a mode change on the MCP he must select the desired mode on the MCP, confirm that the mode is now active by checking the FMA, and then announce this to the monitoring pilot so that he is kept in the loop regarding the automation status of the aircraft. A pilot's mental model of the situation cannot be complete unless he has an accurate representation of the current state of the aircraft automation, particularly when that mental model is used to project the future flight path of the aircraft. The "FMA callouts" made by one pilot to announce a flight mode change to the other pilot are crucial in allowing both pilots to maintain an up-to-date mental model of the situation. Unfortunately, lack of mode awareness appears to be a problem in modern aviation, one that is related but not limited to the phenomenon of automation surprise. A study published in 2006 used eye-tracking to monitor the gaze location of pilots during a series of simulated flights, particularly when automation mode changes either were commanded or occurred spontaneously.[8] A total of 1042 mode changes occurred, of which only 32 (less than 3%) were "called out" by the handling pilot in a manner consistent with the SOPs, that is, after visual verification of the FMAs. Crews sometimes looked at the FMAs but did not say anything, or said something but without looking at the FMAs. Another study was carried out where FMA callouts were analyzed in a real-world setting, during the course of 19 flights across three operators all flying the Airbus A320. This study found that there was a tendency not to make FMA callouts during high-workload phases of flight.[9]

The authors of the first, simulator-based study also refer to guidance published by the Joint Aviation Authority, which states that mode transitions that occur spontaneously (e.g. transitioning from an armed mode to an active mode) should be accompanied by an attention-getting feature such as a flashing FMA. They note that FMAs do not really seem to be attention getting whether the symbols are flashing or not and that such regulatory guidance may not be enhancing awareness in any meaningful way. The argument is one we have seen before, especially when we covered sustained attention (vigilance) in Chapter 2 (Information Processing). To "ensure" that suitable modes are selected, the industry has put the onus on the liveware to confirm this rather than engineering a hardware-based solution. Given that FMA callouts are, by a very large majority, substandard according to the requirements of regulators, perhaps the hardware should be designed to give auditory as well as visual feedback in the event of mode changes. A pilot would then announce his intentional mode change while pushing the appropriate button on the MCP, and would expect this to be immediately followed by the system repeating this when the mode becomes active. The system could also announce when armed modes become active. The combination of auditory and visual feedback may help to keep both pilots more in the loop regarding the flight path of the aircraft. With regard to uncommanded mode changes, a study that looked at 20 Boeing 747-400 pilots found that in 32 cases where an uncommanded mode change was simulated, only one pilot spotted the undesired FMA.[10]

However, given that most automated systems in current airliners rely on FMAs to give mode feedback to the pilots, below are some strategies for best ensuring that both pilots have an accurate mental model of the automation state of the aircraft. Please note that these should not be considered as alternatives to your current company SOPs:

- Because the human brain is very poor at sustaining attention, it is best to assume that the monitoring pilot will not notice a mode change unless he happens to be looking at the FMAs or the handling pilot at the moment the mode is changed.
- The handling pilot should train himself to carry out the following sequence of actions when commanding a mode change:
  1. Select the mode on the MCP.
  2. Visually confirm that the appropriate FMA is present on the primary flight display.
  3. Announce this FMA to the monitoring pilot.
- Avoid the trap of calling out the mode change when selecting the mode on the MCP. It should only be done when the desired FMA is displayed.
- In the event of a spontaneous mode change (e.g. capturing a localizer or capturing an altitude), the new FMA should be announced as well.

9.5   Automation, perception and Newton's laws of motion

An AP/AT engaged in lateral and vertical navigation has the ability to alter the course of an aircraft in potentially complicated ways in all three spatial dimensions. This ability, while very convenient for pilots, limits the pilot's ability to project the flight path of the aircraft into the future. We have some perceptual limitations, which are the result of our genetic programming and our environmental upbringing. Consider the three laws of motion proposed by Isaac Newton[11]:

1. An object is either at rest or moves at a constant velocity unless acted upon by an external force (in our normal environment, gravity and friction are external forces).
2. Force equals mass times acceleration.
3. When one object exerts a force on a second object, the second object simultaneously exerts a force of the same magnitude but in the opposite direction to that of the first object (i.e. every action has an equal and opposite reaction).

Whether we knew it or not, we grew up with these laws. As well as evolving to be able to operate in a physical world where these laws are in force, we learn from an early age that a ball will stay still on a level table until we push it and then it will roll (due to the first law). We also know that a heavier ball will require a stronger push if it is to travel at the same speed as the lighter one (due to the second law) and that when the ball rolls off the table and on to the floor, it will bounce because of the opposite upward force (due to the third law). The laws of motions and the other things we observe in the physical world all go into our perceptual "rulebook" that we can then use to predict how a mental model of a situation will unfold even before it happens. I know, even before I touch the ball, that when I push it, it will roll, fall and bounce. We do not have to calculate the path of every object in our mental model because we have learned these rules from such a young age that they are automatically applied. Thus, we know that we need to duck so that we do not get hit in the face by a football or that standing under a precariously balanced heavy object may result in injury.

What we have not evolved to deal with and what takes some time to learn are the effects of the laws of motion when it comes to aircraft, and how sometimes it is our actions that affect the aircraft's motion and sometimes it is the action of the autopilot. Our perceptual model assumes that objects will move in a straight line unless acted upon by an external force. In this way, our perceptual model struggles with predicting the paths of objects that have a degree of autonomy such as automated aircraft. For this reason, it is harder for us to build a picture of when and how our aircraft is going to maneuver itself unless we update our mental model using instrumentation. Humans are not designed to predict conflicts between two objects that are each changing positions in a non-linear way in three-dimensional space, and so we are forced to rely on automated systems to keep us informed. It is possible that we can overcome these perceptual limitations as our experience grows and this may be why more experienced pilots are credited with more talent in this regard than less experienced pilots. It does not necessarily refer to any natural ability that renders the more experienced pilot superior, but simply that he has had sufficient exposure to objects moving in complicated, three-dimensional trajectories at high speed to be able to factor this knowledge into his perceptual projections of his current mental model. In short, computers are better at predicting when aircraft are going to collide with the ground or with other aircraft than humans are because of our perceptual limitations. This means that we need to trust these protective systems rather than spending time trying to verify potentially complex trajectories using our perceptual system.

9.6   The ironies of automation

In one of the seminal pieces of work on automation, Lisanne Bainbridge was prescient when she wrote a paper about the ironies of automation in 1983.[12] In this paper, which has been cited by other researchers over 1200 times since then, she identified a basic mismatch between the perspectives of the system designer and the human operator. The designer may be a proponent of the Old View of error (i.e. humans are what make safe systems fail) and so wish to design the human out of the system to make it as safe as possible. That gives rise to some ironies. Below is a list of potential automation problems based on Bainbridge's original *Ironies of automation* paper:

1. **Design flaws** – The designer may introduce latent errors (unknowable threats) into the system that the operator will then have to deal with; for example, aberrant, incorrect and unexpected mode changes due to programming bugs. This was certainly the case with the Iberia and Qantas Airbus events we looked at earlier (see Boxes 9.1 and 9.2).

2. **Tasks after automation** – The designer may have to leave the operator with a small range of tasks that he cannot work out how to automate. The operator is then saddled with a group of possibly unrelated tasks that he must manage at the same time as managing the automation. Since 1983, this collection of tasks has grown smaller, as we seem to be able to automate just about everything.

3. **Manual control skills** – If the primary task of manually operating the machine is now replaced by monitoring the automated machine, the manual skills required will degrade. Flying an aircraft is a compensatory tracking task whereby there is continual correction of mismatches between the actual path of the aircraft and the desired path.[13] When learning to fly, students gradually get better at assessing the magnitude of control input required to correct one type of deviation (e.g. a lateral deviation). As their skills develop, they will soon be able to easily correct deviations in two dimensions (e.g. vertical and lateral). Soon their skills will become so good so that there is minimal deviation as any predicted deviation can be corrected before it increases in magnitude. This compensatory tracking skill is less important once a student moves on to an aircraft that has an FD, as the FD will show the direction and magnitude of the input required to give the required flight path. An able student will still be able to judge how much control input force is required to give the required flight path change. Once the student is used to flying with an AP/AT coupled to the FD, this knowledge may fade as well, so that in the event that a pilot has become accustomed to flying with AP/AT and FD and his compensatory tracking skills have degraded, an unexpected loss of both systems will leave the pilot having to carry out a compensatory tracking task in a flight condition that may be very intolerant to the magnitude of errors that would be expected for a novice or a deskilled professional (e.g. high altitude or unusual attitude). Interestingly, a visual approach becomes a compensatory tracking task once the AP/AT and FD are disengaged, but one where the pilot must judge the corrections needed by interpreting the visual environment rather than the instruments. This is no easy feat if the pilot is not used to doing it, especially as there is not a desired flight path superimposed in three dimensions on the outside world. As well as the challenge of the compensatory tracking task, there is the added difficulty of not knowing exactly what the correct flight path is.

4. **Long-term knowledge** – If an operator must control a system constantly, he will inevitably gain knowledge about that system. In an aircraft with a manually operated pressurization system, the operator will be forced to understand pressurization schedules, discharge valve slaving, normal rates of cabin climb and descent, and so on. Knowledge is necessary for task management. As we know from information processing, if chunks of knowledge are used frequently, their base-level activation goes up and they are easier to retrieve when making decisions. If knowledge has not been used for a long period, it may be extremely difficult, even impossible, to retrieve. If a system has been automated, this knowledge may not have been used since an initial type-rating. Should the automation fail, the operator will be a lot less proficient at manually controlling it because the knowledge will have degraded.

5. **Current system state** – For air traffic controllers, whether they are in the tower or covering a section of airspace, when one controller takes over from another, the second controller may sit with the first one for 10 minutes to build up his perceptual model of the state of the system; that is, the air traffic that he will be controlling. If he were to just take over immediately with no handover, he would be forced to build up his mental model of the position and paths of all the aircraft in his sector before he could reliably give instructions. In the same way, a system on an aircraft that must be controlled manually forces the operator to remain engaged with it. He must know the system state, including its recent past state and, more importantly, its future state, in order to control it. If something goes wrong, his engagement with the system means that he has a detailed mental model already which can

be used to quickly fill in the goal module and the imaginal module of his decision-making system, and quickly get to work on solving the problem. If a system is automated but then reverts to manual control because of a system failure, the operator must now correctly direct his selective attention to the relevant system, build a mental model of the system state and then project its future state (i.e. what is going wrong) before he can fill in the goal module and imaginal modules to fix the problem. This process takes time and in the event of a worsening failure or a failure that is leading to other failures, by the time he has filled the goal and imaginal modules, the situation could have changed. The operator is then trying to catch up with the system and, along with the increasing workload that this generates, is more likely to start making rule-based (RB) errors of the type we talked about in Chapter 3 (Error Management and Standard Operating Procedures for Pilots). These RB errors occur under the high-workload conditions that would be associated with an unexpected and deteriorating system failure, are difficult to detect and give rise to a multitude of other potential errors that can threaten flight safety.

6. **The monitoring task** – By automating functions, the operator, while losing some tasks, is given a new one: monitoring. Refer back to Section 2.4.2 on sustained attention in Chapter 2 (Information Processing). In brief, humans are very bad at sustaining attention on a relatively unchanging stimulus in order to detect changes. Research shows that it is surprisingly effortful, cognitively draining and, after 15 minutes, should the person be able to sustain his or her attention that long, deteriorates to the extent that changes are missed.[14] In short, all those complaints that pilots do not monitor the instruments any more don't take account of one of the basic facts about sustained attention, which was uncovered back in the 1940s[15]: humans did not evolve needing to monitor computers and, hence, we are really bad at it. If we cannot change the liveware to solve this problem, we need to change either the procedures or the hardware.

7. **Who is calling the shots?** – The famous line uttered by pilots in the face of an unexpected automated change is "What's it doing now?" An FMS can make decisions about vertical and lateral navigation, thrust and other system settings with little or no input from the pilot. It can make these decisions because it can combine Global Positioning System (GPS) positions, air data, inertial reference data and a myriad of other streams of information to make decisions about how to change the flight path and thrust of the aircraft. The FMS can do this because it is a computer and can handle all these data quickly to make decisions about flight path or speed. Because the pilot is not as connected to the data stream as the computer, if at all, he will be unable to verify the accuracy of the FMS's decision. All he can do is assess whether the commands being issued from the FMS are "acceptable". The computer is making decisions using such a vast array of data and at such a high speed that the human has no way of knowing where the line is between "It knows what it's doing, even if I don't" and "Something's wrong: it shouldn't be doing this". At what stage of uncertainty do we disconnect everything and then make ourselves susceptible to irony number 3 – hand flying an aircraft in an unexpected situation when the computer was probably doing an OK job?

8. **Are we pilots or software managers?** – Bainbridge gave an interesting example in her paper. The management of one processing plant that had recently become automated had to ensure that there was a senior manager present during the night shift or the workers would just switch off the automation. The aviation industry needs to be sensitive to the views of the pilot community with regard to the relationship between the role of the pilot and the nature of the technology in the flight deck. The following lines are copied verbatim from Bainbridge's paper as there is no more eloquent way of summing up this particular dilemma: "One result of skill is that the operator knows he can take over adequately if required. Otherwise the job is one of the worst types, it is very boring but very responsible, yet there is no opportunity

to acquire or maintain the qualities required to handle the responsibilities. The level of skill that a worker has is also a major aspect of his status, both within and outside the working community. If the job is 'deskilled' by being reduced to monitoring, this is difficult for the individuals involved to come to terms with". The design of modern airliners is based around the idea that the automation will be doing the vast majority of the flying, not the pilots. The thing is, that's OK. Modern airliners are technological miracles that allow us to fly higher, faster, safer and more cheaply, so much so that the once rarefied luxury of air travel is now accessible to much of the world's population. Technology and automation made the planet smaller and put new horizons in the pockets of people who could never have afforded them before. We, as an industry, just need to decide how the pilot fits into this equation. If we are to be software managers, we cannot be criticized when we are called on to use our "skills" only to find them degraded by lack of use. If we are to be pilots in the old-fashioned sense of the word, we need to decide when, how and under what conditions we should be allowed to practice our skills, and the industry has to accept that this, in itself, introduces a risk into the system. And so, once again, we find ourselves on the horns of a dilemma: Leave it to the computers altogether, or set standards of manual flying skills that need to be maintained but accept the risks that practicing these skills will involve. There probably is a middle ground but we all need to agree on it in order to move forward.

9. **Flying is safer than it has ever been, thanks to automation** – This is true and could be an argument for saying that we should rely more on the computers given that things so rarely go wrong. Pilots of yesteryear had such good skills because their aircraft were always breaking down. Because that does not happen any more, why do we need to maintain these skills? Bainbridge noted this back in 1983 and said the following. "Of course, if there are frequent alarms throughout the day then the operator will have a large amount of experience of controlling and thinking about the process as part of his normal work. Perhaps the final irony is that it is the most successful automated systems, with rare need for manual intervention, which may need the greatest investment in human operator training."

Fortunately for us, Bainbridge suggests some strategies for dealing with these problems, although most of them are aimed at the aircraft designers, rather than the pilots:

- Monitoring – Forget it. If you want a pilot to notice a change, you have got to have some sort of alarm. You also need to think how the alarms are going to present themselves if several activate at once.
- Compensation is good for the aircraft, but bad for the pilot – Cleverly designed automated systems can compensate for failures or other disturbances without the pilot even knowing. It is only when the system can compensate no more that the automation gives up and suddenly presents the pilot with a serious situation that could have started an hour ago. Consider an aircraft retrimming itself automatically because of ice accumulation on the airframe. The pilot may have no clue that this is happening until the automation gives up and the pilot is left struggling with a grossly out-of-trim aircraft that is covered with ice. If things are going wrong, the system should alert the pilot even if it is compensating. This at least allows him to remain engaged with the developing situation and intervene if necessary.
- There are only two ways of addressing the problem of degradation of manual control skills and degradation of knowledge: frequent real-world practice or frequent simulated practice. The frequency required is based on the skill involved. Are two or three visual approaches every 6 months enough to keep a pilot's compensatory tracking skills up to scratch? I doubt it. Is there a benefit to being in the simulator more frequently but for shorter periods? Do these skills need to be practiced in full-motion simulators? These are all valid questions, but

they are based on the dilemma raised earlier: Do we want to invest time and money in more frequent simulator training (or accept the risks of mandatory hand flying during day-to-day operations) so that pilots have these skills, or would we prefer them to be software managers and let these skills degrade?

In the next four sections, we will look at some of these ironies in relation to aviation operations and expand on the strategies we have available to deal with them.

9.7   Skill fade and automation dependency

Skill fade is the phenomenon of becoming decreasingly less proficient at a skill when it is not practiced regularly. The irony of automation leading to loss of manual control skills is, unfortunately, one with very real consequences. In 2006, a study into the manual flying skills of 110 pilots, some of whom were used to flying modern glass-cockpit aircraft and some who were not, showed that pilots who were accustomed to highly automated flight decks had significantly poorer manual handling skills than pilots who flew less automated aircraft.[16] A review of incidents associated with manual flying between 2000 and 2006 also concluded that there was a subset of incidents attributable to degraded manual flying skills. Ironically, the incidents occurred when pilots had intentionally disconnected the automation in order to practice hand flying prior to a simulator check.[17] Subsequent research from 2009 showed that even pilots who had spent a significant part of their career flying older, less automated aircraft suffered the same deficits as those who had spent most of their career on modern, highly automated aircraft.[7] The key variable in determining whether a pilot will have good manual handling skills is the *recent* hand-flying experience that pilot has accrued, usually within the preceding few weeks. Without this, no matter how many hours a pilot has accrued on less automated aircraft types, skills will fade. As well as deskilled pilots having to put in a great number of control inputs, the magnitudes of the input were also greater (i.e. control was much cruder), showing that there was a decrease in the skills required for this compensatory tracking task. More worryingly, speed control was also impaired and this seems to be one of the primary causes of accidents that occur when flying manually.

We, again, return to the irony of being told by the industry that we need to maintain our skills, of research showing that they need to be practiced regularly during line operations or in the simulator, and being faced with an industry that does not want pilots to hand fly during normal operations and will not commit to more regular simulator sessions. As with so much else, there is no easy answer. Unless regulators stipulate the minimum number of manual approaches and departures that must be flown over a set period for a pilot to remain current, automation-induced skill fade will remain a significant problem in aviation. The alternative to this is mandating planning rules that mean that only specially qualified pilots are allowed to dispatch an aircraft to an airport where a visual approach is expected, in the same manner as aircraft with certain technical failures can only be flown by specially trained crew.

The scientific community is also rising to the challenge, and several universities, research groups, training organizations and aircraft manufacturers are collaborating on a project entitled Manual Operation for 4th Generation Airliners (Man4Gen) with the aim of addressing the difficulties that arise when pilots must transition from monitoring highly reliable systems to taking active control and becoming the main decision maker when automation fails.[18] The hope is that this research will show how salient data are best presented in the flight deck and also the best way to train pilots to maintain control of an aircraft during unexpected events and how to employ strategies to handle the event.

9.8   Automation complacency

This phenomenon is cursed with an unfortunate name. The word "complacency" conjures up negative images of laziness, disregard and general disengagement with the task at hand. The reality is that automation complacency refers to the tendency to trust automation to fulfill the function for which it was designed. The more reliably it does this, the more we trust it. The perceived "negative" effect of this is a decrease in our degree of monitoring over the tasks that the automation is managing. Consider an office manager (the pilot) assigning tasks to his employees. Four out of five employees require constant supervision to ensure that they are satisfying the goals that have been set for them. The fifth employee is well known for his thoroughness and precision. Will you afford him the same attention as you afford the others? No, of course not. He can do his job and he does it reliably and accurately. It is best to direct your resources to where they are needed most, to the employees who require your expert input. Parasuraman found that this was true of our relationship with automation and, rather than its being an indication of negligent entrusting beyond the bounds of what is safe, it reflects a basic program in the human capacity to manage workload: if a task is being adequately managed by another resource, it does not require as much attention as a task that is not.[19] More accurately, automation complacency as defined in the scientific literature comprises three points:

- A human operator is involved in monitoring an automated system.
- The frequency of such monitoring is lower than a perceived standard.
- This perceived monitoring deficit has some observable effect on system performance.

It should be noted that as well as flight path automation, automation complacency extends to other systems in the aircraft, such as pressurization, navigation and engine control. Although this chapter is abound with ironies already, it is another irony that pilots are always being reminded to be constantly vigilant of their automated systems, even in the face of frequent improvements to the reliability of these systems and the preprogrammed rule that says that concentration is best directed to tasks that require supervision. A summary of Parasuraman's findings on the subject of automation complacency is given below:

- Automation that is unchanging in its reliability is more likely to induce complacency than automation where the reliability varies.
- Where automation systems exhibit variable reliability, the human operator is more likely to detect problems in such systems than when the automation is reliable in a more consistent way.

- Complacency still occurs even when the reliability of automation is low.
- In the event of an automation failure, operator trust is reduced but then recovers slowly.
- Automation complacency represents a reallocation of attention to other manual tasks when the workload is high.
- Complacency may be the cost that we must incur for the benefits that automation can provide.

The findings of this research suggest that human operators will naturally develop trust in the reliability of automated systems, particularly if they exhibit high reliability. Without deliberately introducing artificial system failures into aircraft automation, there seems to be little chance of consistently being able to stimulate human operators to monitor trustworthy automated systems to same the degree as they monitor non-automated systems. Automation complacency may be, as Parasuraman says, the price we pay for having high levels of automation (see Box 9.4).

9.9   Automation bias

Those of you who have GPS satellite navigation systems in your car may have been faced with a situation where the system is telling you to go one way, but all the road signs are telling you to go the other way. Which do you believe? The system that relies on a global network of satellites launched at great expense, or a road sign? More often

---

Box 9.4 Accident Case Study – Automation Complacency

Asiana Airlines Flight 214 – Boeing 777-200
San Francisco International Airport, California, USA, 6 July 2013[20,21]
Please note that, at the time of writing, the final accident investigation report has not been published for this accident and this summary is based on the information that the National Transportation Safety Board (NTSB) has released so far. The NTSB released the following information based on an initial reading of the flight data recording, but have stressed that these data will need further validation and cross-referencing with other data sources. The crew of this flight from Incheon, South Korea, were carrying out a visual approach to runway 28L at San Francisco International Airport based on a 14 nautical mile (nm) straight-in final. The calculated threshold speed was 132 knots.[22] The target approach speed (V_{APP}) was 137 knots and the stall speed (V_{SO}) was approximately 101 knots. The following heights and speeds were reported by the NTSB:

- 2400 feet (730 m) – 175 knots (well above normal glidepath).
- 5 nm from runway – 3000 feet (900 m) selected in MCP altitude selector in case of go-around. Speed set to 152 knots (still above glidepath).
- 1600 feet (490 m) (3.5 nm from runway) – FLIGHT LEVEL CHANGE mode was selected. This is not recommended past the Final Approach Fix according to Boeing guidelines. With the autopilot (AP) and autothrottle (AT) engaged, the aircraft

pitched up and attempted to climb the aircraft to 3000 feet (900 m) at 152 knots. The AT increased the thrust in order to permit this. Given that the aircraft was flying an approach, this would have been an undesired scenario and so the AP was disconnected by the pilot and throttles were manually reduced to idle. The flight directors (FDs) were still giving commands consistent with FLIGHT LEVEL CHANGE mode and the AT transitioned to HOLD mode with thrust levels in idle position due to the pilot manually retarding them. In this mode, the AT would not control airspeed to that which was selected.

- 1000 feet (300 m) – 149 knots.
- Air speed selected to 137 knots (V_{APP}) and 5 seconds later left FD switched off but the right FD was left on.
- 500 feet (150 m) – 134 knots (3 knots below V_{APP}). Aircraft descended through desired glidepath.
- The precision approach path indicator (PAPI) showed three reds and so pitch was increased by the pilot through application of back pressure on the control column. However, the thrust remained at idle and aircraft continued to sink below the glidepath.
- 300 ft – 120 knots (17 knots below V_{APP}) . The pilot would have been able to see four red PAPI lights indicating that he was very low on the approach. The aircraft was decelerating through 120 knots with 7 degrees pitch up.
- 200 feet (60 m) – 118 knots (19 knots below V_{APP}).
- 120 feet (35 m) – 114 knots, automated low airspeed caution activated (a quadruple chime).
- 7 seconds before impact, one of the other pilots calls "speed!"
- Just below 100 feet (30 m), throttles were moved fully forward.
- 4 seconds prior to impact, shortly after the radio altimeter called "fifty (ft)", the cockpit voice recorder recorded a sound similar to the stick shaker. This lasted for 2.2 seconds.
- A go-around was called for between 20 and 30 feet (6 and 9 m) above the ground.
- 3 seconds before impact, the speed was 103 knots (29 knots below V_{REF} and 2 knots above stall speed). Thrust was increasing and then engines were at approximately 50% thrust.
- At impact, airspeed was 106 knots.

Although it seems there were many other factors that contributed to this accident, such as culture and training, it is included in this section to highlight the apparent problems with speed control in a flight condition where it was assumed that the AT would still be controlling the speed. If the pilot believed that the AT was controlling the speed, it is unlikely that he would have included this in his scan owing to the effects of automation complacency. Unfortunately, because FLIGHT LEVEL CHANGE had been selected earlier leading to an increase in thrust, the manual retarding of the throttles meant that they would remain at idle and not control the thrust to give the selected airspeed. The NTSB reported that there were several automation mode changes during the final approach, but the exact consequences of these cannot be confirmed until the final report is published.

than not, we tend to believe the computer. Unlike automation complacency, which is a tendency not to monitor automated systems that show a high degree of reliability, automation bias represents another aspect of the trustful relationship between pilot and automation, and focuses on systems other than those used for flight path guidance. In the same fashion as many of the other biases and heuristics mentioned in the decision-making section (Section 2.6) of Chapter 2 (Information Processing), automation bias is the tendency for the pilot to put greater faith in the feedback/diagnosis of the on-board automated computer systems than in other non-automated cues that may be present. The reasons for this should be evident from what you have learned about decision making earlier on in this book. Under high-workload conditions, as would be associated with an on-board technical failure, humans employ cognitive shortcuts to make decisions as quickly as possible. If you happen to have a highly reliable, intuitively designed computer system on board that can gather information from all parts of the aircraft, analyze it and present you with not only the diagnosis of the problem but also the associated checklist required to fix it, there is a high chance that rather than attempting to diagnose the problem yourself, you will trust the automation and go along with its diagnosis. This would indeed be the best course of action, provided the diagnosis of the automated system is correct. In an interesting study from 1992, pilots were tested in a simulator and given an engine fire during which the automated systems on board recommended shutting down the wrong engine. The study found that 75% of pilots followed this erroneous guidance when the system was fully automated, compared with only 25% of pilots who used normal paper checklists.[23] When pilots were tested using a engine indicating and crew alerting system (EICAS) equipped simulator, all the pilots tested shut down an engine in response to an erroneous EICAS alert despite all other parameters being normal.[24] Two-thirds of these pilots said that they saw at least one indication that supported the EICAS warning, even though there was none in reality.

Because automation bias seems to be a sensible cognitive shortcut that will allow the automated system to support pilot decision making in complex, time-critical situations, it can be a difficult phenomenon to overcome. The main opportunity for identifying when the automation has provided an incorrect diagnosis or course of action is during the diagnosis and review steps of TDODAR (time available, diagnose, options, decide, assign tasks, review), the decision-making tool recommended in Chapter 2. In this scenario, automation takes the role of System 1 and is attempting to quickly generate a diagnosis and recommend a course of action, possibly based on incomplete information. It is now up to the pilot to overcome his own System 1 conclusions, which will be partly driven by the diagnosis of the automated systems, and engage System 2 in order to evaluate the non-automated cues and confirm that the diagnosis is correct.

9.10 Automation surprises

Defining what constitutes an automation surprise is easier when we consider how pilots respond to the state of the aircraft and its automated systems. Woods and Sarter listed four questions, any of which would suggest that the pilot has experienced an automation surprise[25]:

- What is it doing now?
- What will it do next?
- How did I get into this mode?
- Why did it do this?

These questions can arise because of one of the ironies of automation that Bainbridge identified: Who is calling the shots? Because the automation can utilize much more data than the human in its autonomous information processing and decision making, it is unlikely that the pilot will be able to keep up, both because of not being fully in the loop when the automation is engaged and because of the inherent processing limitations of the human brain (remember, nerves conduct impulses at 0.000034% the speed of the microchips in the aircraft computers). For this reason, the automated systems may make decisions based on information that is not immediately accessible to the pilot and so the system's decisions may seem confusing. This mismatch between what is observed and what was expected (i.e. something else) is an automation surprise.[26]

Woods and Sarter also noted that automation surprises are most likely to occur when three factors converge:

- spontaneous, autonomous change in the state/mode of the automated system
- gaps in the human operator's mental model of how the hardware is meant to work in different situations
- weak feedback from the automated system to the operator.

Tragically, these three factors converged in the case described in Box 9.5. In a situation characterized by a high degree of confusion and probably a significant amount of fear as the situation deteriorated, the only indication that the aircraft was now operating according to a different set of flight control laws was an amber annunciator on the primary flight display. The accident report noted that the systems were functioning in a degraded manner without the real overall situation being known to the crew. The crew were clearly surprised by two events:

- failure of the angle-of-attack protection due to the frozen sensors
- the inability to lower the nose of the aircraft due to the loss of automatic pitch trim when the aircraft transitioned to direct law, a transition that they were unaware of

Developing strategies to deal with automation surprises is a difficult challenge simply because there are so many ways that automation can surprise us. Unfortunately, the most reliable defense against automation surprises depends on a strategy that pilots are very bad at: checking the FMAs. The old adage of the priorities of the pilot being to aviate, then navigate and then communicate (in that order) is now flawed unless we expand our concept of what it means to aviate to include assessing the automation state (or, indeed, the flight control law state) of the aircraft. Automation surprises are, by definition, unexpected. It may only be after months of painstaking research that we discover why a surprise occurred, and so we need to adopt one crucial strategy when faced with any unexpected situation: check the FMAs at the same

Box 9.5 Accident Case Study – Automation Surprise

XI Airways Germany Flight 888T – Airbus A320
Near Canet-Plage, France, 27 November 2008[27]

Before returning this Airbus A320 to Air New Zealand, XL Airways Germany, who had been leasing the aircraft, needed to carry out an acceptance flight to demonstrate the functional status of various systems. This flight would be observed by a pilot and several engineers from Air New Zealand. The flight departed Perpignan-Rivesaltes aerodrome in the south of France with the intention of completing a series of flight tests. Because of other commercial traffic in the area, air traffic control (ATC) declined permission to carry out some of the tests and so the crew elected to carry out as many as possible on their return flight to Perpignan. Unfortunately, before the flight, the aircraft had been repainted and after this process, a layer of dust remained on the airframe that was hosed off with high-pressure water. It appears that water entered the angle-of-attack sensors on the aircraft and this water froze when the aircraft flew into subzero temperatures. Flight data recordings showed that at FL320, the angle-of-attack sensors stopped moving and remained static until the end of the flight. One of the required flight tests involved testing the angle-of-attack protection that is part of the usual flight control law, known as normal law. To test this, the crew decelerated the aircraft and the automatic pitch trim proceeded to trim the aircraft into a nose-up position. Owing to the malfunction of the angle-of-attack sensors, the normal protection did not occur and the aircraft was now in a nose-high, low-speed condition. This highly unusual condition for an Airbus A320 led to a transition from normal law to direct law, a flight control law where many of the automated envelope protections are lost and where there is a direct relationship between the side-stick control position and the position of the control surfaces. Crucially, there is no automatic pitch trim, which a pilot would expect to have in direct law even when flying the aircraft manually. For this reason, a transition to direct law is accompanied by an amber message on the primary flight display that reads "USE MAN PITCH TRIM". In this case, the crew were at low altitude when they entered this unusual attitude. The handling pilot was pushing forward on the side-stick controller but did not know why this was failing to lower the nose. During the 25 seconds that the aircraft was in direct law, the situation deteriorated rapidly because the aircraft was already trimmed into a nose-high attitude. It seems that none of the three pilots in the flight deck realized that they were now in direct law and that only by manually trimming nose down would they be able to recover. Applying full thrust to try to recover from the impending stall caused the aircraft to pitch up even further and all control of the aircraft was lost. It impacted with the sea, killing all seven people on board.

time as you check the attitude, airspeed and altitude. You may find that the nature of the failure has led to one or more of your flight modes changing, and this can quickly and unexpectedly change the attitude, airspeed and altitude from the mental model you just formed from your scan of the instruments.

Box 9.6 Incident Case Study – Automation Surprise

Flybe Flight 1794 – Bombardier Dash-8 Q400
Exeter, UK, 11 September 2010[28]

On approach to Exeter Airport during the hours of darkness, the crew of this flight encountered an unusual technical malfunction. The crew were attempting to ascertain the cause of this failure but were unaware that it had also caused the altitude select mode to become disengaged, and the aircraft began to descend below its cleared altitude. The descent towards the ground first caused the ground proximity warning system to issue a terrain caution, followed by a terrain warning alert instructing the pilots to pull up. The crew responded to the "PULL UP" instruction and climbed to capture the glideslope. It was only during the recovery maneuver that the captain realized that the failure had led to the altitude selection function dropping out.

The case presented in Box 9.6 illustrates that there may be one further strategy to prevent automation surprises threatening flight safety. Rather than relying on the liveware to detect uncommanded mode changes (given that we have already seen how only one out of 32 uncommanded mode changes was detected when Boeing 747 pilots were tested in the simulator[10]), manufacturers may be able to engineer the hardware to be more resilient to relatively minor technical failures affecting safety-crucial functions such as flight modes. They should, at least, consider how to alert pilots to flight mode changes that have not been commanded. Expecting a pilot to notice the *absence* of an FMA may not be the best way of ensuring flight safety.

Chapter key points

1. The nature of the interaction between the pilot and onboard automated systems is highly complex and is one of the most relevant areas of human factors where it is crucial that evidence, rather than opinion, be used to dictate policy.
2. Saying what a pilot *should* be able to do is nowhere near as important as understanding what a pilot is actually *able* to do.
3. Understanding flight control laws is as fundamental as understanding pitch, roll, yaw and thrust.
4. Different levels of automation can be advantageous in certain situations and disadvantageous in other situations.
5. The highest level of automation is not always the most appropriate.
6. Flight mode annunciators (FMAs) are key to a pilot's understanding of the state of the aircraft.
7. The handling pilot should select a mode, visually check that the required FMA is displayed and then announce this out loud to keep the monitoring pilot in the loop.
8. Difficulties in being able to project the path of an aircraft when the flight path is being controlled automatically (particularly in FMS modes) may be related to our perceptual knowledge about how objects usually move.

9. There are some features of automated systems that have a significant impact on how the human operator can relate to such a system.

10. If an automated system carries out a task that would normally be carried out by a pilot, that pilot's ability to manually carry out that task will deteriorate and he will become more dependent on the automation.

11. The extent of this skill fade is a factor of how recently the manual skill has been practiced and not of how much manual practice has been accumulated in the past; that is, recent experience is more important than past experience.

12. The only way to address this skill fade and automation dependency is through frequent practice, either during normal operations or in the simulator.

13. A degradation in the ability to manually control the airspeed of an aircraft is frequently associated with accidents and incidents that occur when the aircraft is being flown manually.

14. Automation complacency is a heuristic that pilots are almost guaranteed to use to manage workload. It is unrealistic to expect a pilot to monitor an automated system to the same degree as a non-automated system.

15. Humans are very poor at sustaining attention. Expecting a pilot to be able to detect an unannounced change in system state is unrealistic.

16. The more reliable the automation, the less likely it is to be monitored.

17. Pilots will tend to believe what a computer tells them rather than any other cues that may be present.

18. Automation surprise is a symptom of the increasing mismatch between the pilot's understanding of the automated system and how those automated systems will actually perform.

19. Hardware changes are most likely to be able to address the potentially fatal outcomes of automation surprise, but liveware-related strategies, such as always checking FMAs with the same diligence as checking attitude, airspeed and altitude in any abnormal situation, may provide some level of protection.

References

1. National Transportation Safety Board. *Operations group chairman factual report. Interview summaries* (Docket No. SA-537. Exhibit No. 2-B). Retrieved from: <https://dms.ntsb.gov/pubdms/search/document.cfm?docID=408744&docketID=55433&mkey=87395/>; 2013.

2. Airbusdriver.net. (n.d.). *Flight control laws summary*. Retrieved February 24, 2014, from: http://www.airbusdriver.net/airbus_fltlaws.htm.

3. Comisión de Investigación de Accidentes e Incidentes de Aviación Civil. *Accident of aircraft Airbus A-320-214, registration EC-HKJ, at Bilbao Airport on 7 February 2001. (Technical Report A-006/2001)*. Madrid: Comisión de Investigación de Accidentes e Incidentes de Aviación Civil; 2006.

4. Australian Transport Safety Bureau. *In-flight upset 154 km west of Learmonth, WA, 7 October 2008, VH-QPA, Airbus A330-303. (Aviation Occurrence Investigation AO-2008-070 Final)*. Canberra: Australian Transport Safety Bureau; 2011.

5. National Transportation Safety Board Aircraft accident report: Loss of thrust in both engines after encountering a flock of birds and subsequent ditching on the Hudson River, US Airways Flight 1549, Airbus A320-214, N106US, Weehawken, New Jersey, January 15, 2009. (NTSB/AAR-10/03 PB2010-610403). Washington, DC: National Transportation Safety Board; 2010.

6. American Airlines. *Advanced aircraft maneuvering program: Automation dependency* [Video file]. Retrieved from: <http://vimeo.com/64502012/>; 1997.

7. Ebbatson, M. *The loss of manual flying skills in pilots of highly automated airliners.* PhD thesis, Cranfield University. Retrieved from: <https://dspace.lib.cranfield.ac.uk/handle/1826/3484/>; 2009.

8. Björklund CM, Alfredson J, Dekker SW. Mode monitoring and call-outs: an eye-tracking study of two-crew automated flight deck operations. *Int J Aviation Psychol* 2006;**16**(3):263–75.

9. Goteman Ö, Dekker S. Flight crew callouts and aircraft automation modes: an observational study of task shedding. *Int J Appl Aviation Stud* 2007;**6**(2):235–48.

10. Mumaw RJ, Sarter N, Wickens CD. Analysis of pilots' monitoring and performance on an automated flight deck. *11th International Symposium on Aviation Psychology, Columbus, OH*. Columbus, OH: Ohio State University; 2001.

11. Newton I, Motte A, Cajori F, Thompson SP. *Mathematical principles of natural philosophy*, vol. 34. Chicago, IL: Encyclopaedia Britannica; 1955.

12. Bainbridge L. Ironies of automation. *Automatica* 1983;**19**(6):775–9.

13. Harris D. *Human performance on the flight deck*. Farnham, UK: Ashgate; 2011.

14. Warm JS, Parasuraman R, Matthews G. Vigilance requires hard mental work and is stressful. *Hum Factors: J Hum Factors Ergon Soc* 2008;**50**(3):433–41.

15. Mackworth NH. The breakdown of vigilance during prolonged visual search. *Quarterly J Exp Psychol* 1948;**1**(1):6–21.

16. Young JP, Fanjoy RO, Suckow MW. Impact of glass cockpit experience on manual flight skills. *J Aviation/Aerospace Edu Res* 2006;**15**(2):27–32.

17. Ebbatson M. Practice makes imperfect: Common factors in recent manual approach incidents. *Hum Factors and Aerospace Safety* 2006;**6**(3):275–8.

18. Man4Gen. (2013). *Manual operation of 4th generation airliners*. Retrieved February 24, 2013, from: <http://man4gen.eu/about/about-man4gen/>; 2013.

19. Parasuraman R, Manzey DH. Complacency and bias in human use of automation: An attentional integration. *Hum Factors: J Hum Factors Ergon Soc* 2010;**52**(3):381–410.

20. National Transportation Safety Board. (2013). *Chairman Hersman briefs the media on Asiana flight 214 crash in San Francisco, CA. July 8, 2013* [Video file]. Retrieved from: <http://www.youtube.com/watch?v=d9MTLlzf8Co/>; 2013.

21. National Transportation Safety Board. *NTSB Hearing of Asiana Flight 214* [Video file]. Retrieved from: <http://www.youtube.com/watch?v=zwNk8jow5xg/>; 2013.

22. National Transportation Safety Board. *Cockpit Voice Recorder 12 – Exhibit 12A – Factual Report of Group Chairman*. Retrieved from: <http://dms.ntsb.gov/pubdms/search/document.cfm?docID=406200&docketID=55433&mkey=87395/>; 2013.

23. Mosier KL, Palmer EA, Degani A. Electronic checklists: Implications for decision making. In: *Proceedings of the Human Factors and Ergonomics Society Annual Meeting* (Vol. 36, No. 1, pp. 7–11). Thousand Oaks, CA: Sage 1992.

24. Mosier KL, Skitka LJ, Heers S, Burdick M. Automation bias: Decision making and performance in high-tech cockpits. *Int J Aviation Psychol* 1998;**8**(1):47–63.

25. Woods DD, Sarter NB. Learning from automation surprises and "going sour" accidents. In: Sarter NB, Amalberti R, editors. *Cognitive engineering in the aviation domain*. Mahwah, NJ: Lawrence Erlbaum Associates; 2000. p. 327–53.

26. Rankin A, Woltjer R, Field J, Woods D. "Staying ahead of the aircraft" and managing surprise in modern airliners. In: *Proceedings of the 5th Resilience Engineering Symposium, Soesterberg, The Netherlands*. Sophia Antipolis, France: Resilience Engineering Association; 2013.

27. Bureau d'Enquêtes et d'Analyses   Accident on 27 November 2008 off the coast of Canet-Plage (66) to the Airbus A320-232 registered D-AXLA operated by XL Airways Germany. Le Bourget, France: Bureau d'Enquêtes et d'Analyses; 2010.
28. Air Accidents Investigation Branch DHC-8-402 Dash 8, G-JECF (AAIB Bulletin: 6/2012 EW/C2010/09/04). Aldershot, UK: Air Accidents Investigation Branch; 2012.

Conclusion

The goal of this book was to present an integrated framework for considering human factors in aviation. The topics covered in the preceding chapters should not be considered as separate. They all have an impact on the safety and efficiency of flight operations and they all interact with each other. In Chapter 1, the SHEL model (software, hardware, environment and liveware) was adapted to make it more relevant to flight operations. Based on the rest of the content of this book we can now include more detail into this model to show how the topics we have covered interact with each other. As you will see from Figure 1, information processing is at the center of how we interact with our environment. Although not shown in this diagram, problems with information processing are the source of all errors and violations.

Now that you have reached the end of the book, I can probably be honest and explain my true motivation for writing it. I am truly disappointed by what Crew Resource Management (CRM) has become and how we, as an industry, largely ignore the principles on which it is based. I do not think it has lived up to its promise and we need to address this, particularly as other domains (such as healthcare) look to us for guidance. We keep being told that CRM is vitally important but this is not backed up by any meaningful or structured guidance for individuals and organizations about how these principles can be best implemented. Accident reports frequently talk about poor CRM between crew members and "recommend" further training to address a particular issue, with no guidance about what form that training should take or what information it should be based on. Often, the human factors analysis that has gone into such reports is misinformed, flawed or just plain wrong.

As an industry, we focus on training CRM instructors to be able to teach effectively without giving them the *knowledge* they need to really make a difference. As a consequence, CRM training is often hollow and meaningless, a clear sign that the aviation industry is more concerned with the method of delivery than with the message itself. I think that many pilots out there would say the same thing as the training captain I talked about in the Preface to this book: CRM is a waste of time. This is not the fault of the instructor and it is not even the fault of the airline. We are trained and qualified in accordance with the law and it is the lawmakers who must take responsibility for this state of affairs.

This apathy towards human factors is made much worse by the fact that people are dead and will continue to die because of these shortcomings in the industry. The vast majority of aircraft that are crashing are perfectly airworthy. It is the human factor that is the cause of these catastrophes. These deaths are entirely preventable, and the lip-service that we pay to the importance of human factors is an insult to the pilots, cabin crew, passengers and people on the ground who are dead because of our refusal

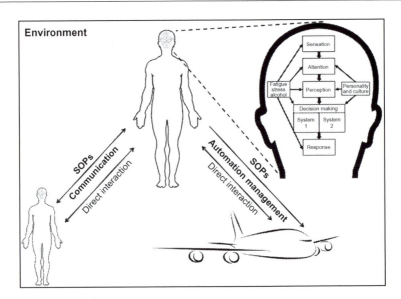

Figure 1 Model of aviation human factors.
Human image copyright vadimmmus and aircraft image copyright mmarius (both from www
.shutterstock.com) SOPs: standard operating procedures.

to engage with this subject in a meaningful way. Three teenage girls, Wang Linjia, Ye
Mengyuan, both 16, and Liu Yipeng, 15 years old, are dead because of human factors
in the crash of Asiana Airlines Flight 214 in 2013. Captain Norbert Kaeppel, First
Officer Theodore Ketzer, Captain Brian Horrell, Murray White, Noel Marsh, Jeremy
Cook and Michael Gyles died in terror and confusion on XL Airways Flight 888T
because vital information about the state of the aircraft systems was not presented in
a way that their information processing systems could take account of. There are still
many questions about how humans interact with other humans and how they interact
with technology, but not only are we not engaging with these questions, we continue
to disregard the decades of scientific research that could give us vitally important
insights into these problems.

The most alarming paradox for me is that although thousands of people have died
in accidents that were entirely preventable, CRM is one of the very few subjects that is
not formally assessed in the classroom. How many more deaths will there have to be
before national and international regulators start to take human factors seriously and
treat CRM training as something more than a "tick-in-the-box" subject? We should
either take human factors seriously or forget about it altogether. Anything in between
is a waste of time.

Perhaps if airlines and regulators realized that effective CRM training is not just
about improving safety we might see more progress. Aircraft accidents are so rare that
there is an argument that says that the industry is as safe as it can ever be. This may
or may not be true but effective human factors training not only decreases the likelihood that an organization will have a catastrophic accident, but should reduce the rate

of incidents and near misses, drive organizational safety culture forward, support the company's training programme and, crucially, improve operational efficiency. Good human factors training can be the silver bullet that turns a shaky, fragile operation with a toxic safety culture into one that is safe, efficient and resilient.

We could make things better. We could have an industry-wide human factors framework that could truly make a difference. Among the things I would recommend would be:

- Dispense with the term CRM. It is tainted by the current sorry state of affairs. Let's call it what it is – *human factors*.
- Set up a Human Factors Regulatory Authority as part of the ICAO. This should be made up of operational experts and, more importantly, human factors scientists. Unless we can apply their findings to our operations, it is just going to be based on individual opinions. We need evidence-based training and policy or we are reduced to guesswork.
- This human factors regulatory authority should be able to carry out the following:
  - Develop an industry-wide human factors training syllabus from zero hours to recurrent human factors training in a multicrew environment. This would cover private pilot licenses, single pilot operations and multicrew operations, both initial, conversion and recurrent.
  - Mandate the course duration and content for any dedicated human factors instructor. The content should give instructors a comprehensive overview of the science of human factors, and they should be the most intensively and highly trained instructors in an airline. Human factors is not an easy subject and requires dedication and hard work on the part of any instructor who wants to teach it.
  - Mandate the course duration and content for any instructor whose job includes instilling human factors principles (e.g., simulator instructors or training captains).
  - All human factors instructors should be required to keep themselves updated with current information regarding human factors and should undergo periodic refresher training.
  - Mandate and set formal end-of-course examinations for instructors.
  - Mandate the course content, duration and frequency for all human factors courses that crew are required to attend, and update the content regularly.
  - Mandate end-of-course examinations for students.
  - Invest in developing a single international standard for non-technical skills assessment to be used in simulator checks and line checks.
  - Mandate that human factors analysis in accident investigations is carried out by suitably qualified human factors professionals and not just investigators who have an interest in the subject.
  - Be able to legislate every aspect of human factors policy across the industry.

Above all, the science must be respected. The dynamic world of aviation raises some truly difficult questions about how humans perform and interact in a variety of complex, high-stakes situations, and these situations also test our capabilities in ways that evolution did not design us for. Fortunately, there are brilliant and insightful people working on these questions and coming up with answers that could really help us. I think we should start listening to them.

Index

Note: Page numbers followed by *"b"*, *"f"* and *"t"* refer to boxes, figures and tables, respectively.

A

Above runway threshold elevation (ARTE), 110
Accident case study
  assertiveness, 181*b*–182*b*
  automation complacency, 260*b*–261*b*
  automation surprises, 264*b*, 265*b*
  availability heuristic, 53*b*
  confirmation bias, 53*b*
  fatigue, 209*b*
  flight control laws, 247*b*, 248*b*
  hindsight bias, 58*b*
  overconfidence heuristic, 56*b*
  perceptual dissonance, 40*b*
  perceptual illusion, 37*b*, 39*b*
  plan continuation bias, 54*b*
  representativeness heuristic, 54*b*
  selective attention, 28*b*
  sensory habituation, 24*b*
  sensory thresholds, 21*b*
  sleep inertia, 200*b*
  toxic captain, 146–148, 146*b*–147*b*
  unstabilized approach, 111*b*
Accident investigation, 271
Adaptive Control of Thought – Rational (ACT-R) model, 43, 44*f*, 47, 134, 136–137
Agreeableness, 135, 142
Airborne conflict, 93–94
Aircraft automation, systems of, 245–246
Air traffic control (ATC), 167–168
Alcohol(ism), 234–240
  abuse, 239
  in aviation, 238–240
  dependence, 239
  incident case study, 236*b*–237*b*
  long-term effects of, 234
  short-term effects of, 234
  units, in common beverages, 235*t*
Allostasis, 228
Allostatic load, 229–230
  due to chronic stress, prevention of, 230

due to chronic stress, management of, 230–233
  at individual, 232–233
  at stressor, 231–232
Alzheimer's disease, 195
β-Amyloid, 195
Antisocial personality disorder, 136
Assertiveness, communication strategies for, 180–186, 181*b*–182*b*
  encouraged by captains, 185–186
  first officer assertiveness
    at threat management stage, 182–183
    at undesired aircraft state management stage, 185
    at unsafe act management stage, 183–184
ASSIST framework, 168
Attention, 26–33
  divided, 32–33
  fatigue effects on, 193
  seeking, 136
  selective, 27–30, 28*b*
  sustained, 30–32
Attentional Control Theory, 45
Attitude direction indicator (ADI), 24, 39–41, 40*f*
Authority gradient. *See* Flight deck gradient
Automation
  bias, 55, 260–262
  complacency, 259–260, 260*b*–261*b*
  dependency, 258–259
  management. *See* Automation management
  surprises, 262–265, 264*b*, 265*b*
Automation management, 243
  aircraft automation, systems of, 245–246
  flight control laws, 245–249, 247*b*, 248*b*
  levels and uses, 249–252, 250*t*
  flight mode annunciators, 246, 252–253
  Newton's laws of motion, 253–254
  ironies of, 254–258
  skill fade and automation dependency, 258–259

Automation management (*Continued*)
  automation complacency, 259–260,
    260*b*–261*b*
  automation bias, 260–262
  automation surprises, 262–265, 264*b*, 265*b*
  manual handling, 258
  monitoring, 243
Autopilot/autothrottle (AP/AT), 245–246
Availability heuristic, 52, 53*b*
Aviation
  alcoholism in, 238–240
  heuristics in, 52
  human factors in, 4–7, 6*f*
Avoidance, uncertainty, 154

B

Balance, 20
Baron, Robert, 146–149
Barriers to communication, 166–167
Basal ganglia, 14
Bedroom, sleep hygiene in, 213
Behavior, 155–156
  and personality, 138–139
Behavioral marker, 8–9
Benzodiazepines, 219
Bescoe, Robert, 149–150
Beverages, alcohol units in, 235*t*
Bias(es), 50–59
  automation, 55
  in aviation, 52
  confirmation, 52, 53*b*
  evolutionary origins of, 51–52
  hindsight, 57–59, 58*b*
  plan continuation, 54, 54*b*
Biological clock, 202
Blood alcohol concentrations (BACs)
  physical effects associated with, 235–236,
    235*t*
Bolt, Usain, 24
Brain
  anatomy of, 13*f*, 14*f*
  structure and function of, 13–18
Brainstem, 14
Briefing
  arrival, 88
  communication strategies for, 178–180
  crew, 88
  departure, 88

C

Cardiovascular disease, stress-induced,
    230
Carter, Jimmy, 2–3
Cerebellum, 14
Cerebral cortex, 14–15
Change blindness, 28–29
Checkerboard, 36*f*
Chernobyl explosion, 1
Chronic stress, 227–234
  allostatic load due to, prevention of,
    230
  allostatic load due to, management of,
    230–233
    at individual, 232–233
    at stressor, 231–232
  critical incident stress management,
    233–234
  incident case study, 231*b*
Chronotypes, 203
Circadian rhythms, 202, 221*t*
Civil Aviation Authority (CAA), 87, 90,
    92–95
Cockpit gradient. *See* Flight deck gradient
Cognitive–behavioral therapy (CBT), 233
  for chronic insomnia, 215
Cognitive effects, of fatigue, 193–195
Collectivism, 154–156
Command, 143–144
Communication, 163
  defined, 163–164
  non-verbal, 164–166
  between pilots, 168–177
    complementary and crossed
      transactions, 170–175, 171*f*, 171*t*,
      172*f*, 173*t*, 175*t*
    ego states, 169–170, 170*f*
    transactional analysis, 168–169
    ulterior transactions, 176–177, 176*f*,
      177*t*
  positive team atmosphere, establishing,
    177–178
  sender–message–channel–receiver model
    of, 163–168, 164*f*
    barriers to communication, 166–167
    NITS briefing, 167–168
  strategies, for assertiveness, 180–186,
    181*b*–182*b*

encouraged by captains, 185–186
threat management stage, first officer
     assertiveness at, 182–183
undesired aircraft state management
     stage, first officer assertiveness at,
     185
unsafe act management stage, first
     officer assertiveness at, 183–184
strategies, for effective briefings,
     178–180
verbal, 164–166
Complementary transactions, 170–175, 171*f*,
     171*t*, 172*f*, 173*t*, 175*t*
Confirmation bias, 52, 53*b*
Conflict resolution styles, 157–159, 157*f*
Conflict-solving strategies, 156–159
individual differences and, 157–158
situational variables and, 158–159
Conscientiousness, 135, 142
Controlled flight into terrain (CFIT), 92–93,
     95, 110
Cooperation, 156–159
Cranial nerves, 15
Crew communication, 99–105, 99*t*, 105*t*
Crew resource management (CRM), 7–9,
     12–13, 28–29, 153, 269–270
Critical incident stress management (CISM),
     233–234
Crossed transactions, 170–175, 171*f*, 171*t*,
     172*f*, 173*t*, 175*t*
Culpability flowchart, 128*f*
Culture, and flight deck gradient, 152–156

D

Decision making, 42–65, 46*f*
anatomy of, 43–46
     goal module, 43
     imaginal (mental manipulation) module,
          43–44
     production (pattern-matching) module,
          44–45
biases, 50–59
     in aviation, 52
     evolutionary origins of, 51–52
fatigue effects on, 194
heuristics, 50–59
     in aviation, 52
     evolutionary origins of, 51–52

memory retrieval heuristics, decision
     making and, 52–55
social, 55–59
operational strategy, 61–65
strategies for, 59–61
     high task-load, managing, 60
     high time-pressure, managing, 60–61
     problem underspecificity, managing,
          61
     System 1 effects, managing, 61
System 1, 47–50, 61
System 2, 48–50
Declarative memory, 44
Divided attention, 32–33

E

Efficiency–thoroughness trade-off (ETTO),
     82, 104, 122–123
Ego states, 169–170, 170*f*
Electroencephalography (EEG), 196, 197*f*
Engine indicating and crew alerting system
     (EICAS), 260–262
Error
     defined, 120
     detection, 82–84
     human, 120, 122–124
     latent, 120
Extraversion, 135, 142
Extravert, sensing, thinking, judging (ESTJ),
     138

F

Fatigue, 31, 192–195
accident case study, 209*b*
cognitive effects of, 193–195
defined, 192
management. *See* Fatigue management
Fatigue management. *See also* Fatigue
risk management strategies, 210–222
     chronic insomnia, 215
     high-quality sleep, 211–213
     jet lag, 219–222
     naps, 213–214
     obstructive sleep apnea, 215–216
     organizational strategies, 222
     risk mitigation, 216–218
     sleep medications, 218–219
sleep, role of, 195–210

Fatigue management (*Continued*)
    homeostatic sleep drive and sleep need,
        200–201
    physiology of sleep, 196–199, 197*f*
    sleep debt, 206–210
    sleep inertia, 199–200, 200*b*
    sleep need, 200–201, 201*f*
    sleep urge, 204–205
Fatigue risk management system (FRMS),
    222
Federal Aviation Administration (FAA),
    215–216
Femininity, 154
Fire, 94–95
First officer assertiveness
    at threat management stage, 182–183
    at undesired aircraft state management
        stage, 185
    at unsafe act management stage, 183–184
Five-Factor Model (FFM), 134–137, 142–145
Flexible culture, 127
Flight control laws, 245–249
    accident case study, 247*b*, 248*b*
Flight data monitoring (FDM) systems,
    243–244, 251
Flight deck gradient, 152–156
    culture and, 152–156
Flight director (FD), 245
Flight guidance system (FGS) modes, 245
Flight Management Attitudes Questionnaire,
    154–155
Flight management computer (FMC), 245–
    246, 251
Flight management system (FMS) modes,
    245–246
Flight mode annunciators (FMAs), 246,
    252–253
Flight safety, leadership and, 144–152
    toxic captain, 146–150
    toxic first officer, 150–152
Flight Safety Foundation, 22
Freezing, fear-potentiated, 66–69, 67*b*
Frequency gambling, 52
Frontal lobe, 15
Fundamental attribution error (FAE), 57–59

G

Gastrointestinal disease, stress-induced,
    229–230

Generic threat management strategies, 97
Global Positioning System (GPS), 256
Goal module, of decision making, 43
Ground handling, 94
Ground proximity warning system (GPWS),
    29, 249–250

H

Halo effect heuristic, 55–57
Heuristics, 50–59
    in aviation, 52
    availability, 52, 53*b*
    evolutionary origins of, 51–52
    halo effect, 55–57
    memory retrieval heuristics, decision
        making and, 52–55
    overconfidence, 55, 56*b*
    representativeness, 54, 54*b*
    social, 55–59
High task-load, managing, 60
High time-pressure, managing, 60–61
Hindsight bias, 57–59, 58*b*
Histrionic personality disorder, 136
Homeostatic sleep drive, 200–201
Housekeeping purpose, of sleep, 195
Human error, 120
    old versus new view of, 122–124
Human factors, 269–271, 270*f*
    in aviation, 4–7
    defined, 3–4
    modern, 1–3
    and non-technical skills, 7–9
    training, 270
Hypnogram, 196, 197*f*

I

Illusion, perceptual, 37*b*
Imaginal (mental manipulation) module, of
    decision making, 43–44
Immune disease, stress-induced, 229
Incident case study
    alcohol, 236*b*–237*b*
    chronic stress, 231*b*
    errors and outcomes, 124*b*
    obstructive sleep apnea, 216*b*
Individualism, 154–156
Information processing, 11
    attention, 26–33
        divided, 32–33

selective, 27–30, 28*b*
sustained, 30–32
decision making, 42–65, 46*f*
anatomy of, 43–46
biases, 50–59
heuristics, 50–59
operational strategy, 61–65
strategies for, 59–61
System 1, 47–50, 61
System 2, 48–50
fear-potentiated startle and freezing,
66–69, 67*b*
overview of, 18–19
perception, 33–42
difficulties with sensory-induced spatial
disorientation, 38–41
mental models, 35–38
perceptual limitations, strategies for
dealing with, 41–42
response, 65–66
sensation, 19–26
sensory habituation, 22–24, 24*b*
sensory limitations, strategies for
dealing with, 25–26
sensory thresholds, 20–22, 21*b*
somatogravic sensory illusions, 24–25
Informed culture, 127
Insomnia, chronic, 215
International Civil Aviation Organization
(ICAO), 270
Interneurons, 17–18
Ironies of automation, 254–258

J

Jet lag, 219–222
Joint Aviation Authority, 252–253
Just culture, 127–129

K

Knowledge-based performance, 79
errors at, 81
checklist for detecting, 101–104
self-detection of, 83
violations, detection of, 104–105

L

Latent errors, 120
Leadership, 140–152
defined, 141

flight safety, 144–152
toxic captain, 146–150
toxic first officer, 150–152
non-technical skills for, 143–144
and personality, 141–143
Learning culture, 127
Life change units, 227–228
Limbic system, 15
Line-oriented safety audits (LOSAs), 87–88,
108–109
L-M-N-O-P-Q-R-S-T-U framework, 96, 97*t*
Loss of control, 90–91

M

Mackworth, Norman, 31
Managing organizational resilience,
principles of, 130–131
Masculinity, 154
Mayday, 168
Melatonin, 203–204, 221–222
Memory consolidation, 195–196
Memory retrieval heuristics, decision making
and, 52–55
Mental illness, stress-induced, 230
Mental models, 35–38
Microsleep, 207–208
Mode control panel (MCP), 245, 252–253
Multitasking, 32–33
Myers–Briggs Type Indicator (MBTI),
137–138

N

Naps, 213–214
National Transportation Safety Board
(NTSB), 58, 208–210
Neurological disease, stress-induced, 230
Neurons, 15, 16*f*
electrical transmission in, 15–16
electrochemical transmission in, 15–16
types of, 17–18
Neuroticism, 135, 142–143
Neurotransmitters, 15–16, 16*f*
Newton's laws of motion, 253–254
NITS briefing, 167–168
Non-rapid eye movement (NREM) sleep,
196–199
Non-technical skills, 7–9
for leadership, 143–144
Non-verbal communication, 164–166

Normal accident theory, 121–122
NOTECHS (non-technical skills) system,
        7–9, 143–146, 156, 163

O

Obstructive sleep apnea, 215–216, 216*b*
Occipital lobe, 14
Openness, 135, 143
Organizations, error management and
        standard operating procedures for, 119
    human error, 120
    managing organizational resilience,
        principles of, 130–131
    resilience engineering, 124–127
    safety culture, 127–130
        just culture, 127–129
        moving from Safety I to Safety II, 129,
            129*t*
    systems thinking, 120–124
        human error, old versus mew view of,
            122–124
        normal accident theory, 121–122
Overconfidence heuristic, 55, 56*b*

P

PACE (probe, alert, challenge, emergency)
        framework, 184–185
PAN-PAN, 168
Parasomnias, 214–215
Parietal lobe, 15
Passive monitoring, 30–32
Perception, 33–42
    difficulties with sensory-induced spatial
        disorientation, 38–41
    fatigue effects on, 194
    mental models, 35–38
    perceptual dissonance, 40*b*
    perceptual illusion, 37*b*, 39*b*
    perceptual limitations, strategies for
        dealing with, 41–42
Performance levels, of pilots, 78–79
    errors and violations at, 79–82
Personal Attributes Questionnaire, 144–145
Personality, 31, 134–140
    and behavior, 138–139
    leadership and, 141–143
    management strategies, 139–140
    structure of, 134–138

Physiology of sleep, 196–199
Pilots
    communication between, 168–177
        complementary and crossed
            transactions, 170–175, 171*f*, 171*t*,
            172*f*, 173*t*, 175*t*
        ego states, 169–170, 170*f*
        transactional analysis, 168–169
        ulterior transactions, 176–177, 176*f*,
            177*t*
    error management and standard operating
        procedures for, 77
        error detection, 82–84
        performance levels, 78–79
        performance levels, errors and
            violations at, 79–82
        Swiss Cheese Model, 84–86, 85*f*, 86*f*
        threat and error management 2. *See*
            Threat and error management 2
Plan continuation bias, 54, 54*b*
Positive team atmosphere, establishing,
        177–178
Post-traumatic stress disorder (PTSD), 229,
        233–234
Power distance, 154–156
Precision approach path indicators (PAPIs),
        112
Problem underspecificity, managing, 61
Procedural memory, 44–45
Production (pattern-matching) module, of
        decision making, 44–45

R

Rapid eye movement (REM) sleep, 198–199
Reporting culture, 127
Representativeness heuristic, 54, 54*b*
Resilience engineering, 124–127
Response, fatigue effects on, 194
Revised NEO Personality Inventory (NEO-
        PI-R), 138–139, 153
Risk mitigation, 216–218
Rule-based performance, 79
    errors at, 80–81
        checklist for detecting, 101–104
        self-detection of, 83
        violations, detection of, 104–105
Runway excursions, 91–92
Runway incursions, 93

S

Safety culture, 127–130
  just culture, 127–129
  moving from Safety I to Safety II, 129, 129*t*
Safety I to Safety II culture, moving from, 129, 129*t*
Selective attention, 27–30, 28*b*
Sender–message–channel–receiver model of communication (SMCR), 163–168, 164*f*
  barriers to communication, 166–167
  NITS briefing, 167–168
Sensation, 19–26
  fatigue effects on, 193
Sensory habituation, 22–24, 24*b*
Sensory illusions, somatogravic, 24–25
Sensory-induced spatial disorientation, perceptual difficulties with, 38–41
Sensory limitations, strategies for dealing with, 25–26
Sensory neurons, 17
Sensory thresholds, 20–22, 21*b*
Serious negative outcomes, threats associated with, 90–95
  airborne conflict, 93–94
  controlled flight into terrain, 92–93
  fire, 94–95
  ground handling, 94
  loss of control, 90–91
  runway excursions, 91–92
  runway incursions, 93
SHEL model, 4–6, 5*f*, 6*f*, 88, 269
Sight, 20
Similarity matching, 52
Sixteen Personality Factor Questionnaire (16PF), 134
Skill-based performance, 79
  errors at, 80
    self-detection of, 82–83
    checklist for detecting, 100–101
Skill fade, 258–259
Sleep, 190–191
  architecture, 196
  debt, 206–210
  homeostatic sleep drive, 200–201
  hygiene, 212–213, 222
    in bedroom, 213
    before sleeping, 212–213

  inertia, 199–200, 200*b*
  initiators, 218–219
  maintainers, 219
  medications, 218–219
  need, 200–201, 205*f*
  physiology of, 196–199, 197*f*
  planning, 211–212
  role in fatigue management, 195–210
  urge, 204–205
Slow-wave sleep, 198
Smell, 20
Social heuristics, 55–59
Social Readjustment Rating Scale (SRRS), 228, 232
Sociotechnical system, 4–5
Somatogravic sensory illusions, 24–25
Sound, 20
Spinal cord, 15
Stage 1 (N1) sleep, 198
Stage 2 (N2) sleep, 198
Stage 3 (N3) sleep. *See* Slow-wave sleep
Startle, fear-potentiated, 66–69, 67*b*
Stimulants, 31
Stress
  acute, 45
  defined, 227–228
  chronic. *See* Chronic stress
Stroke, 169, 177–178
Suprachiasmatic nucleus (SCN), 202–205, 219–222
Sustained attention, 30–32
Swiss Cheese Model, 84–86, 85*f*, 86*f*
Synapses, 15, 16*f*
System 1, 47–50
  characteristics of, 49*t*
  effects, managing, 61
  workload, role of, 49–50
System 2, 48–49
  characteristics of, 49*t*
  workload, role of, 49–50
Systems thinking, 120–124
  human error, old versus mew view of, 122–124
  normal accident theory, 121–122

T

Taste, 20
TDODAR, 61–65, 99, 168, 262

Teamwork, 133–134
Technical sensory organs, 25–26
Temazepam, 219
Temporal lobe, 14–15
Thalamus, 14
Thereat and error management 2 (TEM2),
    86–110, 108*t*
  distinguished from original thereat and
    error management, 108–109
  threat management, 86–97
    generic threat management strategies, 97
    opportunities, 88
    serious negative outcomes, threats
      associated with, 90–95
    threat identification framework, 95–96
    threat management tool for briefings, 97
    types of, 88–90
  undesired aircraft state management,
    106–107
  unsafe act (error and violation)
    management, 97–106
    checklists and crew communication,
      99–105, 99*t*, 105*t*
    prevention, 98–99
    strategies, 105–106
  and unstabilized approaches, 110–115
    undesired aircraft state management for,
      113–115
    unsafe act management for, 113, 113*t*
    threat management for, 111–113
Thomas–Kilmann Conflict Mode Instrument,
    157–158
Threat identification framework, 95–96
Threat management stage, first officer
    assertiveness at, 182–183
Three Mile Island disaster, 1–3, 51–52,
    121–122
Touch, 20
Toxic captain, 146–150, 146*b*–147*b*
Toxic first officer, 150–152
Traffic collision avoidance system (TCAS),
    249
Transactional analysis, 168–169

U
Ulterior transactions, 176–177, 176*f*, 177*t*
Uncertainty avoidance, 154–155
Undesired aircraft state management,
    106–107
  stage, first officer assertiveness at, 185
  for unstabilized approaches, 113–115
Unsafe act management, 97–106
  checklists and crew communication,
    99–105, 99*t*, 105*t*
  prevention, 98–99
  stage, first officer assertiveness at,
    183–184
  strategies, 105–106
  for unstabilized approach, 113, 113*t*
Unstabilized approaches, 110–115
  accident case study, 111*b*
  threat management for, 111–113
  undesired aircraft state management for,
    113–115
  unsafe act management for, 113, 113*t*
US National Aeronautics and Space
    Administration (NASA), 144–146

V
Verbal communication, 164–166
Vigilance, 30–32
  decrement, 31
Violations
  exceptional, 82
  routine, 81–82
  situational, 82

W
Workload, 45

Z
Zaleplon, 218–219
Z-drugs, 218–219
Zolpidem, 218–219
Zopiclone, 218–219

Printed in the United States
By Bookmasters